2nd International Conference on Electronics, Electrical and Information Engineering

AA007202

Journal of Physics: Conference Series Volume 2387

Changsha, China
12-15 August 2022

Editors:

Shengqing Li
Xilong Qu
Bin Hu

ISBN: 978-1-7138-6492-9
ISSN: 1742-6588

Printed from e-media with permission by:

Curran Associates, Inc.
57 Morehouse Lane
Red Hook, NY 12571

Some format issues inherent in the e-media version may also appear in this print version.

Institute of Physics
Dirac House, Temple Back
Bristol BS1 6BE UK

Phone: 44 1 17 929 7481
Fax: 44 1 17 920 0979

techtracking@iop.org

Additional copies of this publication are available from:

Curran Associates, Inc.
57 Morehouse Lane
Red Hook, NY 12571 USA
Phone: 845-758-0400
Fax: 845-758-2633
Email: curran@proceedings.com
Web: www.proceedings.com

TABLE OF CONTENTS

 2387 (2022) 011001 doi:10.1088/1742-6596/2387/1/011001

Editorial

On the behalf of the organizing committee of the 2nd International Conference on Electronics, Electrical and Information Engineering (ICEEIE2022) held during August 12-15, 2022 in Changsha, Hunan, China. Due to the epidemic situation, there were travel restriction for every scholars who were linked with this event by using Zoom conference system, we enjoyed 3 keynote speeches and oral presentations made by the authors which lasted for 15 minutes each, a lot of inspiration sparked during the Q&A parts. The conference has provided a significant platform to encourage researchers, scholars, academician and practitioners to congregate for exchange of ideas and experiences from their work while identifying future directions of relevant areas.

ICEEIE2022 promotes research and developmental activities in the field of Electronics, Electrical and Information Engineering. Researchers who have contributed their work to the conference shared their findings and experience with other researchers and attendee. Many attendees without any contribution of research paper have also been allowed to attend the conference to gain knowledge in their respective areas. With all contributions from the participants, 63 papers are final accepted from 140 submissions. Those accepted ones are well categorized into four sectors as follows:Communication engineering, Electrical engineering, Electronics, Information engineering.

We wish to show our sincere appreciations to all individuals and organizations who have contributed to ICEEIE2022,especially those who did the review of all the submitted papers,dedicated their time and efforts in planning, promoting, organizing, and helping the conference. Also we must extend our thanks to our keynote speakers: Prof.Mohammad S. Obaidat from University of Sharjah,USA, who made speeches titled A New Learning Automata-Based Controller Placement Scheme for Software-Defined Network Systems, Prof. Witold Pedrycz from University of Alberta, Canada, who made speeches titled Developments in Federated Learning and Knowledge Transfer: Pursuits in Green and Granular Machine Learning, Prof. Claudio Cañizares from University of Waterloo, Canada, who made speeches titled Energy Storage Systems.

We wish to show our special thanks to Hunan University of Finance and Economics, Hunan University of Technology, Hunan University of Arts and Science for support, they provided various platform to promote this conference with nice service.

We are pretty sure that readers will benefit and get inspiration for their future work by using this volume, we will feel honorable to have your continuous attention and attendance for our next edition of ICEEIE.

This project is funded by the scientific research project of Hunan Provincial Department of Education(21C1632).

Editors:
Shengqing Li
Xilong Qu
Bin Hu

COMMITTEES

General Chair

Prof. Dr.Shengqing Li , Hunan University of Technology,China

Shengqing Li was born in 1961 in Hunan province, China. He is a Doctor of Engineering, professor, doctoral supervisor and the expert entitled to government special allowance（GSA）. Professor Li is the dean in the School of Electrical and Information Engineering at Hunan University of Technology. Meanwhile, he is serving as the Chairman of Hunan Engineering Research Center, the academic leader of provincial key discipline as well as the permanent member of Hunan Electrotechnical Society and Hunan Electrical Engineering Society.

Professor Li's extensive experiences with electricity and engineering lead to his broader interest in Power Quality Control Strategy for Distribution Network, New Energy Grid-Connected System Power and Electrical Energy-saving Technology. He currently hosted and participated in 20 important research projects of the Provincial Natural Science Foundation, the National Natural Science Foundation and National Key R&D Program of China etc. He received 9 ministerial Second Prizes and honored prizes of Progress in Science and Technology, Science and Technology Award of China Electrotechnical Society etc. with his papers, researches, 14 Authorized Patents and Software Copyright.

Professor Li is the author of 6 book including 2 of them were published by China Science Press. He has published numerous papers, 45 of which are included by SCI/EI consisting of Proceedings of the CSEE and International Journal of Robotics and Automation etc.

Program Chair

Prof.Dr.Mufti Mahmud, Nottingham Trent University, UK

Dr Mufti Mahmud is a Senior Lecturer of Computer Science at the Nottingham Trent University, UK with a PhD degree in Information Engineering (with a specialisation in Bioengineering) from the University of Padova – Italy, in 2011. A recipient of the prestigious Marie-Curie postdoc fellowship, he has served various positions in the industry and academia in India, Bangladesh, Italy, Belgium, and the UK in the last 17 years.

Dr Mahmud has a vision towards leaving behind a healthy, secure, smart, and better world to live in. Combining computational intelligence and big data technologies, Dr Mahmud's research aims to convert the ever-expanding amount of raw data to smart data to be able to build predictive, secure and adaptive systems aiming personalized services towards improving quality of life. Dr Mahmud aims to develop secure and intelligent tools to advance healthcare access in low-resource settings.

Dr Mahmud serves as Vice Chair of the Intelligent System Application Technical Committee of IEEE Computational Intelligence Society, Member of the IEEE CIS Task Force on Intelligence Systems for Health, Co-Chair of the IEEE P2733 Standard on Intelligent systems design, with AI/ML, Member of the IEEE R8 Humanitarian Activities Subcommittee, Project Liaison Officer of the IEEE UK and Ireland Special

Interest Group on Humanitarian Technology, and Publication Chair of the IEEE UK and Ireland Industry Applications chapter.

Dr Mahmud is also actively involved in dissemination activities through the organisation of workshops, conferences, and specialised short duration schools (e.g., Local organising chair of IEEE-WCCI2020; General Chair of BI2020 and BI2021; and Program Chair of IEEE-CICARE2020 and CICARE2021).

Prof. Dr.Jianqi Li, Hunan University of Arts and Science, China

Professor Li Jianqi obtained a doctorate degree in control science and engineering. He is currently the dean of the School of Computer and Electrical Engineering of Hunan University of Arts and Science, and the vice chairman of Hunan Institute of Instrumentation. He is mainly engaged in the teaching and research of computer control and detection technology. He is also an off-campus postgraduate tutor for Hunan University and Xiangtan University, and an expert on the major administrative decision-making advisory committee of Changde City.

He has published more than 30 academic papers, 12 of which are collected by SCI/EI, 2 invention patents, 8 utility model patents, and 10 software copyrights. He presided over the completion of 1 National Natural Science Foundation project and 2 provincial self-research fund projects. He actively carried out the work of serving local areas and once served as a special commissioner for science and technology in Changde City. During his tenure, he completed the development of more than 10 horizontal projects for the enterprise and achieved good economic and social benefits. Among them, the high-precision steel strip parameter measurement system was the first in China, and the scientific research results were specially reported by the Hunan Provincial Public Channel and other media.

Professor Li Jianqi instructed students to have won more than 10 provincial competition awards, published more than 10 papers, and won 1 third prize of provincial teaching achievement.

Prof. Dr.Jinhui Zeng, Hunan University of Technology, China

Prof. Jinhui Zeng is the associate dean of College of Electrical and Information Engineering, Hunan University of Technology, a master's Supervisor. He gets both his doctor degree and master degree in Hunan University. He is the Executive Director of Hunan Electrotechnical Society and holds more than 10 research projects funded by National Natural Science Foundation of China, Ministry of Education Cooperative Education Project, Natural Science Foundation of Hunan Province and so on. He hold numerous patents and copyrights during participating in the research projects like China Railway Construction Corporation and has academic cooperation with some senior conferences and forums which proceedings are published by Springer Elsevier and IEEE, he publishes more than 20 articles in journals and most of them are SCI or EI-indexed, his research interests include Power Electronics Converter, Power quality control, Inductive Filtering Technology in Multi-infeed HVDC Transmission System.

Organising Chair

Prof.Dr. Xilong Qu, Hunan University of Finance and Economics,China

Prof. Xilong Qu is the dean of School of Information Technology and Management, Hunan University of Finance and Economics, he is also the master degree student advisor for both Xiangtan University and Hunan Institute of Engineering. Prof. Xilong Qu received his Ph.D on Southwest Jiaotong University, he finished his post-doctor study in Post-doctoral Research Center of Computer Science and Technology, South China University of Technology. Prof. Xilong Qu is the young core teacher of Hunan and a member of "121" talents program of Hunan. Prof.Xilong Qu joined 4 projects from the National High Technology Research and Development Program of China (863 Program) and the National Natural Science Foundation of China, also he led the National Natural Science Foundation of Hunan, the Research Foundation of Education Bureau of Hunan Province, China and the National Science Foundation for Post-doctoral Scientists of China for 3, 1 and 1 times respectively. Prof. Xilong Qu published 4 books 35 articles (5 articles are indexed by SCI and the rest are indexed by EI). Prof. Xilong Qu is also invited for chairs and keynote speakers of numerous conferences and forums in China.

Technical Program Committee

Prof. Vladicescu Popentiu, City University, UK
Prof. Guangzhi Qu, Oakland University, USA
Prof. V.S.S. Yadavalli, University of Pretoria, South Africa
Prof. Bruno Apolloni, Università degli Studi di Milano, Italy
Prof. Harry Bouwman, Delft University of Technology, Netherlands
Prof. Shyi-Ming Chen, National Taiwan University of Science and Technology, Taiwan
Prof. Yahaya Coulibaly, University Technology Malaysia, Malaysia
Prof. Ing Kong, RMIT University, Australia
Prof. Gerald Midgley, Centre for Systems Studies, Business School, University of Hull, UK
Prof. Khubaib Ahmed, Hamdard University, Pakistan
Prof.Dr. Tongliang Fan, China Maritime Police Academy, China
Prof. Junjie Lv, University of Wuhan Polytechnic, China
Prof.Dr. Hao Wang, State Key Laboratory of Rolling and Automation, Northeastern University, China
Prof.Dr. Imran Memon, Zhejiang University, China
Dr. Liang Zong, Shaoyang University, China
Dr. Xilang Tang, Airforce Engineering University, China
Prof. Jun Tao, Jianghan University, China
Dr. Wei Wei, Xi'an University. of Tech, China
Dr.Xu Zezhong, Changzhou Institute of Technology, China
Dr.Jian-wei Liu, University of Petroleum, China
Dr.Jun Guo, East China Normal University, China
Dr.Ing. Kuan Chee, University of Nottingham Ningbo China , China

The contact details

Full: name:Xilong Qu
Affiliation:Hunan University of Finance and Economics
Institutional email address: quxilong@126.com

Peer Review Statement

All papers published in this volume have been reviewed through processes administered by the Editors. Reviews were conducted by expert referees to the professional and scientific standards expected of a proceedings journal published by IOP Publishing Publishing.

- **Type of peer review:** Single Anonymous
- **Conference submission management system:** Morressier
- **Number of submissions received:** 140
- **Number of submissions sent for review:** 140
- **Number of submissions accepted:** 39
- **Acceptance Rate (Submissions Accepted / Submissions Received × 100):** 27.9
- **Average number of reviews per paper:** 1
- **Total number of reviewers involved:** 4
- **Contact person for queries:**
  Name: Shengqing Li
  Email: 11697@hut.edu.cn
  Affiliation: Hunan University of Technology

Overview of ship image recognition methods based on computer vision

Guoming Chen[1,3], Qinghua Liu[2,*]

[1] School of Computer Science, Jiangsu University of Science and Technology, ZhengJiang, JiangSu, 212000, China

[2] School of Computer Science, Jiangsu University of Science and Technology, ZhengJiang, JiangSu, 212000, China

[3] Email: 211710701105@stu.just.edu.cn

*Corresponding author's e-mail: liuqh@just.edu.cn

Abstract. This paper introduces the research background of water traffic problems, analyzes the Marine equipment in the process of shipping the deficiencies, on the basis of introducing the concept of image recognition, computer vision and computer vision image recognition are briefly introduced the development of the status situation, sums up and summarizes the recent years research direction such as neural networks based on machine learning, convolution algorithm, For ship image recognition method, and from static image recognition to the complex environment of image recognition, and machine learning, convolutional neural networks based algorithm was analyzed, and the advantages and disadvantages and applicable environment, finally put forward to build a multilateral participation information fusion detection framework to implement the idea of the safe navigation of the ship.

1. Introduction

China has a long history of inland waterway shipping, as early as 400 years BC, China began to realize the transport of goods through the Beijing-Hangzhou Grand Canal, the Yangtze River and other waterways. China is accessible by water. The Yangtze River basin alone can be divided into six systems, among which the Yangtze River Delta waterway network of Shanghai, Nanjing and Hangzhou is the largest, with a total navigable length of over 20,000 KM.

With the development of economy, railway, highway, air and other modes of transportation are developing rapidly, but shipping occupies a stable place in the freight market by virtue of its unique advantages such as low transportation cost, large cargo capacity and dependence on natural environment. As a branch of shipping, the economic benefit of inland waterway shipping has always maintained vitality, and the quantity and quality of inland waterway vessels have maintained a steady growth. In the early 20th century, the transportation department has accelerated the waterway capacity by structure adjustment, promote inland ship type standardization work, hang pulp vessel because of its low pollution, high energy consumption, safety factor gradually withdrew from the stage of shipping, can only see the figure of tug and single ship in inland, and inland water transport ship's load is growing, As a result, the inland waterway has been subjected to great navigable pressure, which is also an important reason for the collision and grounding of inland vessels with the development of inland waterway navigation. In order to study ship collision rate, DUAN Aiyuan [1] calculated the average change of ship collision rate in port area with ship tonnage based on maritime statistics data as follows:

Collision rate =0.0014GT+0.0009 GT - Total tonnage.

This shows that the greater the tonnage of the ship, the higher the probability of collision.

In the early 20th century, shipping ships were basically equipped with VHF communication systems, a set of radio communication equipment for close range. In the same period, China Maritime Safety Administration explicitly required that transport vessels with a load of more than 200 gross tons must be equipped with AIS (Automatic Identification System of Ships), which provided a guarantee for navigation safety to a certain extent. Today, in the environment of technology software and hardware constantly updated and iterated, the number of equipment to ensure safe navigation continues to increase. It is mainly divided into three modules. The first, navigation communication system: Very High Frequency; The second, Navigation scanning system: Ship borne Navigation Radar, Echo Sounder, video Circuit Television; The third, Navigation positioning System: Automatic Identification System and BeiDou Navigation Satellite System. However, in the harsh weather environment, no matter how accurate the detection instrument is, it only transmits information to people through images, and then people make decisions, which cannot avoid accidents caused by human causes.

In order to realize the safe navigation of ships in frost and fog weather and at night, the image recognition technology of ships, piers and other objects in the waterway based on computer vision technology has become the research focus. Through the existing Marine equipment, the real-time image data is acquired and transmitted to the image recognition system to identify the information of the object to be intersecting. Combined with the radar equipment to obtain the distance, so as to make hedging countermeasures. In this paper, the relevant literature is collected and sorted out, and on the basis of learning, analysis and induction, all kinds of image recognition in waterway by computer vision is discussed.

2. Computer vision image recognition

2.1. Development status of target detection and recognition

Image target detection and recognition is an important part of image processing. With the continuous development of technology, target detection and recognition technology has been combined with various fields. The basic steps of image target detection and recognition are shown in Figure 1. Compared with traditional target recognition algorithms, researchers are more focused on convolutional neural network algorithms at present.

Convolutional neural network originated from the relevant research of the visual cortex of the brain by scientific researchers. The classic convolutional neural network is a forward neural network, which is evolved on the basis of the development of neural network. Convolutional neural network has three characteristics: local connection, weight sharing and spatial up-down sampling. These three characteristics make the convolutional neural network with translation, scaling and distortion invariance, which is conducive to extracting more representative features of the target image.

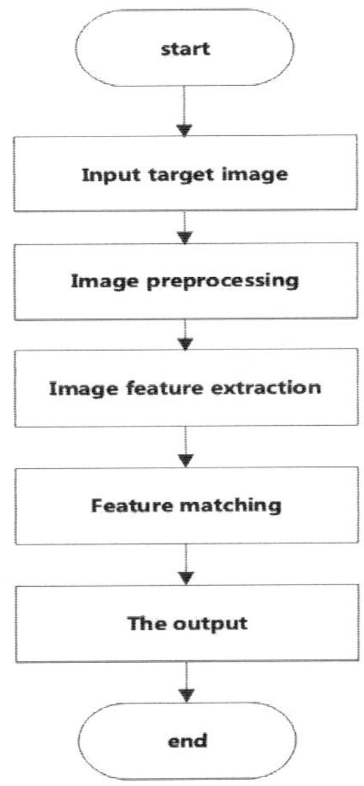

Figure 1. The steps of image target detection and recognition

2.2. Convolutional neural network algorithm

Convolutional neural network algorithms can be roughly divided into three categories: R-CNN algorithm, YOLO series algorithm and SSD target recognition algorithm.

ROSS[2] et al. proposed to use high-capacity convolutional network for target detection. R-cnn is a simple and extensible detection algorithm. It is the first one to explore the application of convolutional neural network to target detection, and it is a typical two-stage target detector. R-cnn firstly applies high-capacity volume and neural network to bottom-up regions to locate and segment target objects; secondly, pre-training auxiliary tasks when training data is scarce; finally, region-specific micro-adjustment is carried out to improve performance effects. Although the computational speed of R-CNN has made great progress compared with the traditional machine learning target detection algorithm, researchers hope to make greater use of the target detection speed of deep convolutional neural network. Ross et al. once again proposed an extremely fast but phase detector -- YOLO.

YOLO is to directly divide the image into multiple sub-modules, and then extract features from the images in the sub-modules. Although this algorithm improves the computational speed, it ignores the location information, which is particularly important for small targets. Therefore, YOLO algorithm is not suitable for small target detection.

Compared with YOLO, SSD[3] is another single-stage target detector. It adopts the pyramid network structure for target detection, ensures the improvement of accuracy through the structure and large-scale data expansion, and improves the accuracy by setting the ratio of positive and negative targets at 1:3.

DING Peng [4] made a detailed description of the advantages and disadvantages of the three algorithms of convolutional neural network, and classified the target detection methods based on deep convolutional neural network into single-stage detector and two-stage detector. The former is typical of YOLO and SSD, while the latter is typical of RCNN. Comparatively speaking, the single-stage detector has more advantages in detection speed, so its performance in accuracy is not satisfactory accordingly. In particular, the YOLO algorithm is not ideal in the detection accuracy of small targets. Although the two-stage detector is slow, the detection accuracy is due to the single-stage detector.

3. Computer vision image recognition

3.1. Identification of multi-scale ship targets

The multi-scale phenomenon of ship targets in imaging is caused by the difference in visual distance between ships in the process of moving, especially in the case of small ships with less image feature information, and as mentioned above, the recognition accuracy of YOLO algorithm for small targets cannot reach the expected value. ZHANG Yu et al. [5] proposed a new loss function MLoU based on the minimum closure region between the target prediction frame and the real frame and the normalized distance of key points. The loss function can significantly improve the regression speed and accuracy of multi-scale ship target prediction box, and the ship identification accuracy of the six different ship types involved in the experiment reaches more than 97%. The Yolov3-MIOU algorithm improved based on YOLOv3[6] algorithm has a high recognition degree for ships of different scales and different types. In particular, the identification accuracy of fishing boats in inland rivers is improved by more than 3% compared with other algorithms, which is suitable for inland river shipping applications. However, it also has some limitations. There is a lack of research on complex environment in the algorithm data, and the influence of weather, illumination and other complex environmental factors is not taken into account.

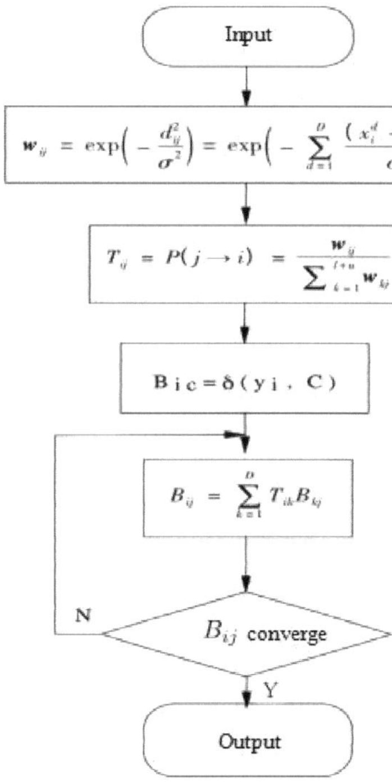

Figure 2. LPA algorithm description

3.2. Target recognition based on improved SSD algorithm

LIU Shuo [7] et al. proposed a maritime identification method for unmanned ships based on improved single-lens and multi-box detectors to solve the problem of inaccurate and low identification accuracy of ships in pictures acquired by the vision system during sea navigation. L2 regularization is added to the seventh layer of the traditional SSD network to improve it, so that the parameters of each convolution layer of SSD are more balanced, and the sensitivity of the model to local features is reduced, so as to improve the confidence of ship identification. The improved experimental results show that the proposed method has better maritime ship identification effect, and the accuracy of the improved algorithm is higher than that of the traditional SSD algorithm. In addition, LIU Wei et al. [8] showed in their research on SSD algorithm that SSD has better recognition effect for large objects. As mentioned in the previous paper, the identification accuracy of SSD, a single-stage target detector, is difficult to meet the actual demand for small targets. However, the identification effect of SSD placed in this paper for large targets such as freight ships is very encouraging.

3.3. Target recognition based on GA-LPA algorithm

Label Propagation algorithm[9] is a machine learning method for data classification, which is suitable for dealing with the labeling and tracking of ship images. XIONG Yong et al. [10] analyzed the parameter determination of LPA algorithm. In order to solve this problem, the LPA algorithm based on GA is proposed, and the method of ship image recognition using GA-LPA algorithm is established. The feasibility and efficiency of using GA-LPA algorithm for ship image labeling and recognition are verified by an example. However, the quality of the label algorithm not only depends on the algorithm itself, but also is affected by factors such as the degree of correlation between data. Therefore, its demand for data is large, and the carrying capacity and processing capacity of the computer are higher, otherwise it cannot achieve the desired higher efficiency.

4. Target recognition in complex environment

4.1. Target recognition based on improved YOLOv5 algorithm

LI Nan et al. [11] proposed an AAM-YoloV5 algorithm to solve the problem of missing detection in image ship detection under complex scenes. By introducing the adaptive attention module into the feature prediction structure of YOLOv5, the feature learning ability of ship target region is improved. At the same time, the loss function is optimized by referring to the attention mechanism, so that the model will pay priority to important information when updating parameters in the reverse direction, and finally improve the target detection effect in complex scene regions. The experimental results show that the improved AAM-YoloV5 detection algorithm significantly improves the recall rate and the quality of the output prediction box. From the perspective of detection efficiency, this algorithm adds additional pooling layer and convolution layer. Compared with the original YOLOv5 detection algorithm, the detection speed is slightly lower, but the average accuracy is significantly improved compared with the original YOLOv5 algorithm and Faster R-CNN algorithm, reaching 79.8%.

4.2. Target recognition based on shallow machine learning algorithm

In order to study the effective detection of ships using shallow machine learning methods in complex environments, BI Zhenbo et al. [12] believe that it is difficult to accurately achieve only one detection method, while current ship detection methods are mostly limited to the application of a method in video surveillance, satellite remote sensing, radar, AIS and other equipment. In addition, more than 20 derivative algorithms, such as Hough transform [13], straight line fitting [14], edge detection [15] and image segmentation [16], among the three categories of water boundary method, visual attention mechanism and motion area detection, were analyzed and studied. Bi Zhenbo et al. believed that the research of single algorithm was restricted by complex environment. At the same time, ship detection based on shallow machine learning relies on artificial design features, and the selection of features affects the final detection results. In particular, dynamic water background, weather conditions, platform shaking and natural light changes affect the detection effect, which increases the difficulty of the detection algorithm. Therefore, the ship target fusion detection should not only combine several kinds of target attribute information collected by other equipment, but also find out invariable features for reliability ship target detection, and combine several invariable features for classification detection. It can be combined with the features obtained by the same detection equipment for comprehensive detection, or based on the features obtained by different types of detection equipment for fusion detection, such as the comprehensive extraction of ship radar image features and background knowledge for comprehensive detection.

5. Comparison of different identification methods

Based on the characteristics of the above mentioned ship identification methods, it can be found that the research of automatic identification algorithm is developing towards large data volume and high identification target. See TABLE 1 for details.

Table 1. Comparison of different identification methods.

Recognition method	adaptive	date	accuracy
R--CNN	big	big	medium
YOLO	medium	medium	medium
SSD	big	medium	medium
Yolov3 -MIou	big	medium	high
Yolov5 -AAM	medium	big	high
GA -LPA	medium	big	high

6. Conclusion

In academic circles at present, for the ship image recognition, are presented in a single algorithm was improved to increase the recognition rate, but this premise is on the basis of strict requirements for other indicators, may be a large demand for data, for the bearing capacity of computer processing power and has certain requirements, may be a need to establish a test under ideal conditions, Applicability to complex environments will be reduced. This point was also mentioned in the study of LUO Mingming et al. [17].

In general, the research of ship image target detection is still a challenging task. It has developed from traditional artificial feature search to shallow machine learning and then to the target detection algorithm based on deep learning which is hot in recent years. The ship target detection method based on convolutional neural network will still exist and develop for a long time.

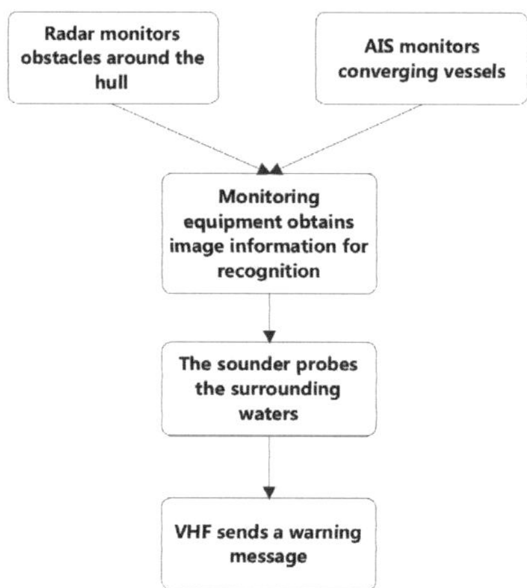

Figure 3. Multi-party platform framework

Based on the whole discussion, it is envisaged that a multi-party platform can be built in the future to ensure the safety of ship navigation, as shown in Figure3. Externally, radar, AIS, sounder and other equipment can be combined with computer vision processing to build a complete multi-information fusion detection framework to further improve the accuracy and supervision ability of ship detection. Internally, the dangerous driving behaviors proposed by LIU Changrong et al. [18] can be well detected by detecting and identifying the dangerous driving behaviors of drivers so as to give early warning. If such a comprehensive platform is built, it will also be of great help to the improvement of shipping security.

References

[1] Duan, A. Y. (2006) Application of Formal Safety Assessment(FSA) to the Shipping Traffic Safety Management of the Port and its Adjoining Waters[D].Huazhong University of Science and Technology.
[2] Girshick, R., Donahue, J., Darrell, T., Malik, J. (2015) Region-Based Convolutional Networks for Accurate Object Detection and Segmentation.[J]. IEEE transactions on pattern analysis and machine intelligence, 38(1): 142-158.
[3] Wang, S., Fei, S. M. (2019) SSD(Single Shot MultiBox Detector) Research and improvement of target detection algorithm [J].Industrial control computer, 32(04):103-105.
[4] Ding, P. (2019) Research on Object Detection Technology in Optical Remote Sensing Based on Deep Convolutional Neural Networks [D].University of Chinese Academy of Sciences .
[5] Zhang, Y., Kang, Z., Ma, J., Li, B. (2021) Multi-scale ship image target recognition method for inland navigation safety Monitoring [J/OL].Journal of Dalian Maritime University: 1-11.
[6] Liu, S., Ge, Y., Li, Y. Y., Hu, J. X., Ye, G. (2021) Study on Improvement of YOLOv3 Algorithm[J]. Journal of Physics: Conference Series, 1884(1).
[7] Liu, S., Ge, Y., Li, Y. Y., Hu, J. X., Ye, G. (2021) Recognition algorithm of unmanned ship at sea based on improved SSD[J].Sensors and Microsystems, 40(02):120-123.
[8] Liu, W., Anguelov, D., Erhan, D. (2016) SSD: Single shot multi- box detector[C] Proceedings of the European Conference on Computer Vision, Amsterdam, 21-37.
[9] Zhu, X. J., Ghahramani, Z. (2002) Learning from labeled and unlabeled data with label propagation: Technical report CMU- CALD-02-107[R]. Pittsburghers, USA: Carnegie Mellon University,1-8.
[10] Xiong, Y., Zhang, J. L., Huang, L. W. (2016) Study on ship image recognition based on GA-LPA algorithm[J].Journal of Guangxi University (Natural Science Edition), 41(02):554-561.
[11] Li, N., Ye, X. D., Wang, H., Huang, X. Y., Tao, S. F. (2021) A Ship Detection Method for SAR Image in Complex Scene based on Improve YOLOv5 [J/OL].The signal processing:1-16 [2021-11-22].
[12] Bi, Z. B., Zhang, S. Y., Yang, H., Wu, Y. H. (2021) Survey of Ship Detection in Video Surveillance Based on Shallow Machine Learning[J/OL].Journal of System Simulation:1-17[2021-11-22].
[13] Fefilatyev, S., Goldgof, D., Shreve, M., et al. (2012) Detection and tracking of ships in open sea with rapidly moving buoy-mounted camera system[J]. Ocean Engineering (0029-8018) , 54: 1-12.
[14] Ma, T., Ma, J., Fu, W. (2016) Sea-sky line extraction with linear fitting based on line segment detection[C].2016 9th International Symposium on Computational Intelligence and Design (ISCID). IEEE, 1: 46-49.
[15] Lu, J. W., Ren, J. C., Lu, Y., et al. (2006) A modified canny algorithm for detecting sky-sea line in infrared images[C].Sixth International Conference on Intelligent Systems Design

and Applications. IEEE, 2: 289-294.
[16] Liu, S. J., Jiang, M., Zhuang, L. (2011) A Fast and Effective Algorithm for Sea-sky-line Extraction in Infrared Image [J].Infrared technology, 33(4):230-232+240.
[17] Luo, M. M., Zhu, F., Wang, D. S. (2018) Overview of Automatic Identification of Ship Image Based on computer vision[J].Software Tribune, 17(07):5-8.
[18] Liu, C. R., Zhao, X. H., Liu, Q. H., Jin, W., Du, C. W., Zhang, Z. R. (2020) Design of Shipboard Dangerous Driving Behavior Detection System Based on Deep Learning SSD Model[J]. Ship Electronic Engineering, 40(12):139-143.

Journal of Physics: Conference Series **2387** (2022) 012002 doi:10.1088/1742-6596/2387/1/012002

A high precision phase extraction method in heterodyne interferometer based on FPGA

Wen Deng[1,]*, Xifeng Peng[1,2]

[1]The center of Metrology and Testing ,China Academy of Engineering Physics, Mianyang, China
[2]e-mail:pxf33@163.com

*Corresponding author e-mail: dwayne1213@163.com

Abstract. The high-precision phase extraction method directly determines the final measurement performance of the heterodyne laser interferometer. This paper proposes a digital quadrature phase-locked (DQPL) method with sub-nanometer precision based on FPGA hardware and compiled on the LABVIEW FPGA platform. DQPL method can accurately extract the phase difference of the two optical signals in the heterodyne laser interferometer. The validity and stability of the method have been verified by simulations and experiments. The experimentally obtained measurement resolution is 0.074 nm, and the measurement standard deviation obtained by the Monte Carlo method is 0.22 nm.

1. Introduction

In the fields of ultra-precision machining equipment and high-precision position metrology, heterodyne laser interferometer systems have been widely researched and applied due to their advantages of high resolution, strong anti-interference ability, and non-contact measurement [1-2]. The principle of heterodyne laser interferometer is a high-precision measurement technology derived from the Michelson interference optical path. Nowadays, the measurement technology based on the principle of laser interference has been able to achieve precise measurement of displacement, angle, velocity, etc [3-5]. In the field of laser heterodyne interferometry, in addition to the nonlinear error caused by the manufacturing process of each optical device in the interference optical path, the extraction accuracy of the heterodyne interferometric phase in the signal processing process is also another key factor that determines the measurement performance of the system [5-6].

Commonly used heterodyne interference signal processing methods mainly include frequency demodulation and phase demodulation. The frequency demodulation method essentially converts the Doppler frequency shift generated in the interference optical path into a pulse signal through an analog device, and then the measurement result is obtained through counting and processing. The resolution which can be achieved is about 0.31nm using this method [7-9].

The phase demodulation method uses analog or digital means to calculate and demodulate the phase of the interference signal, which is a general term for integer phase counting and fractional phase calculation. Terry G. McRae et al. proposed a high-speed signal acquisition and processing method based on the principle of quadrature lock-in amplifier for the linearization and minimization of interferometer cyclic error [10]. The lock-in amplifier and subsequent signal processing are all completed within the FPGA, achieving a cyclic error of about 100 pm for a commercial interferometer. Similarly, P Kochert et al. realized a heterodyne interferometer for static and quasi-static measurements with the static resolution of 1pm, at a maximum speed of 10mm/s [11].

In order to ensure the precise displacement measurement of sub-nanometer level, a simple and high-precision laser heterodyne interference phase extraction method is implemented in this paper. The method is based on FPGA hardware and developed on the Labview FPGA platform. Through model simulation and experimental test, within the range of Doppler velocity of ±0.5m/s, the measurement resolution obtained by the experiment is 0.074nm, and the measurement accuracy is 0.22nm obtained with Monte Carlo method, which can fully meet the requirements of ultra-precision machining equipment and high-precision position metrology.

2. Principle

2.1. Optical path

The optical path of heterodyne interferometer is shown in Fig. 1. The light, emitted from the linearly polarized laser source, is spilt into two beams by the first beam splitter (BS1). A beam of light passes through the first acousto-optic modulator (AOM1), the polarizer (POL) and the half-wave plate (HWP) successively. POL and HWP are used to guarantee the linear polarization direction coincides with the fast axis of the polarization beam splitter (PBS). Another beam, with the same polarization direction as the slow axis of PBS, is obtained using the same method. Thus, the two beams of polarized light converge at the second beam splitter (BS2). Under the action of the analyzer, the reference interference signal is detected by the first photodetector (PD1). Similarly, the measured interference signal is detected by the second photodetector (PD2) with the quarter-wave plate (QWP) and the retro-reflector (RR).

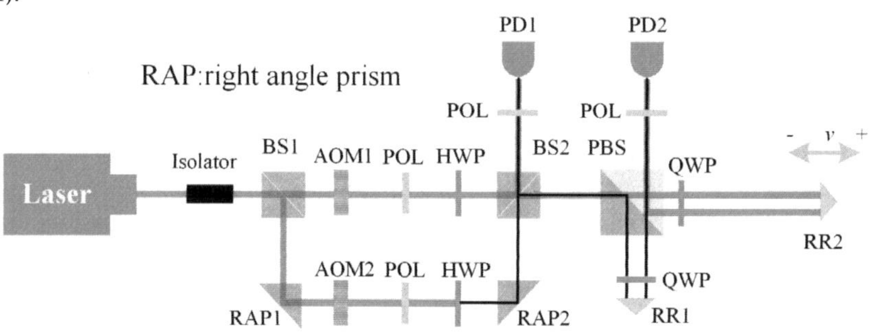

Figure 1. Schematic diagram of the optical path of heterodyne interferometer.

While the second retro-reflector (RR2) moves at the speed v towards the optical axis direction, according to the laser Doppler effect, the optical frequency of the measured arm can be expressed as:

$$f' = f_0 + f_1 + \frac{2v}{\lambda}, \tag{1}$$

where f_0 is the initial frequency of the light source, f_1 is the frequency shift amount of AOM1, λ is the wavelength.

Finally, the frequency of the measured interference signal detected at the photodetector PD2 is:

$$\begin{aligned} f_m &= f' - (f_0 + f_2) \\ &= (f_1 - f_2) + \frac{2v}{\lambda}, \end{aligned} \tag{2}$$

where f_2 is the frequency shift of AOM2.

According to the relationship between frequency and phase, the phase difference between the measured interference signal and the reference interference signal can be obtained as:

$$\begin{aligned}\Delta\varphi &= 2\pi\int_{t_1}^{t_2}[(f_1-f_2)+\frac{2v}{\lambda}]dt - 2\pi\int_{t_1}^{t_2}[(f_1-f_2)]dt \\ &= \frac{4\pi\Delta L}{\lambda},\end{aligned} \tag{3}$$

where L is the displacement in time $[t_1,t_2]$. It can be further rewritten as:

$$L=\frac{\lambda}{2}\cdot N, \tag{4}$$

where N is the number of phase difference fringes in time $[t_1,t_2]$. For heterodyne interferometer, the displacement is the product of half wavelength and the number of phase difference fringes. Since the laser wavelength is stable and known, the displacement variation can be obtained by counting the fringes of the interference phase difference.

2.2. Digital quadrature phase-locked method

The measurement of the displacement by the heterodyne interferometer is essentially the measurement of the phase. A digital phase demodulation method is applied in this paper, which is called the digital quadrature phase-locked method. This method no longer needs to convert the signal into a square wave and then perform pulse counting. Moreover, both the time and amplitude components of the signal are taken into account in DQPL method, with high resolution and anti-interference ability. The basic principle of the FPGA-based DQPL method is shown in Fig. 2.

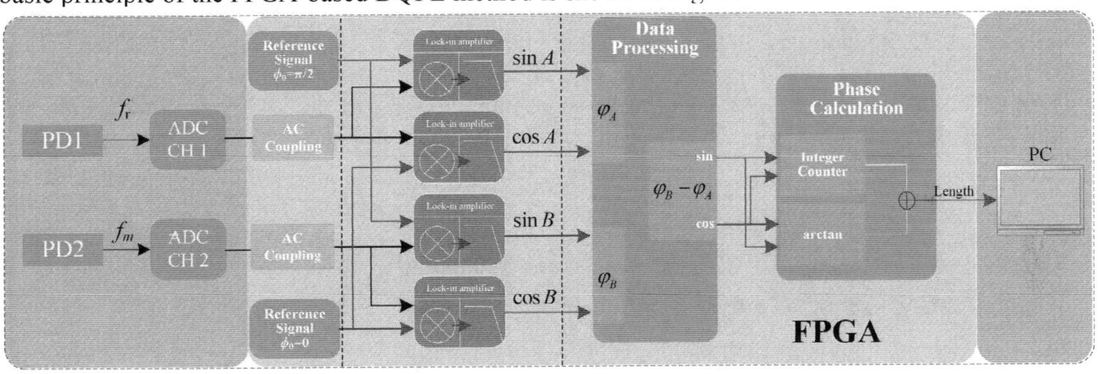

Figure 2. Principle of the FPGA-based DQPL method.

The reference interference signal I_R and the measured interference signal I_M are respectively detected in the reference arm and the measurement arm, which can be expressed as:

$$\begin{cases} I_R = r\cdot\sin[2\pi(f_2-f_1)t+\varphi_r] \\ \quad = r\cdot\sin(\omega_r t+\varphi_r) \\ I_M = m\cdot\sin[2\pi(f_2-f_1\pm\Delta f)t+\varphi_m] \\ \quad = m\cdot\sin(\omega_m t+\varphi_m), \end{cases} \tag{5}$$

where r, m are the amplitude of the interference signals; Δf is the Doppler frequency shift amount; φ_r, φ_m are the initial phase of the signals, respectively.

A quadrature reference signal with a phase difference of 90° is generated inside the FPGA chip, which is defined as:

$$\begin{cases} I_{\sin} = \sin(2\pi f_L t) = \sin(\omega_L t) \\ I_{\cos} = \cos(2\pi f_L t) = \cos(\omega_L t), \end{cases} \tag{6}$$

where f_L is the frequency of the reference signal generated by the FPGA. Generally it is selected as the difference between f_1 and f_2.

After the optical signal and the reference signal are processed by a lock-in amplifier, that is, after mixing and low-pass filtering, the signals of each channel are:

$$\begin{cases} I_A = \sin A = 0.5r\cos[(\omega_r - \omega_L)t + \varphi_r] \\ I_B = \cos A = 0.5r\sin[(\omega_r - \omega_L)t + \varphi_r] \\ I_C = \sin B = 0.5m\cos[(\omega_m - \omega_L)t + \varphi_m] \\ I_D = \cos B = 0.5m\sin[(\omega_m - \omega_L)t + \varphi_m] \end{cases} \tag{7}$$

The required phase difference is the phase difference between the measured interference signal I_M and the reference interference signal I_R:

$$\begin{aligned} \Delta\varphi &= (\omega_m t + \varphi_m) - (\omega_r t + \varphi_r) \\ &= [(\omega_m - \omega_L)t + \varphi_m] - [(\omega_r - \omega_L)t + \varphi_r] \end{aligned} \tag{8}$$

Formula (8) can be further written as:

$$\begin{cases} I_x = \sin(\Delta\varphi) = \dfrac{4(I_A \cdot I_D - I_B \cdot I_C)}{m \cdot r} \\ I_y = \cos(\Delta\varphi) = \dfrac{4(I_A \cdot I_C + I_B \cdot I_D)}{m \cdot r} \end{cases} \tag{9}$$

In the phase counting process, the integer part of the heterodyne interference phase difference is subdivided by four times in a complete cycle interval according to the relationship between I_x and I_y, which is called the four-subdivision counting method. Furthermore, the movement direction is defined by the change of I_y compared to I_x. The specific principle is shown in Fig. 3.

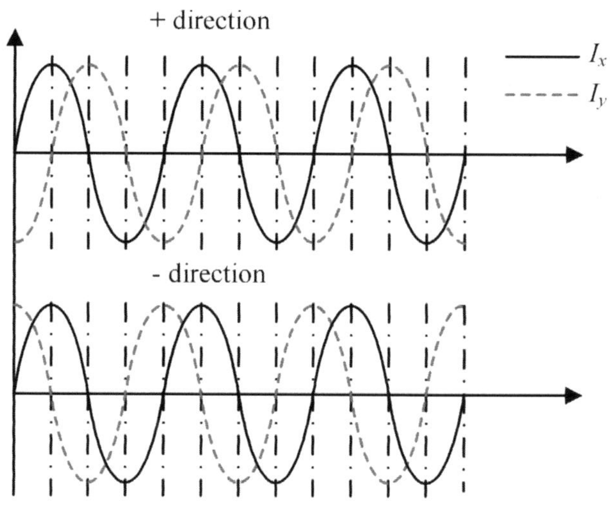

Figure 3. Integer phase counting principle.

In addition, the fractional part can be obtained by using the arctangent formula:

$$\Delta\varphi = \arctan\frac{I_x}{I_y} = \arctan\frac{I_A \cdot I_D - I_B \cdot I_C}{I_A \cdot I_C + I_B \cdot I_D} \tag{10}$$

The counting principle of the fractional part including the direction is shown in TABLE 1. in more detail, and the displacement L is finally calculated by formula (4).

Table 1. Counting criteria for the fractional part

Direction	Fractional part count value ($\Delta\varphi / \pi$)	
	$I_x \cdot I_y \leq 0$	$I_x \cdot I_y > 0$
+	$\arctan\left\|I_x / I_y\right\|$	$0.5 - \arctan\left\|I_x / I_y\right\|$
-	$\arctan\left\|I_x / I_y\right\| - 0.5$	$-\arctan\left\|I_x / I_y\right\|$

3. Simulations

In order to verify the correctness of the above method, a model is built in the Matlab Simulink module. In addition, in order to simulate the different signal quality in the actual situation, the signal-to-noise ratio (SNR) of the optical signals is changed, which is used to test the anti-interference ability of the method.The simulation parameter settings of the constructed Simulink model are shown in TABLE 2. Set the Doppler frequency shift amount at 2MHz, 1Hz, and 0 respectively, and the simulation results obtained are shown in Fig. 4. It can be seen that the final test results can perfectly measure the displacement change. Meanwhile, owing to the input optical signal has no noise signal, the obtained I_x and I_y signals both have good signal quality.

Table 2. Simulation parameters of the simulink model

Parameters	Describtion	Value
f1、f2	Frequency shift amount	150MHz、160MHz
fL	Reference signal frequency	10MHz
fs	Sampling rate	62.5MHz
r、m	Amplitude of optical signals	1V
φ_r、φ_m	Initial phase of the optical signals	0、3π/4
A	Amplitude of reference signals	1V
fpass、fstop	Pass frequency and cutoff frequency of low-pass filter	2MHz、5MHz
Δf	Maximum Doppler shift amount	2MHz

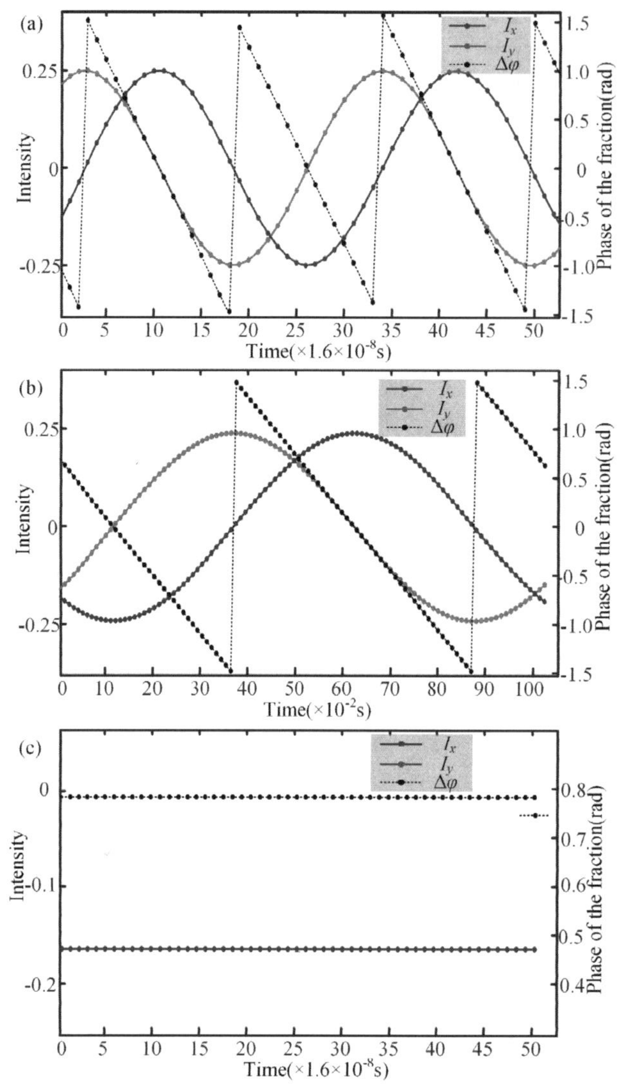

Figure 4. Simulation results at different Doppler shifts. (a) Δ f=2MHz; (b) Δ f=1Hz; (c) Δ f=0.

As shown in Fig. 5, in the case of different SNRs of the input optical signals, the error of the fractional part phase relative to the theoretical value under actual conditions is simulated while the Doppler frequency shift amount is 2MHz. It can be seen that, under the condition of (40~60) dB SNRs, the maximum phase error of the fractional part does not exceed 0.04rad. Since the phase of the integer part is obtained by the changing characteristics, it can be considered that there is no error in the counting of the integer part. So the phase difference of the fractional part obtained by the simulation is the error of the phase measurement. If a 532nm solid-state laser is used as the laser source, and the double-subdivided optical path shown in Fig. 2 is applied in the heterodyne interferometer, the maximum error of the displacement measurement is 1.7nm. The above simulation results not only prove the correctness of the method, but also verify its excellent anti-interference ability. Even in the case of normal signal quality, it can still obtain nanometer-level measurement accuracy.

IOP Publishing
2387 (2022) 012002 doi:10.1088/1742-6596/2387/1/012002

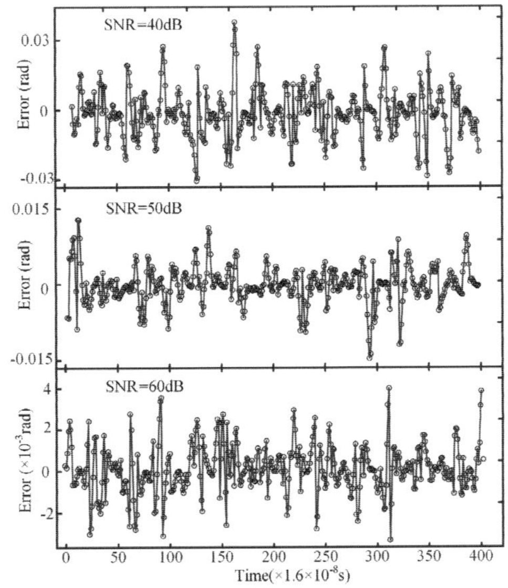

Figure 5. Phase extraction errors with different SNRs.

4. Experiments and analysis

4.1. Experimental results

The DQPL processing flow is compiled and tested based on the Labview FPGA module. In the experimental design, a signal generator (VICTOR 2060H, internal frequency accuracy: ±1ppm) is selected to simulate two optical signals in the interference optical path, and experiments are performed on the entire test system under static and dynamic conditions.

Under the static conditions, the frequency of the two analog signals of the signal generator is 10MHz. The initial phases of I_R and I_M are 0° and 20°, respectively. Doppler frequency shift amount is 0, and the experimental results are shown in the Fig. 6. In terms of the obtained phase variation trend, it can be seen that there are obvious random errors. Therefore, in order to effectively eliminate random errors in the measurement results, a moving average filtering algorithm is used for the phase measurement results, and the filter width is set to 1024 points. As can be seen from the Fig. 6, the filtering effect is very obvious, which effectively reduces the standard deviation of the measurement results. The standard deviation before filtering is 0.08nm, while the standard deviation after moving average filtering reaches 3.5pm.

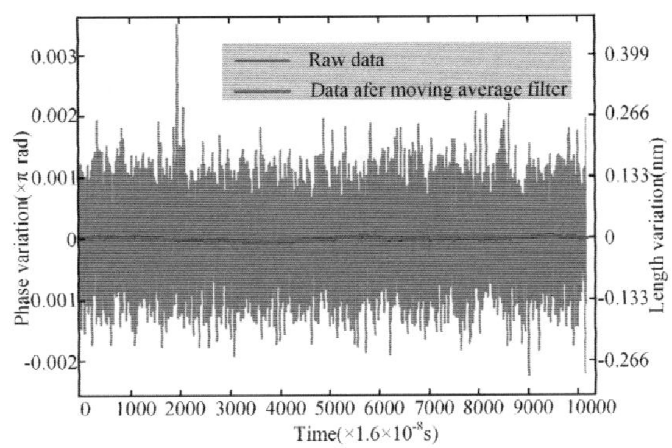

Figure 6. Phase measurement variation at static.

In order to test the measurement resolution of the system, keep the frequency of the two analog signals unchanged, and fine-tune the phases of I_M. The fine-tuning range is 20°~21°, and the step size is 0.1°. As can be seen from the Fig. 7, the measurement results can sensitively resolve 0.1° of angular adjustment. At the same time, the variation of the measurement results is compared with the theoretical value, and the maximum measurement accuracy obtained is 0.053nm. Therefore, the resolution of this method is better than 0.074nm, and the measurement accuracy is 0.053nm.

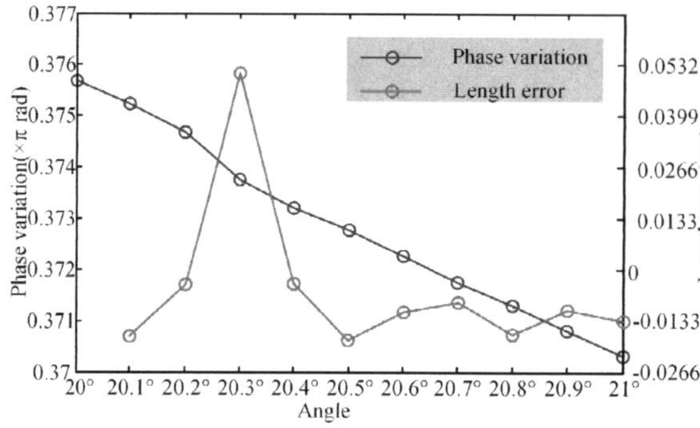

Figure 7. Resolution test results.

In the dynamic case, since the Cordic function has multiple cycles of delay in the FPGA implementation, it is impossible to compare the measured results with the theoretical values in real time. In order to reflect that the system can effectively extract the signal of the corresponding frequency within the Doppler frequency shift range of ±2MHz. In our test, the frequency of the reference optical signal was fixed at 10MHz, and the frequencies of I_M were adjusted to 12MHz, 11MHz, and 10.5MHz. Thus the Doppler frequency differences were 2MHz, 1MHz, and 500kHz. After each measurement, the spectrum of the obtained I_x is analyzed to obtain its spectrogram as shown below. It can be observed from the Fig. 8 that the frequency of the I_x signal is exactly the same as the theoretical Doppler frequency, indicating that the method can still accurately distinguish the heterodyne signal under dynamic conditions.

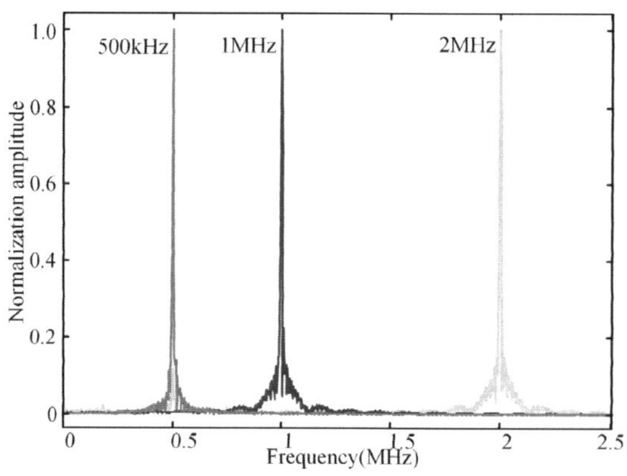

Figure 8. Spectrogram of Ix under different doppler frequency.

4.2. Uncertainty analysis

Under the premise of ignoring the local error of the signal generator, the main sources of uncertainty of the system are: the amplitude error of the data acquisition system, the synchronization error between the two channels, and the quantization error. Due to the complexity of the model, the uncertainty cannot be directly calculated by the GUM method, so the Monte Carlo method is used to analyze the measurement uncertainty. Under the condition that the input optical signal amplitude is 1V, the specific parameters are shown in TABLE 3.

Table 3. Parameters of Monte Carlo Uncertainty Analysis

Errors	Value	Type of probability distribution
Amplitude error of data acquisition system	±2.9mV	Uniform distribution
Synchronization error between two channels	120ps	Uniform distribution
Quantization error	±0.031mV	Uniform distribution
Number of samples	1000000	/

As shown in the figure, the standard uncertainty of this method is 0.0052 rad, and the 95% probability interval is (-0.0098rad, 0.0098rad). The corresponding standard uncertainty of length measurement is 0.22nm, which also confirms the measurement capability of the system with sub-nanometer precision.

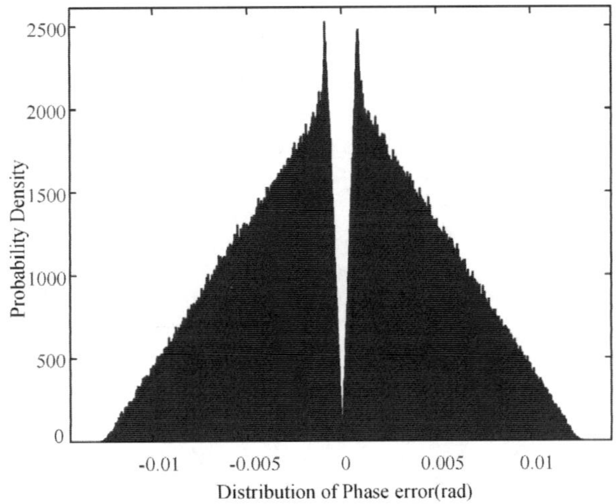

Figure 9. Spectrogram of Ix under different Doppler frequency.

5. Conclusion

In this paper, we adopt a DQPL method which is applied to a heterodyne laser interferometer. Whether in static or dynamic conditions, the method can accurately extract and distinguish the phase difference of two optical signals, with a good anti-interference ability. The measurement resolution obtained by the experiment is about 0.074nm, the measurement standard deviation obtained by the Monte Carlo method is 0.22nm. The confidence interval with 95% probability of the phase measurement error is (-0.098rad, 0.098rad). In addition, the method can easily implement a multi-channel heterodyne interferometric signal processing system on one FPGA card, which can be further applied to multi-wavelength interferometry and other applications requiring multi-channel scenarios.

References

[1] Zhang, Z., Cheng, Z., Qin, Z and Zhu, J., (2007) "Micro-Gal laser absolute gravimeter based on

high precision heterodyne interferometer," Chinese Optics Letters, vol. 5, no. 6, pp. 344–346.

[2] Liu. S., Zhong, X., (2021) " Study and simulation of heterodyne optical phase-locked loop for inter-satellite laser interferometer," Journal of Physics: Conference Series, vol. 1952, p. 022013.

[3] Lou, Y., Li. Z., Yan, L., Chen, B and Jian, J., (2021) "A phase differential heterodyne interferometer for simultaneous measurement of straightness error and displacement," Optics Communications, vol. 497, p. 127195.

[4] Wang, X., Su, J., Yang, J., Miao, L and Huang, T., (2021) "Investigation of heterodyne interferometer technique for dynamic angle measurement: error analysis and performance evaluation," Measurement Science and Technology, vol. 32, no. 10, p. 105016.

[5] Straube, G., Calderón. J. S. F., Ortlepp, I., Fül, R and Manske, E., (2021) " A heterodyne interferometer with separated beam paths for high-precision displacement and angular measurements," Nanomanufacturing and Metrology, vol. 4, no. 3, pp. 200–207.

[6] Hou, W., (2006) "Optical parts and the nonlinearity in heterodyne interferometers," Precision Engineering, vol. 30, no. 3, pp, 337–346.

[7] Oldham, N. M., Kramar, J. A., Hetrick, P. A and Teague, E. C., (1993) "Electronic limitations in phase meters for heterodyne interferometry," Precision Engineering, vol. 15, no. 3, pp. 173–179.

[8] Yim, B., Eom, C. I., Kim, S. W., (2000) " Dual mode phase measurement for optical heterodyne interferometry," Measurement Science and Technology, vol. 11, pp. 1131–1137.

[9] Demarest, C., (1999) "High-resolution, high-speed, low data age uncertainty, heterodyne displacement measuring interferometer electronics," Measurement Science and Technology, vol. 9, pp. 1024–1030.

[10] McRae, T. G., Hsu, M. T., Freund, C. H., Shaddock, D. A., Herrmann, J., Gray, M. B., (2012) "Linearization and minimization of cyclic error with heterodyne laser interferometry," Optics letters, vol. 37, no. 13, pp. 2448–2450.

[11] Köchert, P., Flügge, J., Weichert, C. H., (2012) "Phase measurement of various commercial heterodyne He–Ne-laser interferometers with stability in the picometer regime," Measurement Science and Technology, vol. 25, no. 7, p. 074005.

Filtering method of airborne gravity gradiometer signal based on EMD

Xuewu Qian*[1,2], Liye Zhao[1,3]

[1]School of Instrument Science and Engineering, Southeast University, Nanjing, China
[2]School of Electronic Engineering, Jinan Vocational College, Jinan, China;
[3]e-mail: liyezhao@seu.edu.cn

*Corresponding author e-mail: sdqxwu@163.com;

Abstract. Gravity gradiometer sensor performance, platform instability and electronic circuit noise can cause serious influence on the weak gravity gradiometry signal, and the noise level is usually one or several times of gravity gradient signal. This paper put forward a method to resolve this problem according to the characteristics of the signal. First of all, filtrate the gravity gradient signal with a band-passing filter. Then, demodulate the output signal of the band-passing filter, at last, remove the gravity gradiometry noise with the empirical mode decomposition (EMD) method. Then the gravity gradiometer signal was tested on the hardware-in-the-loop simulation platform. The experiment results verify that the various harmonic noise can be removed with the method proposed above, and the gravity gradient signal accuracy and gravity gradient space resolution is further improved, the good properties of the method is of high practice and maneuverability.

1. Introduction

Gravity gradient reflects the spatial differential of gravitational acceleration, which can reflect more details of the field source body and has a higher resolution than gravity. Gravity gradient measurement is of great significance to the development of space science, earth science, geological science and inertial navigation, etc[1-2]. At the same time, gravity gradient measurement is considered to be one of the most effective means of resource exploration. It has important application prospects in the fields of research, geodesy, and exploration of oil and gas resources. Aerial gravity gradient measurement has important application value for basic data acquisition in mountainous areas, uninhabited areas, islands and coastal continental shelves [3-4]. In recent years, the domestic gravity gradiometer instrument(GGI) has been developed rapidly, and the aviation gravity gradiometer has already possessed the ability of aviation gravity flight measurement. After several years of continuous experiments, the domestic aviation gravity gradiometer will certainly be able to be competent for practical application tasks[5].

Rotating accelerometer gravity gradiometer is composed of a variety of mechanical devices, accelerometer performance, accelerometer installation accuracy, rotating mechanism stability, temperature control performance, platform stability and other factors will have a greater impact on the gravity gradient signal[6-8], these influencing factors are directly reflected in the output signal of the GGI. The noise contained in the gravity gradient signal will have a serious impact on the demodulation of the gravity gradient, and will also affect the real-time adjustment of the accelerometer scale factor. Therefore, it is very important to study the gravity gradient signal processing method. Because the gravity gradient signal captured by the aviation gravity gradiometer contains various frequency gravity gradient information, the ordinary digital filter has a smoothing effect, which will affect the high-frequency information of the gravity gradient. The wavelet transform method is widely used in signal denoising[9-12], but it is difficult to determine the wavelet basis function and the number of decomposition times, so it can only be done by trial and error method.

In this paper, according to the characteristics of gravity gradient signal, the FIR zero-phase band-pass digital filter is used to filter the output signal of the GGI, a and then the filtered signal is demodulated to obtain the gravity gradient signal. Finally, the EMD method is used to extract the gradient signal from the gravity gradient data. , through hardware-in-the-loop simulation experiments, it shows that the gravity gradient signal processing method proposed in this **paper** can effectively improve the gravity gradient demodulation accuracy and has good application value.

2. Processing method of airborne gravity gradiometer

In the actual working environment, due to various factors, the output signal of the GGI contains various components of interference noise, which has a serious impact on the high-precision gravity gradiometry. Gravity gradient signal is filtered out by FIR digital band-pass filter, and the gravity gradient information is decomposed by EMD method, and finally the gravity gradient component is obtained. The specific process of gravity gradient signal processing is shown in Fig. 1.

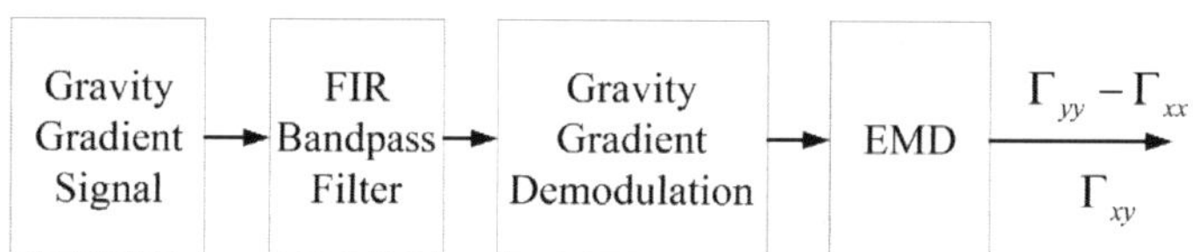

Figure 1. Block diagram of gravity gradiometer signal

2.1. FIR digital bandpass filter

The frequency impulse response function of an ideal digital bandpass filter as defined below:

$$H_d(e^{j\omega})=\begin{cases} e^{-j\omega\tau} & \omega_{c1}\le\omega\le\omega_{c2} \\ 0 & 0\le\omega<\omega_{c1} \text{ and } \omega_{c2}<\omega\le\pi \end{cases} \tag{1}$$

where, ω_{c1} is the low cutoff frequency of the bandpass filter, ω_{c2} is the high cutoff frequency of the bandpass filter, and τ is the sampling delay. FIR filters can be divided into four forms according to whether the unit impulse response $h(n)$ is parity symmetric and whether the length N is an odd or even number. When the impulse response $h(n)$ is even symmetric and the length N is odd, various filter functions can be realized, so this paper chooses the impulse response $h(n)$ to be even symmetric and the length N of the impulse response is odd. The unit impulse response function of the filter is:

$$\begin{aligned} h_d(n) &= \frac{1}{2\pi}\int_{-\pi}^{\pi} H_d(e^{j\omega})e^{j\omega n}d\omega \\ &= \frac{\sin\left(\omega_{c2}(n-\tau)\right)}{\pi(n-\tau)}-\frac{\sin\left(\omega_{c1}(n-\tau)\right)}{\pi(n-\tau)} \end{aligned} \tag{2}$$

In order to ensure the linear phase, take $\tau=(N-1)/2$, use the window function $w(n)$ with finite length N to intercept $h_d(n)$ at N points, and the intercepted impulse function $h(n)$ can be expressed as:

$$h(n)=h_d(n)w(n) \quad ,n=1,2,\dots,N-1 \tag{3}$$

2.2. Gravity gradient signal demodulation method

According to Newton's second law, the ideal output signal expression of the GGI can be expressed as:

$$E_{out} = 2RKK_I\left(\Gamma_{yy} - \Gamma_{xx}\right)\sin 2\omega t + 4RKK_I\left(\Gamma_{xy}\right)\cos 2\omega t \tag{4}$$

where, R is the distance from the center of the accelerometer to the center of the disc, K is the signal amplification gain, K_I is the scale factor of the accelerometer, and ω is the rotational angular frequency of the disc. Gravity gradient component demodulation formula:

$$\begin{cases} \left(\Gamma_{yy} - \Gamma_{xx}\right) = \dfrac{\int_0^T E_{out}\sin 2\omega t dt}{TRKK_I} \\ \left(\Gamma_{xy}\right) = \dfrac{\int_0^T E_{out}\cos 2\omega t dt}{2TRKK_I} \end{cases} \tag{5}$$

where, E_{out} is the output signal of the gradiometer, and T is the rotation period of the disc of the gravity gradiometer. In practical engineering applications, hardware demodulation and software demodulation are generally used. Hardware demodulation is mainly carried out by phase-sensitive detection circuit, and software demodulation is mainly carried out by digital signal processing method. Software demodulation needs to discretize the gravity gradient analog signal, and then the signal at the fixed frequency is processed by continuous addition and summation. Therefore, the integral form of the gravity gradient demodulation formula (5) is transformed into the continuous addition and summation demodulation form. The demodulation formulas of the two gravity gradient components are similar, only given Digital demodulation expression for gradient component ($\Gamma_{yy} - \Gamma_{xx}$):

$$\left(\Gamma_{yy} - \Gamma_{xx}\right) = \frac{\sum_{i=1}^{N} E_{out}(i)\sin\left[2\omega\Delta t(i-1)\right]}{NRKK_I} \tag{6}$$

where, E_{out} (i) is the signal of the gradiometer after sampling, $\triangle$t is the sampling time interval, and N is the number of sampling data when the disk of the gravity gradiometer rotates for one cycle.

2.3. Empirical mode decomposition method

The EMD method is a new signal analysis and processing method proposed by Huang et al. It can deal with dynamic data with non-stationary and nonlinear characteristics. EMD adopts the "sieve" method to decompose the original signal into a series of eigenmode functions (IMFs) and residual terms with different frequencies (different feature scales) arranged from high to low, and the IMF component with the lowest frequency represents the original signal. Change trend or mean value, with good adaptability. In order to decompose different signals quickly and efficiently, the eigenmode function IMF must satisfy two conditions: ① In the whole signal sequence, the difference between the number of extreme points and the number of zero-crossing points is not greater than 1. ② At any time, The mean value of the upper envelope formed by the local maximum points and the lower envelope formed by the minimum points is 0.

Using the EMD method to decompose the gravity gradient data calculated by equation (6), we get:

$$\Gamma = \sum_{i=1}^{N}\Gamma_i + \Gamma_r \tag{7}$$

where, Γ_i is the IMF component of the i_{th} group of gravity gradient signals, representing the signal components of different characteristic scales in the original gradient signal Γ; Γ_r is the residual component of the gravity gradient, which reflects the gravity gradient change in the actual gravity gradiometer measurement. Therefore, after EMD decomposition of the gravity gradient signal Γ containing large noise, the sum of the stationary components of N groups of different frequency components and a set of trend terms can be obtained, and the trend term reflects the size of the real gravity gradient signal.

3. Computer simulation verification

Due to the influence of various factors, the output signal of the gravity gradiometer is very complex, and the signal-to-noise ratio(SNR) of the gravity gradient signal is generally below 0dB. In order to verify the effectiveness of the processing method in this paper, the SNR is set to 0dB, -10dB, -20dB, -30dB , in the case of -40dB, compare the processing effects of 100s smoothing, 200s smoothing and EMD methods on the signal.

The rotation frequency of the gravity gradiometer is 0.25hz. According to the working principle of the gravity gradiometer of the rotating accelerometer, the signal frequency containing the gravity gradient component is 0.5Hz. In order to reflect the authenticity of the output signal of the gravity gradiometer, harmonic signal components of 0.25hz, 1Hz and 1.2hz are added, and the signal sampling rate is 10Hz. The output signal form of the gravity gradiometer is set as follows:

$$\begin{aligned} s(t) &= 100\sin 0.5\pi t + 400\sin \pi t + \\ & 50\sin 2\pi t + 20\sin 2.4\pi t + n(t) \end{aligned} \tag{8}$$

where, $n(t)$ is gaussian white noise, and the noise energy is changed by setting the SNR. It can be seen from the above signal expression that to extract the 0.5Hz signal, it is necessary to use a band-pass filter for filtering, and then perform amplitude demodulation on the filtered signal. The amplitude results of demodulation are processed by 100s smoothing, 200s smoothing and EMD methods respectively. The processing results are shown in Fig. 2 and Table 1. It can be seen from the results in Fig. 2 and Table 1 that with the decrease of the SNR, the processing effect of the above methods gradually deteriorates, but the processing effect of the EMD decomposition method is obviously better than that of smooth filtering, and the original gravity is also obtained. At the same time, it can also be obtained that after the output signal of the original gravity gradiometer is processed by band-pass filter, the gravity gradient data is more stable and more accurate.

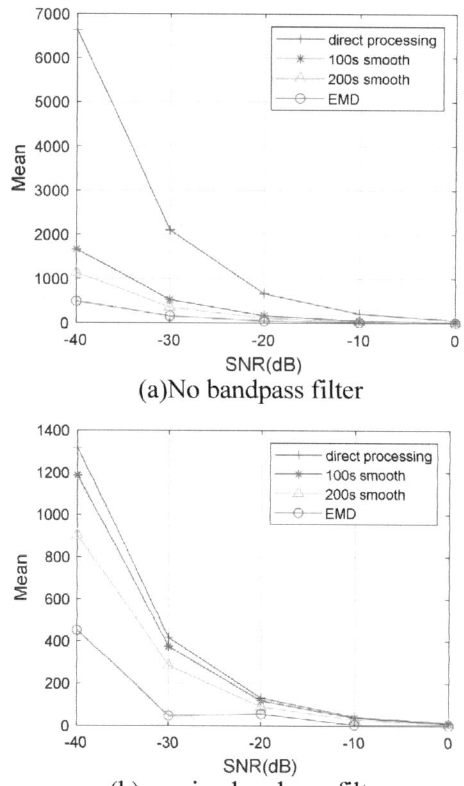

Figure 2. The effect comparison of different processing methods

Table 1. The effect comparison of different processing methods

whether to pass zero-phase bandpass filtering	signal to noise ratio	original signal	100s smooth	200s smooth	EMD
NO	0dB	66.397	16.621	11.347	4.421
	-10dB	209.97	52.559	35.884	16
	-20dB	663.99	166.21	113.47	49.116
	-30dB	2099.7	525.59	358.84	155.32
	-40dB	6639.9	1662.1	1134.7	487.67
YES	0dB	14.276	11.744	8.7557	1.579
	-10dB	42.238	37.332	28.251	4.8861
	-20dB	132.14	118.48	89.976	56.194
	-30dB	416.94	375.15	285.19	47.83
	-40dB	1317.7	1186.8	902.51	452.12

4. Hardware-in-the-Loop Simulation

According to the working principle of the rotary accelerometer gravity gradiometer, a high-performance computer, high-precision programmable power supply, ultra-low-noise current amplifier, ultra-low-noise voltage amplifier, multiplex switch and high-precision programmable digital voltmeter are used to build a gravity gradiometer. Gradient signal hardware-in-the-Loop simulation analysis system, sending the simulation data to the semi-physical simulation analysis system through the computer can simulate the gravity gradient signal in all aspects, which is helpful for further detailed research on the signal characteristics of the gravity gradiometer, the accelerometer scale factor feedback adjustment algorithm and the signal processing method. For the physical diagram of the hardware-in-the-Loop simulation analysis system, please refer to Reference [13].

The hardware-in-the-Loop Simulation parameters of the gravity gradient signal are as follows: the density of the sphere is 18000kg/m^3, the mass is 1000kg, the coordinates of the center of the sphere are (0.8, 0, 0) m, the rotation period of the disc is 4s, and the baseline distance of the accelerometer pair is 0.2m, the sampling rate of the simulation analysis system signal is 10Hz, and according to the law of universal gravitation, the four accelerometers on the disc sense the gravitational acceleration of the ball to the accelerometer, and this value is the ideal accelerometer signal. The theoretical value of the gravity gradient component produced by the above parameter sphere is: $(\Gamma_{yy} - \Gamma_{xx}) = -393$Eu. The accelerometer data sampled from the gravity gradiometer prototype is used as the signal source of the semi-physical simulation system of gravity gradient, and the output signal of the simulation system is sampled and processed. Fig. 3 is the original signal and its signal spectrum of the output signal of the semi-physical simulation analysis system of the gravity gradiometer.

After SNR estimation, the SNR of the original gravity gradient signal is about -22dB. It can be seen from Fig. 3 that the gravity gradient signal is completely submerged by noise. It can be obtained by spectrum analysis that the 0.5Hz gravity gradient signal is more obvious. In terms of distribution, the noise is not completely white noise, and the low-frequency noise has a large amplitude, which is mainly caused by the electronic circuit, mainly $1/f$ noise, and the low-frequency interference cannot be completely removed by the low-pass filter. Since the gravity gradient signal is sensitive to the phase, in order to keep the phase of the filtered signal unchanged, a zero-phase digital band-pass filter is used to process the original gravity gradient signal.

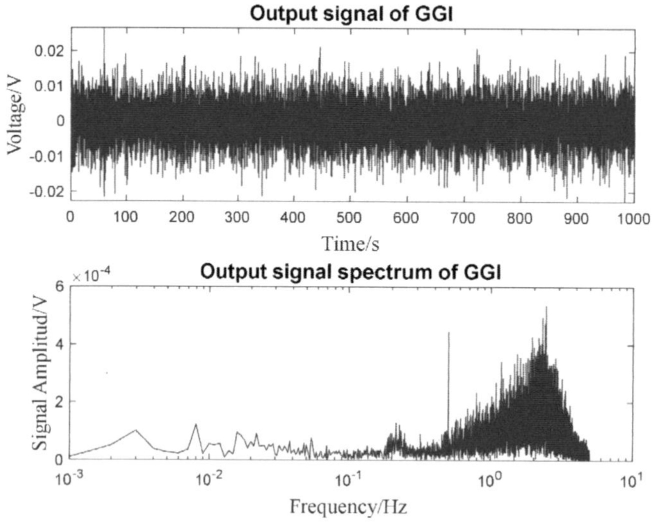

Figure 3. The output signal and its spectrum on the hardware-in-the-loop simulation platform for GGI

The output signal of the gravity gradient hardware-in-the-Loop simulation system is first filtered by a FIR zero-phase digital band-pass filter, and the gravity gradient signal after filtering is subjected to gravity gradient demodulation. Different methods are used to process the gravity gradient data. The processing results are shown in Fig. 4. It can be obtained from the processing results, compared with the traditional smoothing filtering effect, the EMD processing method can significantly reduce the mean square error of the gravity gradient information and significantly improve the accuracy of the gravity gradient signal.

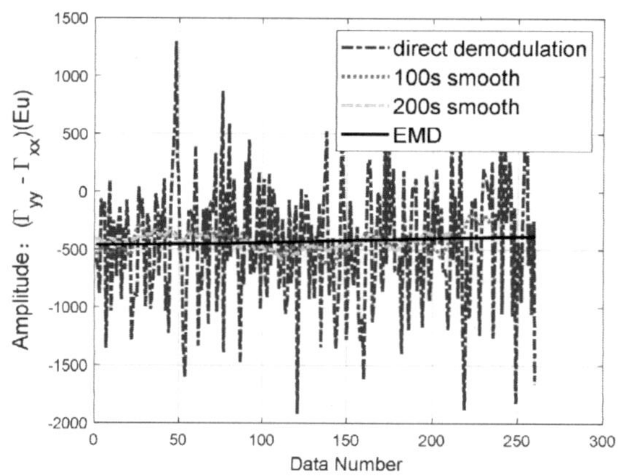

Figure 4. The comparison of the filtering results of gravity gradiometer signal

5. Conclusions

The output signal of the airborne gravity gradiometer contains various noises, and the gravity gradient signal is very weak and completely submerged by the noise, which brings great challenges to the high-precision aviation gravity gradient measurement. In this paper, according to the characteristics of gravity gradient signal, the FIR zero-phase digital band-pass filter is used to filter the output signal of the gravity gradiometer, and then the filtered signal is subjected to gravity gradient demodulation. Through computer simulation and hardware-in-the-loop simulation experiments, it is verified that the proposed processing method can effectively improve the accuracy of gravity gradient measurement and has high engineering application value.

Acknowledgment
This work was supported by the China Postdoctoral Science Foundation funded project (Grant No. 2020M671287), Jiangsu Planned Projects for Postdoctoral Research Funds (Grant No. 2020Z387), Shandong provincial Natural Science Foundation, China (Grant No. ZR2020MF089)

References

[1] Cai, T. J., Zhou, B. L., (1999) Status and prospects of gravity gradiometers. Journal of Chinese Inertial Technology, vol. 7, pp. 39-42.
[2] Dransfield, M. H., Christensen, A. N. (2013). Performance of airborne gravity gradiometers. The Leading Edge, 32(8), 908-922.
[3] Galder, C. V. (2014) Improved resolution of fixed-wing airborne gravity gradiometer surveys. Seg Technical Program Expanded Abstracts, 5183.
[4] Yamanaka, H. (2013) HeliFALCON® Airborne Gravity Gradiometer Data Acquisition in Rugged Terrain. Global Meeting AbstractsSociety of Exploration Geophysicists, 580.
[5] Zhang, L. D., Sun, F. P., (2021) The Key Technologies Development and Analysis of Rotating Accelerometer Grabity Gradiometer. Geomatics Science and Technology, vol. 9, pp. 26-39.
[6] Qian, X., Zhu, Y. (2019) Self-gradient compensation of full-tensor airborne gravity gradiometer. Sensors, vol. 19, pp. 1950-1963.
[7] Lee, J. B. (2001) FALCON gravity gradiometer technology. Exploration Geophysics, vol. 32, pp. 247-250
[8] Martinez, C. D. (2015) Quantitative interpretation of airborne gravity gradiometry data for mineral exploration. Dissertations & Theses - Gradworks.
[9] Lyrio, J. C. S. O. (2001) Wavelet denoising of gravity gradiometry data. Seg Expanded Abstracts, vol. 1, 1474.
[10] Birkelo, B. (2014) Noise reduction procedures for gravity gradiometer data. Seg Technical Program Expanded Abstracts, 5183.
[11] Cesar, J., De, S., Lyrio, O. (2004) Efficient automatic denoising of gravity gradiometry data. Geophysics, vol. 69, pp. 772-782.
[12] Pajot, G., Viron, O. D., Diament, M. (2008) Noise reduction through joint processing of gravity and gravity gradient data. Geophysics, vol. 73, I23.
[13] Qian, X. W., Cai, T. J. (2016) Data processing method for rotating accelerometer gravity gradiometer. JOURNAL SOUTHEAST UNIVERSITY(Natural Science Edition), vol. 46, No.4, pp. 708-711.

Safety DV-hop positioning algorithm of underwater acoustic sensor networks based on Nesterov accelerated gradient descent method

Qinyi Zeng[1], Ziyu Zhou[1], Ye Chen[1,2,*] and Tian Ji[1]

[1]School of Applied Science and Technology, Hainan University, Haikou 570228, China
[2]School of Electrical Automation and Information Engineering, Tianjin University, Tianjin 300072, China

First author: Qinyi Zeng (zengqinyi@hainanu.edu.cn)

*Correspondence: Ye Chen (chenye@hainanu.edu.cn)

Abstract. Before sensing and transmitting data, the underwater nodes need to calculate and estimate their location information through anchor nodes with known locations. The ranging-free positioning algorithm DV-hop can be applied to the underwater wireless sensor network in a relatively poor environment, which is not required high positioning accuracy. However, some anchor nodes may be destroyed and utilized to deter unknown nodes' precise positioning in the underwater confrontation scenario. Therefore, we propose a computationally efficient algorithm for determining the location of sensor nodes in harsh and confrontational underwater environments and resisting non-cooperative network attacks. This algorithm integrates Nesterov accelerated gradient descent (NAG) into the DV-hop positioning algorithm and prunes the data by selecting the measurement inconsistency to get better positioning accuracy and higher security. The simulation experiment confirms that our proposed algorithm has better measurement accuracy in the underwater environment than the existing algorithms.

1. Introduction

Underwater Wireless Sensor Network (UWSN) [1] is widely available for underwater environment detection, marine animal monitoring, marine resource development and protection, military confrontation, and navigation monitoring. Usually, ground control stations, surface buoys, underwater sensor nodes, bottom nodes, and underwater vehicles are deployed in UWSN to perform collaborative monitoring tasks. However, the existing algorithms of wireless sensor networks on the ground cannot be directly applied to UWSN, even though the localization of wireless sensor networks on the land has been widely researched. Positioning is a basic problem in UWSN [2]. According to our acknowledgment, the impact of security attacks has not been considered in the underwater DV-hop algorithm.

In the current research, the location of unknown underwater nodes is often determined by anchor nodes with known location information. The underwater positioning methods can be divided into ranging-based and ranging-free positioning according to the different technical means[3]. In this paper, we focus on the latter. More importantly, some anchor nodes may be destroyed and utilized to provide false information in confrontation scenarios, thus hindering the effective positioning of unknown nodes and damaging the operation of underwater sensor networks. Therefore, there should be necessary to design a safe underwater DV-hop localization algorithm. The secure location algorithm should be robust to malicious underwater attacks and can still measure the position of underwater nodes even under attack. Meanwhile, the designed underwater mobile localization algorithm should be safe and efficient due to the limited computing resources in the sensor nodes.

2. Related works

In [4], Niculescu and Nath proposed a DV-hop algorithm for positioning to locate unknown nodes. In this algorithm, the average receiving distance per hop and the minimum hop number of anchor nodes is proposed, which can be used to estimate the position of the nodes. Many improvement measures for the DV-hop algorithm are proposed to acquire more reasonable positioning accuracy.

Given the positioning process of the traditional DV-Hop localization algorithm, [5] pointed out that the insufficient estimation of average hop distance between anchor nodes and unknown nodes that appear in the process of the localization and the unknown node coordinate beyond the existing bound of a given area leading to the positioning error.

[6] proposed a distributed active multi-hop localization algorithm, which extends the false anchor to help position and collects the average multi-hop distance in the whole network as the estimated distance between the unknown and anchor node to improve the positioning accuracy. Then, the algorithm uses maximum likelihood estimation to calculate the position coordinates of unknown nodes.

[7] proposed an improved DV-hop algorithm, which improves the distance estimation based on the error of beacons in the estimation stage of the DV-hop algorithm and revises the distance estimated by the multiple beacon error weighted method. In [8], the algorithm improved the average distance by using the inverse distance weighting method based on [7] to improve the accuracy.

[9] proposed another improved DV-hop algorithm: anchor nodes refine the average hop-size by minimum mean square error and modify it by error factor. Then, a dynamic weight coefficient modifies the average hop size between the unknown node and the anchor node. Finally, the coordinates of unknown nodes are estimated by the 2D hyperbolic localization algorithm.

[10] proposed an improved DV-hop localization algorithm that corrects the estimated distance between the unknown node and different anchor nodes based on fractional hop count information and relatively accurate coordinates of the anchor nodes information and uses the improved Differential Evolution algorithm to get the estimated location of unknown nodes to further reduce the localization error.

An enhanced variant of the DV-hop algorithm [11] is utilized to estimate the distance between the unknown node and the anchor node to give the algorithm better stability and faster convergence speed based on high precision. Then the position estimation stage is formulated as the minimization problem solved by the recently developed squirrel search algorithm (SSA).

In [12], the improved cosine similarity parameter is used to measure the similarity between path pairs. The path from the unknown node to the target anchor point is selected to independently calculate the average jump distance of the node anchor point path. Then, the hybrid algorithm of improved particle swarm optimization and simulated annealing is used to improve the initial positioning accuracy of unknown nodes.

According to my acknowledgment, the existing literature does not consider the safe location of the DV-hop algorithm underwater in attack scenarios.

3. Problem description

The scenarios we envisaged may involve either the movement and reconfiguration of the network or the deployment in an unreachable area where the UWSN can operate in a fully distributed manner. In this case, nodes can use the network hop-by-hop propagation ability to obtain distance information. In general, the nodes actively exchange information and contact in one hop. Assuming that the nodes can obtain depth information, so the localization of the three-dimension is transformed into a two-dimension. Once any node estimates the distance information of at least three anchor nodes, it can obtain its position in the plane. Therefore, the DV-hop algorithm [13] is used here.

In this paper, a challenging situation is considered. Specifically, assuming that there exist malicious attack nodes in the network, which affect the unknown node localization by a non-cooperative attack. Since the signal strength of the underwater acoustic signal can be acquired through software access, the distance between the adjacent nodes can be estimated. However, errors caused by the underwater environment and network attacks will affect these measured values.

Therefore, we adjusted the effective NAG method, found the confidence hop count by gradient descent method, and then determined the relative position of each node in the whole sensor network, which reduced the positioning error caused by network attacks. Hence, we apply the algorithm in a distributed way. In this algorithm, each unknown node obtains and updates the current estimated value of the relative position of the anchor node and can receive the hop count information between itself and the anchor node.

4. System model

4.1. Location model

The proposed algorithm first calculates the minimum number of hops between unknown nodes and beacon nodes and then estimates the average distance of each hop. Therefore, the distance between the unknown node and the anchor node is obtained by estimation, that is, the minimum number of hops multiplied by the average distance of each hop. Finally, the NAG method is used to calculate the localization of the unknown nodes.

The process is as follows: the anchor node broadcasts a beacon to the UWSN, and the beacon includes a parameter with an initial value of 1, which is used to indicate the hop count and the location information of the anchor node. This beacon is propagated in a flooding manner in underwater sensor networks, and the hop count increases by one each time the beacon is forwarded. On the other hand, the receiving node saves the beacon with the smallest hop value among all beacons. In addition, the receiving node collects the information of the anchor node and abandons the beacon of the same anchor node with a significant hop value. Through this mechanism, all nodes in UWSN get the minimum hop count of each anchor node. Besides, each node maintains a vector $\{X_n, Y_n, h_n\}$ and exchanges updates only with adjacent nodes. Once a node obtains the distance from other nodes, it will estimate the average size of one hop and then use it as a correction for the entire network deployment. When a correction is received, any nodes can estimate the distance to the anchor node in meters, which can be used for triangulation. The correction of position coordinates (X_n, Y_n) is :

$$C_i = \frac{\sum \sqrt{(X_n - X_m)^2 + (Y_n - Y_m)^2}}{\sum h_m}, n \neq m, all\ landmarks\ m \tag{1}$$

The flow chart of the DV-Hop localization algorithm is as follows:

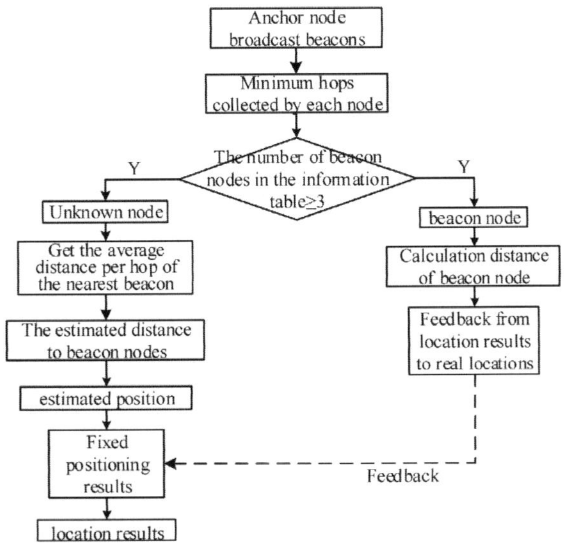

Figure 1. Flow chart of DV-hop localization algorithm.

The unknown node obtains updated information from neighboring anchor nodes, usually the latest updated information. This process depends on the deployment strategy and when the correction stage of DV-hop starts from each anchor node. In addition, the correction is controlled by feedback. Once a node obtains and forwards a correction, it will delete all subsequent corrections. This way, it can be ensured that most nodes only receive the correction once from the nearest anchor node.

4.2. Communication attack model

The malicious nodes should be excluded first to avoid using tampered location information and hops for localization Considering network attacks in the DV-hop location process.

Assuming that L is the number of anchor nodes at known locations. If j is a malicious node, then (X_j, Y_j) is the false location information and f_j is the mendacious hop information. Therefore, the unknown node needs to receive not only the relative position information of each anchor node (X_k, Y_k), but also the hop count information between the unknown node and the anchor node and the estimated hop distance (C_i) from the unknown node to the anchor node. In the attack scenario, malicious nodes can provide false information to other nodes that need to be located by broadcasting false location information to the network, thus changing the average hop count and hop distance from unknown nodes to anchor nodes. Therefore, they can increase the positioning error to achieve the purpose of the attack. We model the measurement error caused by Na on-cooperative attack as independent and uniformly distributed noise and the error caused by the underwater environment as additive Gaussian noise (variance σ^2 and zero mean).

$$d_k^{(nc)} = \begin{cases} dist_k + n_k + u_k, & If\ k\ nodes\ are\ malicious \\ dist_k + n_k, & Other\ cases \end{cases} \quad (2)$$

In which u_k is an independent zero-mean uniform random variable with variance σ^2_{attack}, n_k is an independent Gaussian variable $N(0, \sigma^2)$ and $dist_k$ is the estimated value of average communication distance per hop.

Attackers need more information to initiate coordinated attacks in underwater sensor networks successfully. The attack, as mentioned above, is applied to the uncoordinated attack of malicious nodes. This paper only studies uncoordinated attacks in underwater sensor networks. In future work, we will consider the issue of launching intelligent, cooperative attacks in underwater sensor networks.

5. Nesterov accelerated gradient descent algorithms

5.1. Rationale

We define the network attack as a noise parameter model. To this end, the NAG descent algorithm is modified to this setting, such as below.

Every underwater node i stochastic initializes the estimated value of its current hop $\hat{P}_i(m)$. In each subsequent single hop m, the i_m node obtains the measured value $\{\hat{P}_k(m-1), d_{ik}^{(nc)}(m)\}k = 1,2,\cdots, N; i \neq k$ from the rest nodes. The following is the cost function of i_m node for a single hop m:

$$f_i^{(m)}(P_i(m)) = \sum_{k=1\cdots Nk\neq i}\left(\left\|P_i(m) - \hat{P}_k(m-1)\right\| - d_{ik}^{(nc)}(m)\right)^2 \tag{3}$$

Node i estimates the cost function gradient in (3) at evaluating its current hop $\hat{P}_i(m-1)$. Then the evaluation is updated by increasing a step in the negative direction of the gradient:

$$g_i'(m) = rg_{m-1}(m) + \eta\nabla_P\left(f_i^{(t)}(P - rg_{m-1}(m))\right)\Big|_{P=\hat{P}_i(m-1)} \tag{4}$$

$$\hat{P}_i(m) = \hat{P}_i(m-1) - g_i'(m) \tag{5}$$

After the amplitude of the NAG gradient falls below the threshold, the selection phase will be activated by pruning some items in the cost function with a large gradient range.

5.2. Nesterov accelerated gradient equivalence

In this section, an ergodic underwater node secure localization algorithm is proposed, which combines NAG gradient descent and selective clipping to filter out the localization measurements caused by malicious nodes underwater.

Localization Estimation of Underwater Nodes. We first acquire the position of underwater nodes estimated by the DV-hop algorithm. When no malicious underwater node exists, and the distance measurement noise is a Gaussian function, the likelihood measurement value of the real position of the underwater node to be located is:

$$P_r\left(\{d_k\}_{k=1}^N \middle| P, \{P_k\}_{k=1}^N\right) = \frac{1}{(2\pi\sigma^2)^{N/2}} exp\left\{-\frac{1}{2\sigma^2}\sum_{k=1}^N (\|P_k - P\| - d_k)^2\right\} \tag{6}$$

Then, by minimizing the exponential negative value, *i.e.*, the sum of weighted mean squared residuals ξ_p, we can find the maximum likelihood (ML) estimate $\hat{P}$ for the real location of underwater nodes.

$$\hat{P} = \arg\max_P P_r\left(\{d_k\}_{k=1}^N \middle| P, \{P_k\}_{k=1}^N\right) = \arg\min_P \frac{1}{2}\sum_{k=1}^N (\|P_k - P\| - d_k)^2 = \arg\min_P f(P) \tag{7}$$

Where $f(P)$ represents the cost function in (3), therefore, the maximum likelihood estimation is the same as the least square estimation of solution (1) in the sense of least square.

When malicious underwater nodes exist, the minimum hop from the beacon to unknown nodes should be counted first. (x_i, y_i, i_d, h) is a grouping of beacon nodes broadcast to the neighbor nodes, where (x_i, y_i) is the coordinate location of beacon node i, h is the jump value, and i_d is the identifier of node i. Each node receives this packet, if not repeated, the information is recorded in the table, discarded, and then broadcast to its neighbors. After this broadcast round, all nodes know that all beacon nodes are calculated.

Secondly, formula (2) calculates the average distance per hop. The beacon node broadcasts the location information C to the network, and the unknown node records the firstly received location information and relays it to the adjacent node. Therefore, $C \times h$ can calculate the hop distance d of each beacon node.

According to the trilateral measurement method, Equation (8) calculates the coordinates of unknown nodes. The unknown node uses each hop d obtained in the previous phase to calculate its coordinates using the trilateral measurement method. It is assumed that the coordinates of the three beacon nodes are (x_a, y_a), (x_b, y_b),(x_c, y_c), and the estimated distances from the unknown nodes are d_a, d_b, and d_c, respectively.

$$\begin{cases}(x_a - x)^2 + (y_a - y)^2 = {d_a}^2 \\ (x_b - x)^2 + (y_b - y)^2 = {d_b}^2 \\ (x_c - x)^2 + (y_c - y)^2 = {d_c}^2\end{cases} \tag{8}$$

5.2.1 Nesterov accelerated gradient descent secure localization algorithm. We can set the security positioning problem when the DV-hop positioning is available. It is assumed that there are anchor nodes with known location coordinates, where some are malicious nodes and launch un-collaborative attacks. Malicious nodes can be concatenated to arrest the accurate location of other nodes. It is assumed that attackers cannot successfully launch attacks by modifying hops individually using tamper-proof trusted nodes to assist located nodes. In reverse, the attacker's strategy is to intelligently modify the minimum hops of nodes to $P_k^{'}$. Suppose the minimum hop between two nodes is P_k. So, we have the following relationship:

$$U_k = C \cdot P_k \tag{9}$$

$$U_k^{'} = C \cdot P_k^{'} \tag{10}$$

Among them, U_k is the actual distance between nodes, and $U_k^{'}$ is the distance after malicious modification between two nodes.

We can extend the selective pruning method based on gradient descent, as the previous section illustrated, to implement security positioning in this instance. This algorithm initiates with the random initialization of LS estimation $\hat{P}(m)$. In the primary step of the iteration, the gradient of the cost function $f_{td}(P)$ is estimated according to the current evaluation $\hat{P}$(i-1), and the estimated value is renewed by moving $g_{td}(i)$.

$$g_{td}(i) = rg_{td-1}(i) + \eta\nabla_P(f_{td}(P - rg_{td-1}(i)))|_{P=\hat{P}_i(i-1)} \tag{11}$$

5.2.2 Nesterov accelerated gradient descent secure localization algorithm flow. At every turn of iteration, the NAG descent algorithm will obtain a new estimated value, which has a greater possibility of the real location of the node. The algorithm finally converges to the ML estimation, the same as the least square estimation without underwater malicious nodes. As mentioned earlier, in adversarial scenarios, the measurement error of node hops increases due to malicious attacks from adversaries, and the least squares estimation may have a large error. Therefore, once the NAG descent algorithm converges to the least square solution, we will switch to a selective clipping stage, where some force vectors are trimmed.

Selection clipping stage: In the circumstance of a non-cooperative attack, the independent disturbance increased by various underwater malicious nodes tends to be averaged, and the least squares solution is close to the real position. We update the cost function to exclude the distance measurement values with large residuals, likely to correspond to the measurement of underwater malicious nodes. The proposed algorithm cuts a small part of the large force vector and uses the residual force vector to calculate the gradient to achieve this point. The estimated value is updated by moving one step in this modified gradient direction at each iteration. Fig. 2 shows the final algorithm, and the following is the flow chart of the algorithm.

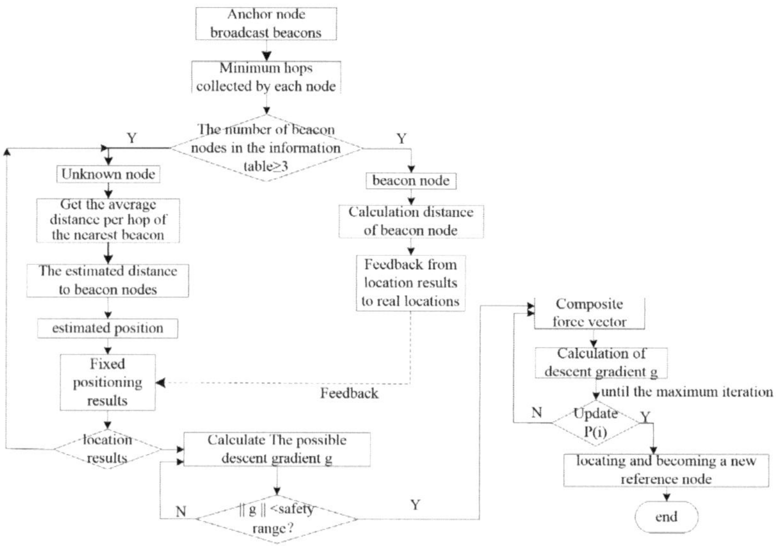

Figure 2. Flow chart of NAG descent safe location algorithm.

6. Simulation

6.1. Simulation parameter

This section demonstrates the accuracy of the proposed method in sensor networks under uncoordinated attacks through experiments. Initially, 75 sensors are randomly deployed in the region of 500 meters × 500 meters. To illustrate the positioning accuracy, we compare our DV-hop NAG descent algorithm with the other three estimation methods, including the Maximum likelihood estimation, Quadratic curve, and least square method.

6.2. Influence of coverage area on positioning error

As the area increases, the positioning errors of Maximum likelihood estimation and Quadratic curve methods become more and more obvious. As depicted in Fig.3, when the UWSN area is the same, the average positioning error of the NAG descent algorithm is better than the Maximum likelihood estimation and Quadratic curve methods and slightly better than the Least square method. When the area of UWSN is $5\times10^5 m^2$, the average positioning error of the NAG descent algorithm is only slightly higher to 5.33m, while the quadratic error is higher to 16.61m, the maximum likelihood estimation is higher to 9.22m, and the least squares is close to 5.84m. Similarly, the average positioning error of the NAG descent algorithm is lower than other estimation methods even when the sensor network area is $1\times10^5 m^2$.

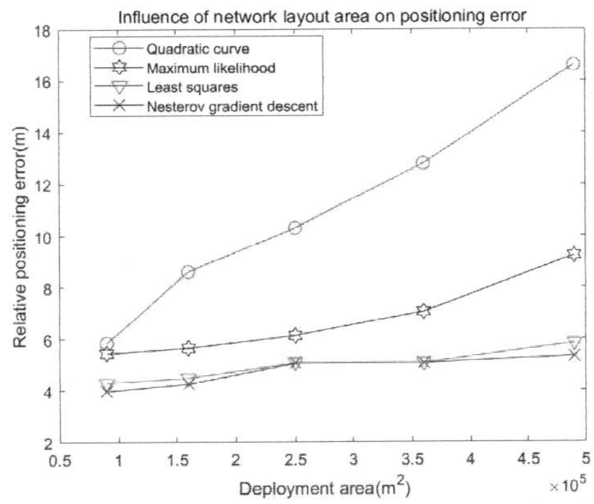

Figure 3. Influence of coverage area on positioning error.

6.3. Influence of sensor node number on positioning error

In this section, we change the number of nodes. As depicted in Fig.4, location error reversely reduces for all algorithms with increases in the nodes. Besides, it can be seen that in the case of the same node number, the average positioning error of NAG descent is significantly better than that of the other algorithms. Particularly in the case of sparse nodes, the advantage is more obvious. When there are only 25 nodes in the test area, the average error of NAG descent is 6.72m, while the quadratic curve error is as high as 13.75m, the maximum likelihood error is 10.91m, and the least square error is slightly higher to 7.12m.

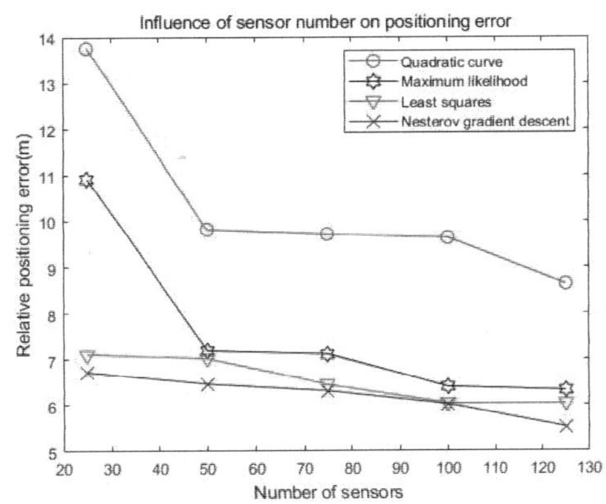

Figure 4. Influence of sensor node on positioning error.

6.4. Influence of different anchor node ratios on positioning error

In this section, we change the ratio of anchor nodes to analyze the location error. As the ratio of anchor nodes change from 0.2 to 0.5, the average positioning error of the four methods is significantly reduced, and the NAG descent algorithm is more accurate than the other algorithms. When the proportion of the anchor node is 0.2, the average positioning error of NAG descent is 10.70m. Besides, the error of Maximum likelihood and least squares is 11.01m and 10.76m, respectively, which are close to the NAG descent method. When the ratio of the anchor node is 0.5, the average positioning error of NAG descent decreases to 8.29m, which is still lower than the others.

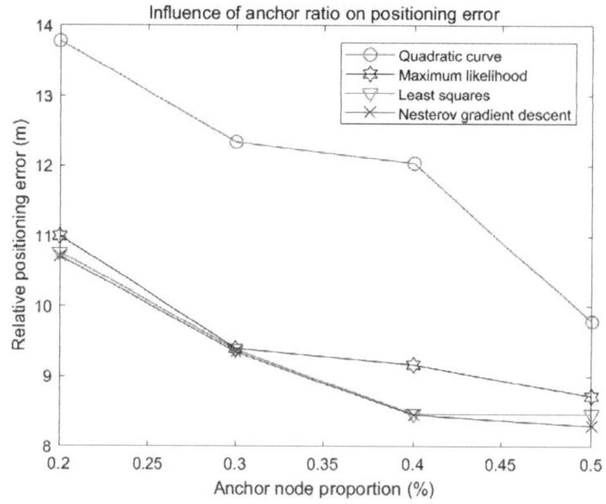

Figure 5. Influence of different anchor node ratios on positioning error.

6.5. Influence of different communication radius on positioning error

Fig.6 depicts the relationship between the node communication radius and the average positioning error. As the node communication radius varies from 200m to 400m, the average error of the four algorithms is simultaneously descending. When the communication radius goes up to 400m, the average error of the four algorithms is almost the same. On the other hand, the average error of NAG descent is 2.43m when the communication radius is 300m, which is slightly better than 2.66m for Least squares, better than 4.84m for Quadratic curve, and much better than 13.95m for Maximum likelihood. Peculiarly, the decline of NAG descent and Least squares are gentle compared to the other two methods.

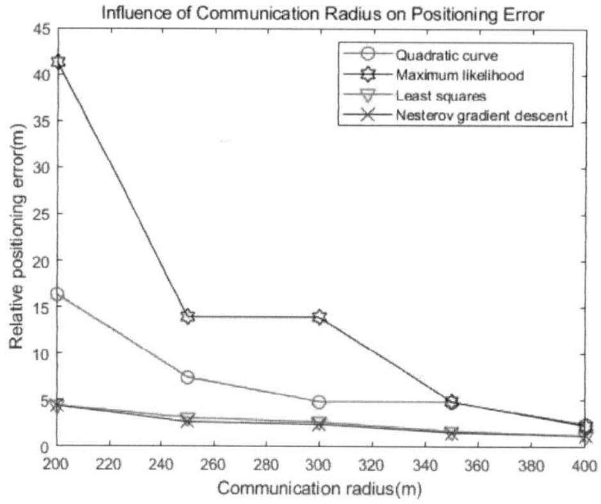

Figure 6. Influence of different communication ratios on positioning error.

7. Summary

In this paper, we propose a computationally effective and secure localization algorithm for UWSN to reduce the error caused by network attacks. The proposed algorithm combines the NAG descent method with the pruning stage, which filters out the inconsistent measurements to acquire the location of nodes.

We prove the availability of the NAG descent algorithm for location errors. Furthermore, the simulation results show that this method has better positioning accuracy than existing methods under coordinated attack. For non-cooperative attacks in the size of a $500m \times 500m$ deployment area, the positioning accuracy of 125 nodes is improved by about 0.5m when there are Gaussian measurement noise and uniformly distributed noise. The calculation requirements and running time in the displayed method are lower than those in the existing method.

Acknowledgments
This work has been funded by the Natural Science Foundation of Hainan Province with grant number (RZ2100000416) and partially supported by the critical project of Hainan Province under Grant (ZDYF2020199).
The authors thank all the anonymous reviewers for their valuable comments and suggestions.

References

[1] Ahmad, I., Ullah, U., Masood, J., et al. (2021) A SURVEY ON UNDERWATER WIRELESS SENSOR NETWORKS REQUIREMENTS[C]// First International Two-Day Conference (Online) on Advanced Research and Development in Computer Science and Information technology (ARDCSIT - 2021) 13 and 14 JULY 2021.
[2] Sahana, S., Singh, K., Kumar, R.,Das, S. (2018). A review of underwater wireless sensor network routing protocols and challenges. Next-generation networks, 505-512.
[3] Chen, D. H., Zhang, X. M., Ke, H. E. (2009) Research on Node Localization in Underwater Acoustic Sensor Networks[J]. Audio Engineering.
[4] Niculescu, D., Nath, B. (2001). Ad hoc positioning system (APS). In GLOBECOM'01. IEEE global telecommunications conference (Cat. No. 01CH37270) (Vol. 5, pp. 2926-2931). IEEE.
[5] Hu, L and Evans, D. (2004) Localization for mobile sensor networks, in Proc. 10th ACM Annu. Int. Conf. Mobile Comput. Netw. (MobiCom), Philadelphia, PA, pp. 45–57.
[6] Baggio, A. (2008) Monte-Carlo Localization for Mobile Wireless Sensor Networks[J]. Ad Hoc Networks, 6(5):718-733.
[7] Zeng, Y., Cao, J., Hong, J., Zhang, S and Xie, L. (2009) SecMCL: A secure monte carlo localization algorithm for mobile sensor networks, in Proc. IEEE 6th Int. Conf. Mobile Adhoc Sens. Syst. (MASS), Macau, China, pp. 1054–1059.
[8] Liu, D., Ning, P., Liu, A., Wang, C and Du, W. K. (2008) Attack-resistant location estimation in wireless sensor networks, ACM Trans. Inf. Syst. Security, vol. 11, no. 4, pp. 1–39.
[9] Misra, S., Bhardwaj, S and Guoliang, X. (2006) ROSETTA: Robust and secure mobile target tracking in a wireless ad hoc environment, in Proc. IEEE Military Commun. Conf., Washington, DC, pp. 1–7.
[10] Michael, N., Zavlanos, M., Kumar, V and Pappas, G. (2008) Distributed multi-robot task assignment and formation control, in Proc. IEEE Int. Conf. Robot. Autom., Pasadena, CA, pp. 128–133.
[11] Mauve, M., Widmer, A and Hartenstein, H. (2001) A survey on position-based routing in mobile ad hoc networks, IEEE Network, vol. 15, no. 6, pp. 30–39.
[12] Ramanathan, S and Steenstrup, M. (1996) A survey of routing techniques for mobile communications networks, Mobile Netw. Applicat., vol. 1, no. 2, pp. 89–104.
[13] Han, F., Abdelaziz, I., Liu, X., et al. (2020) A Survey on DV-Hop localization Techniques in Three-Dimensional Wireless Sensor Networks. International Journal of Online and Biomedical Engineering (iJOE), 16(10):23.

A Cancelling-Blocking structure for decoupling of microstrip antenna array

Di Gao[1,*], Wenwu Zhang[1, 2]

[1]Research Institute for Special Structures of Aeronautical Composite AVIC, The Aeronautical Science Key Lab for High Performance Electromagnetic Windows, Ji'nan, China.
[2] Email:1119452858@qq.com

*Corresponding author's e-mail:1119452858@qq.com

Abstract. In this work, a cancelling-blocking structure (CBS) for decoupling of microstrip antennas was proposed, which can reduce the coupling between arbitrary elements of a one-dimensional E-plane antenna. Results demonstrate the optimized active VSWR and improved radiation efficiency. The cancelling-type decoupling structure consists of a plurality of electrically small metal sheet arrays printed on the dielectric above the antenna. The blocking-type decoupling structure means that each element uses independent dielectric and ground plane to cut off the coupling current propagation path. Firstly, the decoupling principle of the cancelling-type decoupling structure was deduced, and the decoupling conditions were obtained. Then, the mechanism of the surface wave in the antenna array was qualitatively analyzed, and the applicable conditions of the blocking-type decoupling structure were deduced. To verify the proposed structure, a 1×8 E-plane antenna with the designed CBS was modeled. Simulation results manifested that a mutual coupling reduction of more than 10 dB in 1.2-1.3 GHz was achieved, and the active VSWR was below 2.0 over the range of $\pm60^{\circ}$ scan angle.

1. Introduction

As a transceiver for radio signals, antennas have become an indispensable part of wireless systems [1]. In an antenna array, the mutual coupling implies the energy coupled to other elements when one element is working. A strong mutual coupling would significantly affect the matching characteristics and far-field radiation characteristics of the antenna element and array [2-4]. Therefore, optimized mutual coupling of antenna arrays is urgently needed and would be critical for wireless communication applications.

In the early 1966, the MIT Lincoln Laboratory published a research report on the mutual coupling of antenna arrays [5]. Since then, numerous efforts have been devoted to reduce the mutual coupling between antennas, e.g. defect ground structure [6], electromagnetic band gap structure [7], decoupling network [8], and metamaterial structure [9]. In 2020, Cao et al [6] reported a new type of defected ground structure to decrease the coupling of microstrip antennas. This structure was artificially etched on the ground plane, surrounded by each antenna element and parallelly placed to the radiating side of the antenna. In 2017, Wu and colleagues designed the innovative array-antenna decoupling surface

(ADS) [10] to reduce the mutual coupling in base station antennas. Their results demonstrated that the structure was effective to decrease the coupling of adjacent elements in a 2 ×2 array.
However, previous decoupling methodologies are merely restricted in two-element coupled antennas or small array antennas. Here, based on the above pioneer works, an eight-element E-plane high-isolation antenna with a center-to-center spacing of 0.5 λ_0 based on the CBS decoupling structure was proposed. This work will focus on the following unique and attractive features:
1) The mechanism of surface wave in the E-plane antenna array was qualitatively analyzed, and the applicable conditions of a blocking-type decoupling structure were deduced.
2) A mutual coupling decline exceeding 10 dB in the frequency band of 1.2-1.3 GHz was achieved.
3) The active VSWR was less than 2.0 over the range of ±60° scan angle in 1.2-1.3 GHz.
This paper was constructed as follows. In Section II, the CBS decoupling structure and its underlying mechanism for decoupling was elucidated. Besides, performance of the CBS structure on mutual coupling decrease was demonstrated. Finally, the conclusion was drawn in Section III.

2. The Hign-Isolated E- plane Antenna with CBS

2.1. Decoupling Mechanism of the Cancelling-Type Structure
The decoupling mechanism of the typical cancelling-type decoupling structure, ADS, is schematically depicted in Fig.1. The ADS is a slender surface consisting of multiple electrical small metal patches, which is located in front of the array antenna. The partially diffracted waves from the ADS could be tuned to counterbalance the undesired coupled waves. By reasonably optimizing the geometrical size and spatial location of the ADS, the out of phase and the equal intensity conditions of the diffracted waves, as well as the coupling waves, could be achieved simultaneously. As a result, most of the undesired mutual coupling of antenna elements could be canceled out.

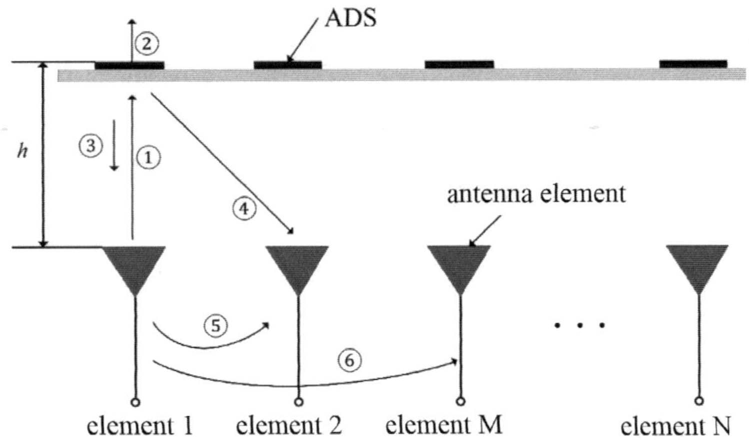

Figure 1. Schematic illustrations of the antenna array and the ADS with spacing distance h

2.2. Decoupling Mechanism of the Blocking-Type Structure
Unfortunately, ADS only decreases the coupling of adjacent elements. To reduce the coupling between non-adjacent array elements, a blocking-type decoupling structure was designed, i.e. each element uses independent dielectric and ground plane to cut off the coupling current propagation path.
Firstly, the strength of the mutual coupling field was compared in three cases:① continuous substrate and ground plane, ② discontinuous substrate and continuous ground plane, and ③ discontinuous substrate and ground plane. In case ①, the coupling field includes three parts: far-field radiation, near-field radiation and surface wave propagation. In cases ② and ③, the surface wave propagation is cut

off, but the surface wave on the substrate would be converted into the near-field coupled wave. The amplitude of the near-field coupled wave varies radially by $1/\rho^2$ (ρ is the center-to-center element spacing). When the spacing is relatively large, the attenuation of the near-field would be faster than that of the surface wave, since the amplitude of the surface wave varies radially by $1/\rho^{1/2}$. Therefore, if the distance between the elements is relatively far, the coupling in case ① would be larger than that in cases ② and ③.

To qualitatively compare the magnitude of the near-field coupled wave in cases ② and ③, as shown in Fig.2, the propagation modes of the conductor covered with substrate in cases ② and ③ were analyzed. No tangential E-field on the z=0 plane in the propagation mode of the conductor covered with substrate. They are the TMn modes (when the wavefunction is odd about the z-axis), n=0, 2, 4, ..., and the TEn modes (when the wavefunction is even about the z-axis), n=1, 3, 5, ... For any kind of wave, Ψ, its arbitrary TM field can be expressed as:

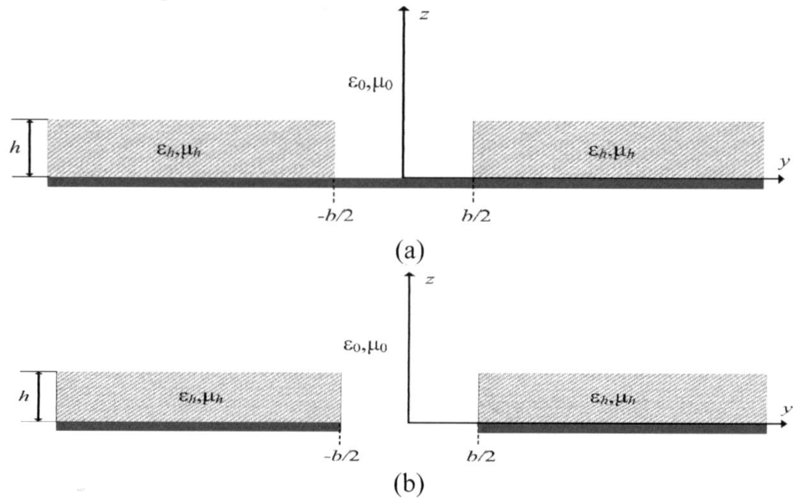

Figure 2. Schematic of conductor covered with substrate in the case of (a) discontinuous substrate; (b) discontinuous substrate and conductor.

$$E_x = \frac{1}{\hat{x}}\frac{\partial^2\Psi}{\partial x\partial y} \qquad H_x = -\frac{\partial\Psi}{\partial z}$$
$$E_y = \frac{1}{\hat{x}}\left(\frac{\partial^2}{\partial y^2}+k^2\right)\Psi \qquad H_y = 0$$
$$E_z = \frac{1}{\hat{x}}\frac{\partial^2\Psi}{\partial z\partial y} \qquad H_z = \frac{\partial\Psi}{\partial x} \tag{1}$$

The electromagnetic wave is subjected to two-dimensional processing. Assuming there is no change along the x-axis, for the y-axis TM mode, equation (1) can be simplified to

$$E_y = \frac{1}{j\omega\varepsilon}\left(k^2-k_y^2\right)\Psi; E_z = \frac{-k_y}{\omega\varepsilon}\frac{\partial\Psi}{\partial z}; H_x = -\frac{\partial\Psi}{\partial z} \tag{2}$$

Because Ψ is an odd function of z, in the substrate region, the wavefunction is:

$$\Psi_h = A\sin uz e^{-jk_y y}, \quad 0 < z < h \tag{3}$$

In the vacuum region, the wavefunction is:

$$\Psi_0 = Be^{-vz}e^{-jk_y y}, \quad z > h;$$
$$\Psi_0 = -Be^{vz}e^{-jk_y y}, \quad z < 0; \tag{4}$$

To simplify the signs in the formula, one can use $k_{zh}=u$ and $k_{zh}=jv$. The separation parameter equation in each region could be written as:

$$u^2 + k_y^2 = k_h^2 = \omega^2 \varepsilon_h \mu_h;$$
$$-v^2 + k_y^2 = k_0^2 = \omega^2 \varepsilon_0 \mu_0; \tag{5}$$

By calculating the field component with the vacuum-substrate interface, one can obtain:

$$\left.\begin{aligned} E_y &= \frac{A}{j\omega\varepsilon_h} u^2 \sin(uz) e^{-jk_y y}, \\ H_x &= -Au\cos(uz) e^{-jk_y y}, \end{aligned}\right\} \quad 0 < z < h$$
$$H_x = Bve^{-v|z|}e^{-jk_y y}, \qquad z<0, z>h$$
$$E_y = \frac{-B}{j\omega\varepsilon_0} v^2 e^{-vz} e^{-jk_y y}, \qquad z > h$$
$$E_y = \frac{B}{j\omega\varepsilon_0} v^2 e^{vz} e^{-jk_y y}, \qquad z < 0 \tag{6}$$

In the region of -b/2<y<b/2, the boundary conditions in case ② are: (a) when z=0, E_y=0, and (b) E_y is symmetric about the y-axis. According to the boundary conditions, the E-field distribution in case ② can be obtained:

$$E_y = \frac{-B}{j\omega\varepsilon_0} v^2 \sin(vz) \cos\left(\frac{m\pi}{b} y\right) \quad m = 0,1,2... \tag{7}$$

In the region of -b/2<y<b/2, the boundary conditions in case ③ are: (a) when z=0, $E_y \neq 0$ and $E_y \neq$ maximum, (b) Ey is symmetrical about the y-axis. According to the boundary conditions, the E-field distribution in case ③ can be obtained:

$$E_y = \frac{-B}{j\omega\varepsilon_0} v^2 e^{-vx} \cos\left(\frac{m\pi}{b} y\right) \quad m = 0,1,2... \tag{8}$$

Comparing equations (7) and (8), in case ②, the E-field is a standing wave along the z-axis. In case ③, the E-field is a fading wave along the z-axis. Therefore, it could be drawn that the strength of E-field in case ② is stronger than that in case ③.

2.3. The Design and Decoupling Effect of CBS

On the basis of the above analysis, the CBS was designed to decrease the E-plane coupling. The cancelling-type decoupling structure is the ADS above the antenna. The blocking-type decoupling structure means that each element utilizes independent dielectric and ground plane to cut off the coupling current propagation path. The schematic illustration of the designed highly isolated 1×8 E-plane antenna with CBS was exhibited in Fig. 3, with parameters labeled in the figures. Note that each element was fed with separate SMA connectors. The dielectric constant and loss tangent of the used Taconic substrate in this work are 2.55 and 0.006, respectively. Simulations were performed via the commercial software High-Frequency Structure Simulator. The parameter values of the antenna with CBS are summarized in Table 1.

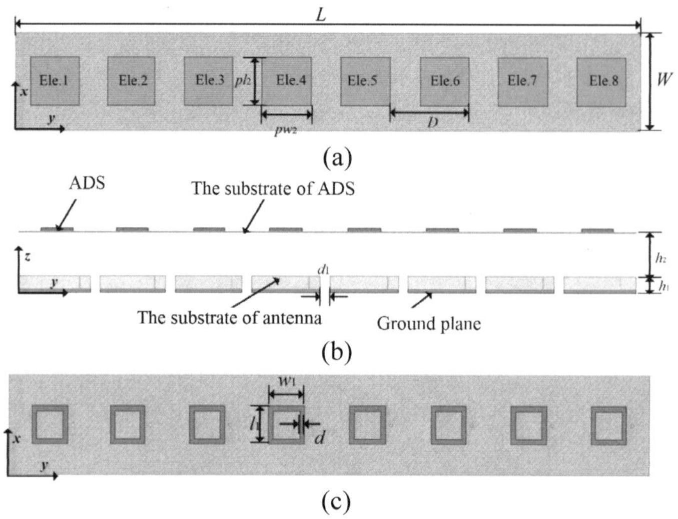

Figure 3. Schematic illustrations of the antenna array with CBS. Top (a) and side view (b) of the antenna. (c) Top view of the ADS.

Table 1. Parameters of the Antenna with CBS

Parameters	Value	Parameters	Value	Parameters	Value
L	1000	W	150	D	120
pl_2	75	pw_2	75	w_1	50
l_1	52	d	5	h_1	20
h_2	69	d_1	15		

The S-parameters of the eight-element E-plane antenna with and without ADS were shown in Fig. 4. Because of the symmetry in this array, only the S-parameters of typical elements 1-5 are presented here. As displayed in Fig. 4(a), the return loss of each element hardly changes no matter with or without ADS. The impedance curves of different elements almost overlap, indicating the good consistency of the array. As exhibited in Fig. 4(b), the coupling coefficient between adjacent elements without ADS resides in the range of [-16 dB, -19 dB]. Specifically, the coupling coefficient between adjacent elements decreased by more than 10 dB with ADS. However, the size and profile height of the ADS are adjusted according to the intensity and phase of the coupled waves between adjacent elements. Therefore, as shown in Fig. 4(c), there is no significant decrease in the coupling between non-adjacent array antennas. Besides, the coupling coefficient between elements 1 and 3 increased from -30 dB to -20 dB with ADS.

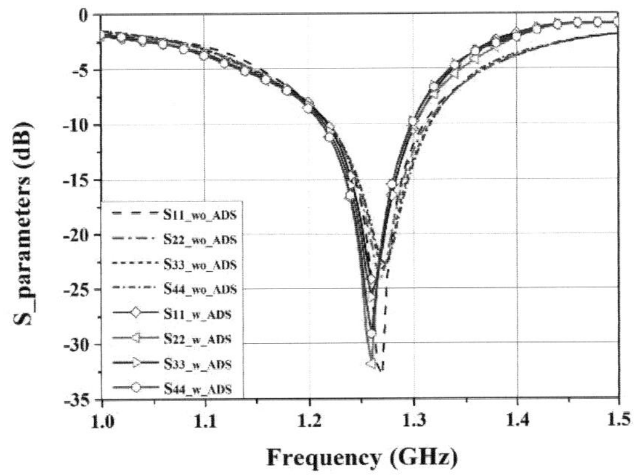

(a)

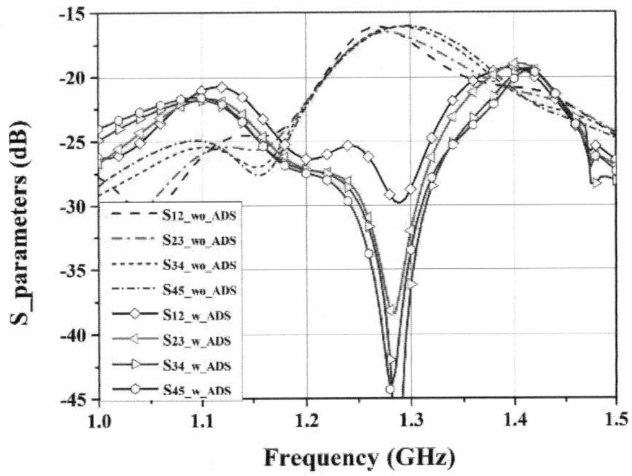

(b)

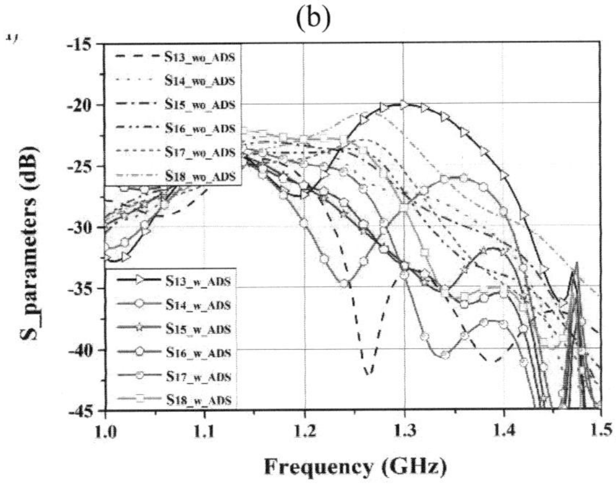

(c)

Figure 4. The return loss (a) and mutual coupling coefficient (b) between adjacent elements with and without ADS. (c) The mutual coupling coefficient between non-adjacent elements of the antenna array with and without ADS.

To reduce the coupling of non-adjacent array elements, as shown in Fig. 3, each element uses independent dielectric and ground plane. The S-parameters of the antenna with and without CBS were illustrated in Fig. 5. As shown in Fig. 5(b), the coupling between adjacent elements was still decreased by more than 10 dB with the independent substrate and ground plane. At the same time, as displayed in Fig. 5(c), the coupling between non-adjacent elements is also reduced to varying degrees. All the coupling coefficients between non-adjacent elements are less than -30 dB in the 1.2-1.3 GHz frequency band.

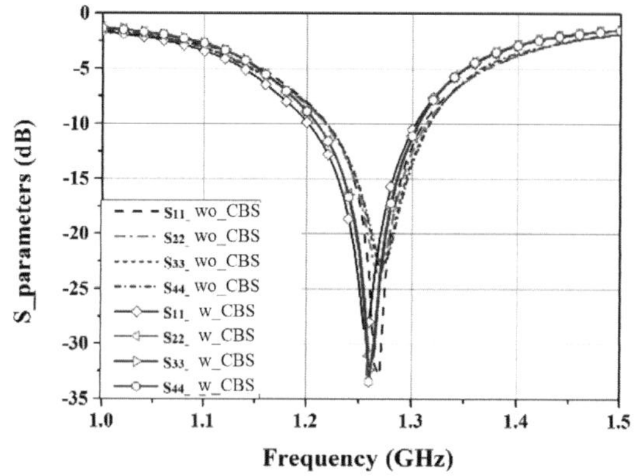

(a)

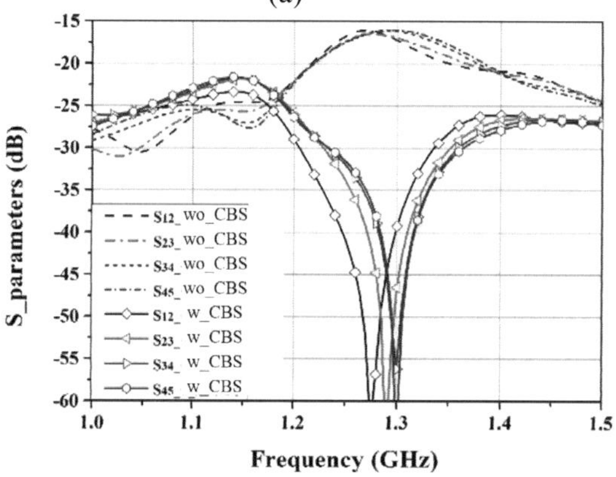

(b)

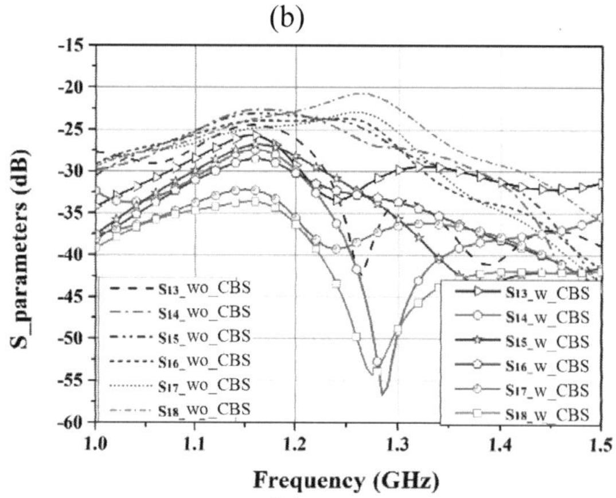

(c)

Figure 5. The return loss (a) and mutual coupling coefficient (b) between adjacent elements, and mutual coupling coefficient (c) between non-adjacent elements of the antenna array with and without CBS.

The active VSWR of the antenna will also be influenced by the CBS. The calculation of the active VSWR is shown in Equation 9.

$$Active(S_{mm}) = \sum_{i=1}^{N} \frac{|a_i|}{|a_m|} e^{j(\phi_{ai}-\phi_{am})} |S_{im}| e^{j\phi_{im}} \tag{9}$$

Table 2 exhibited the active VSWR maximum of the original antenna, the antenna with ADS, and the antenna with CBS over the 0°-60° scanning angle in the 1.2-1.3 GHz frequency band. Compared with the original antenna, one can find that the active VSWR of the antenna with ADS is generally improved, but the active VSWR is still larger than 2.0 at some scanning angles. Besides, the active VSWR in individual elements is larger than that of the original antenna. However, the active VSWR of the antenna with CBS is less than 2.0 over the scanning angle of 0°-60° in the 1.2-1.3 GHz frequency band.

Table 2. The Active VSWR of the Original Antenna, Antenna with ADS, Antenna with CBS and Decoupling Antenna.

Antenna type	Scan angle	Active VSWR			
		Element 1	Element 2	Element 3	Element 4
Original Antenna	0°	2.0	2.6	2.4	2.5
	20°	2.0	2.0	2.0	2.0
	40°	1.8	2.4	2.6	3.5
	60°	2.3	2.4	2.3	1.9
Antenna with ADS	0°	1.9	1.9	2.5	2.1
	20°	2.3	2.0	2.1	2.2
	40°	1.8	1.5	1.7	2.0
	60°	2.1	1.7	2.0	2.1
Antenna with CBS	0°	1.9	1.9	2.0	1.9
	20°	1.9	1.9	2.0	1.9
	40°	1.8	1.7	1.7	1.9
	60°	1.7	1.6	1.8	1.9

Fig. 6 displays the far-field radiation patterns of the antenna with and without CBS. The working state of the antenna is set as only the calculated element was active, while other elements are terminated with a 50 Ω matching load. As shown in Fig. 6, in contrast with the original antenna, the radiation pattern of the antenna with CBS does not change significantly. However, the main polarization gain of the antenna with CBS was improved by 0.5-1 dB. This indicates that the proposed decoupling CBS can effectively decrease the mutual coupling among arbitrary elements without sacrifice of the antenna pattern.

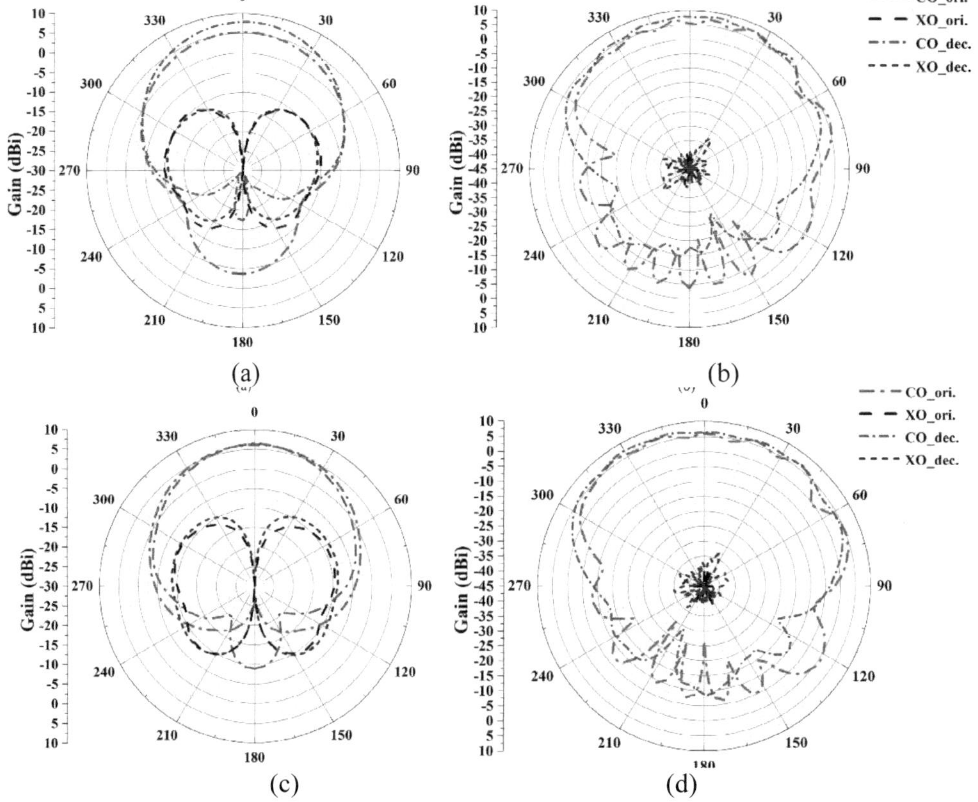

Figure 6. The radiation pattern of the element in the original antenna and decoupling antenna. (a) H-plane of element 1, (b) E-plane of element 1, (c) H-plane of element 2 and (d) E-plane of element 2.

3. Conclusion

In summary, a decoupling CBS for a 1×8 E-plane microstrip antenna was designed and demonstrated in this work. The cancelling-type decoupling structure consists of the ADS above the antenna. The blocking-type decoupling structure means that each element uses independent dielectric and ground plane. By optimizing the geometrical size and spatial position of the ADS, the mutual coupling of adjacent elements could be decreased. Using optimized size and position of the independent substrate and ground plane, the mutual coupling between non-adjacent elements is reduced. Simulation results demonstrated that the enhanced isolation exceeding 10 dB between adjacent elements was achieved, and all coupling coefficients between non-adjacent elements were less than -30 dB in the 1.2-1.3 GHz frequency band. Furthermore, the active VSWR of the antenna with CBS is below 2.0 over the scanning angle of 0°-60° in the 1.2-1.3 GHz frequency band. Our results indicate that the CBS design might be promising in the field of wireless communication and radar engineering.

References

[1] Asplund, H, et al. (2020) Advanced Antenna Systems for 5G Network Deployments[M]. USA: Academic Press, pp. 639-676.

[2] Abdalla M. A and Ibrahim, A. A. (2013) Compact and Closely Spaced Metamaterial MIMO Antenna with High Isolation for Wireless Applications[J], IEEE Antennas and Wireless Propagation Letters, vol. 12, pp.1452-1455.

[3] Hwangbo, S., Yang, H. Y and Yoon, Y. K. (2017) Mutual Coupling Reduction Using

Micromachined Complementary Meander-Line Slots for a Patch Array Antenna[J], IEEE Antennas and Wireless Propagation Letters, vol. 16, pp.1667-1670.

[4] Yin, J., Wu, Q., Yu, C., Wang, H and Hong, W. (2017) Low Sidelobe Level Series-Fed Microstrip Antenna Array of Unequal Inter-Element Spacing[J], IEEE Antennas and Wireless Propagation Letters, vol. 16, pp.1698-1698.

[5] Allen J. L and Diamond, B. (1966) Mutual Coupling in Array Antennas[R], Lexington: MIT Lincoln Lab, 1-58.

[6] Gao, D., Cao, Z., Fu, S., Quan, X., Chen, P. (2020) A Novel Slot Array Defected Ground Structure for Decoupling of Microstrip Antenna Array[J], IEEE Transactions on Antennas and Propagation, vol. 68, pp. 7027-7038.

[7] Kumar N and Kommuri, U. K. (2018) MIMO Antenna Mutual Coupling Reduction for WLAN Using Spiro Meander Line UC-EBG[J], Progress in Electromagnetics Research C, vol. 80, pp.65-77.

[8] Cheng Y and Cheng, K. K. M. (2020) Compact Wideband Decoupling and Matching Network Design for Dual-Antenna Array[J], IEEE Antennas and Wireless Propagation Letters, vol. 19, pp.791-795.

[9] Jafargholi, A., Jafargholi, A and Choi, J. H. (2019) Mutual Coupling Reduction in an Array of Patch Antennas Using CLL Metamaterial Superstrate for MIMO Applications[J], IEEE Transactions on Antennas and Propagation, vol. 67, pp.179-189

[10] Wu, K. L., Wei, C., Mei, X and Zhang, Z. Y. (2017) "Array-Antenna Decoupling Surface," IEEE Transactions on Antennas and Propagation, vol. 65, pp. 6728-6738.

A novel Ultra-wideband Wide-angle scanning Dual-polarized antenna with multilayer dielectric

Huaimin Zhou[1], Xin Quan[2], Zhenxin Cao*, Zihao Xu[3]

State Key Laboratory of Millimeter Wave , Southeast University, Nanjing, China

*caozx@seu.edu.cn; [1] zhdde6@163.com; [2]quanxin@seu.edu.cn; [3]xuzihao@seu.edu.cn

Abstract. In this paper, a dual-polarized planar ultra-wideband modular antenna (PUMA) with wide-scanning angle matching layers is proposed. The multiple dielectric layers and frequency selective surfaces (FSS) are loaded on the dipoles. Each frequency selective surface is etched on the dielectric plate as one wide-angle -matching layer. By applying three wide-angle-matching layers, the wide-scanning angle performance is guaranteed. As for the ultra-wideband performance, a metal cylinder is added to the lower space between adjacent dipoles, and the upper end of the cylinder has a certain distance from the dipole. This forms a "ridge" structure which shift problematic common-mode resonances caused by circulating current out of band. A dual-polarized 256-port array (16×16) is set and simulated. The simulation result indicates that the proposed antenna operating over 4.79GHz -20.15GHz (4.20:1) achieves VSWR<3 while scanning to 60° in E-plane and D-plane, and to 45° in H-plane respectively.

1. Introduction

With the maturity of space exploration activities, the demand for space TT&C antennas is also increasing. The frequency band commonly used in space TT&C antennas includes C/X/Ku band. Traditional separated antenna receiving systems are usually composed of multiple antennas, one of which only receives signals in a specific frequency band. Therefore, if there is an ultra-wideband antenna that covers all these bands meanwhile ensures a wide-scanning angle, the cost of this system can be greatly reduced and the overall weight of the antenna can be reduced.

In the field of ultra-wideband antenna, Vivaldi antenna has been a very classic antenna structure for a long time. It was characterized with simple structure and large impedance bandwidth, so it has been favoured by researchers for a long time. However, Vivaldi antenna has a high profile and a large array scale, which results in poor matching performance in low frequency band. To take count of UWB performance in low frequency, tightly coupled dipole antenna has gradually become the focus of research. The concept of tightly coupled dipole antenna was first put forward by professor Munk's team in Ohio state university. Based on the ideal model of continuous current sheet proposed by Wheeler, professor Munk broke the electrical connection between the adjacent dipole units and introduced coupling capacitors at the end of the dipoles. This kind of coupling capacitors just helps to off-set the inductive effect of the floor at low frequency, thus effectively expanding the low frequency bandwidth, and this kind of antenna is called TCDA (tightly coupled dipole antenna).[1]

The PUMA antenna proposed by Vouvakis in 2010 is a novel form of tightly coupled antenna. [2] Compared with the TCDA antenna originally proposed by Professor Munk's team, this antenna does not apply any external baluns, and has a simple feed structure. It suppresses the common-mode phenomena caused by the unbalanced feed out of the pass band by connecting a short-circuit probe to the dipole arm. [3][4] Domestic researchers have also made many explorations on the improvement and development of wideband tightly coupled array antennas.

The theory of wide-angle matching layer was first proposed by Wheeler. [5] In the design of tightly coupled antenna arrays, one or two layers of uniform and isotropic dielectric loads are often arranged above the array to help achieve broadband matching performance through impedance transformation and counteracting the effect of floor reactance. [6] Furthermore, when the tightly coupled array is scanned to θ in the E-plane, the impedances of equivalent transmission lines representing air and dielectric layers are shown in the following formula respectively

$$Z_0^E = Z_0 cos\theta \quad (1)$$

$$Z_{r1}^H = Z_0 cos\theta_{r1}/\sqrt{\varepsilon_{r1}} \quad (2)$$

When the tightly coupled array is scanned to θ in the H-plane, their impedances are shown in the following formula respectively

$$Z_0^E = Z_0/cos\theta \quad (3)$$

$$Z_{r1}^H = Z_0/(\sqrt{\varepsilon_{r1}} cos\theta_{r1}) \quad (4)$$

θ_{r1} is refraction angle of θ in the dielectric layer, according to Snell's refraction theorem

$$\theta_{r1} = sin^{-1}(\sqrt{\varepsilon_{r1}} cos\theta_{r1}) \quad (5)$$

Therefore, $\theta_{r1} < \theta$. The impedance of equivalent transmission line of the dielectric layer changes with the change of the scanning angle, which is slower than that of the air. The greater ε_{r1}, the more obvious the contrast. However, the height of the dielectric layer should not be too large. When the thickness of the dielectric layer reaches a certain value, the real part of the impedance will be 0 at high frequency. This is actually a "blind zone" phenomenon of the array, which is caused by the high-order mode surface wave in the dielectric layer. Its occurrence depends on the dielectric constant and thickness of the dielectric layer.

Based on the improved PUMA antenna, this paper proposes a design of wide-band and wide-scanning angle antenna which can cover three bands of C/X/Ku. The antenna adopts a tightly coupled form to obtain a wide bandwidth. Three dielectric matching layers and frequency selective surfaces are added to achieve wide angle scanning performance in a wide bandwidth range without grating lobe. Under the periodic boundary condition, the impedance bandwidth (VSWR<3) of the designed wideband array antenna element is 4.20:1 (4.79GHz to 20.15GHz) while scanning to 60° at E-plane and D-plane or to 45° at H-plane.

2. Antenna structure design

The antenna structure is shown in Fig.1. This dual-polarized UWB and wide-scanning angle antenna unit is composed of two linearly polarized antenna units placed orthogonal quadrature, which is designed based on the improved PUMA antenna. The antenna is fed by the coaxial line, and two dipole patches are placed horizontally and connected to the inner and outer conductors of the coaxial line respectively by the conductor column. These two dipole antennas are placed orthogonally to each other to achieve dual polarization.

The overall profile height of the antenna is 5.334mm, and the aperture size of the antenna unit is 7.3mm*7.3mm. In this paper, the master-slave boundary conditions are applied to simulate the tightly coupled array with 8*8 excited units

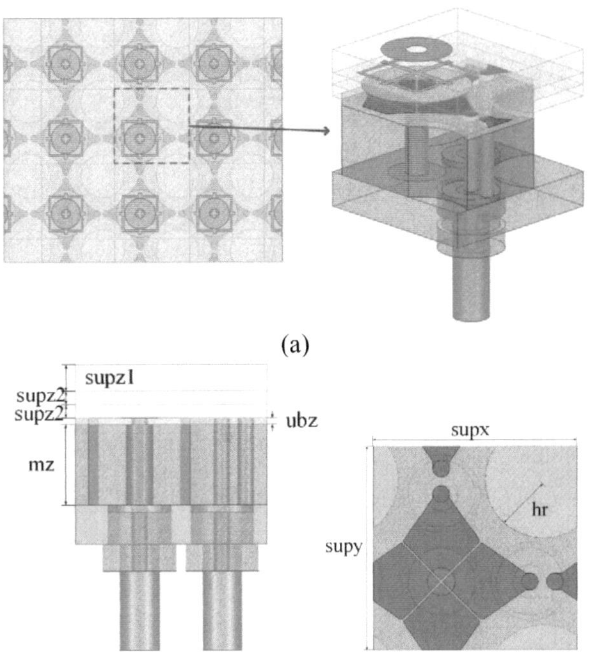

Figure 1. Proposed wideband and wide-scanning angle dual-polarization PUMA. (a) Overall structure of the antenna array and unit cell. (b) Side view of the 4.11:1 PUMA unit. (c) Top view of the Antenna dipole without multiple dielectric layers.

There is a cylindrical conductor below the intersection of the ends of the orthogonal dipole arms. The upper end is not directly connected to the dipole patch, and is isolated by a dielectric layer between them. This structure forms a coupling capacitance between the upper end of the cylindrical conductor and the dipole patch, [7] which strengthens the coupling effect between adjacent elements. This structure also forms a "ridge" structure in the space below the dipole [8], which lengthens the path of the current and makes the common-mode resonant point drift to the lower frequency, thus expanding the overall bandwidth.
At the lower end of the cylindrical conductor, there is a circle patch connect it to the substrate

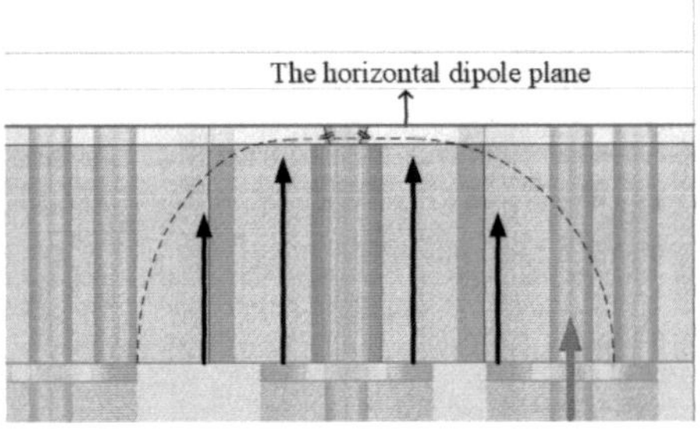

Figure 2. Schematic of "ridge" waveguide structure and equivalent coupling capacitance under the dipole.

There are three dielectric matching layers above the dipole patch, and above each dielectric matching layer there is a patch with different patterns. They can improve the performance of impedance matching

at particular frequency points to a certain extent.[9] This structure is also used to improve the performance of wide-scanning angle, and strengthen the coupling between adjacent elements. [10]
The space below the horizontal dipole is filled with two dielectric matching layers, while the dielectric layer below the dipole patch is perforated to reduce the overall weight.
Rogers 5880LZ dielectric plates with thickness of 0.508mm, relative dielectric constant of 1.96 and loss angle tangent of 0.0021 are applied for the three wide-scanning angle matching layers above the dipole patch. In the space below the dipole patch, the portion below the top of the cylindrical conductor is filled with Rogers 5880LZ with a thickness of 3.048mm. Rogers RT/duroid 5880 (TM) with a thickness of 0.254mm, relative dielectric constant of 2.2, and loss angle tangent of 0.0009 is used to fill the portion above the top surface of the cylindrical conductor. A circular hole with a radius of 2.1mm was drilled in the center of the four adjacent orthogonal di-poles to reduce the weight of the antenna element. The wide-scanning angle matching layers of the antenna is shown in Figure 3.

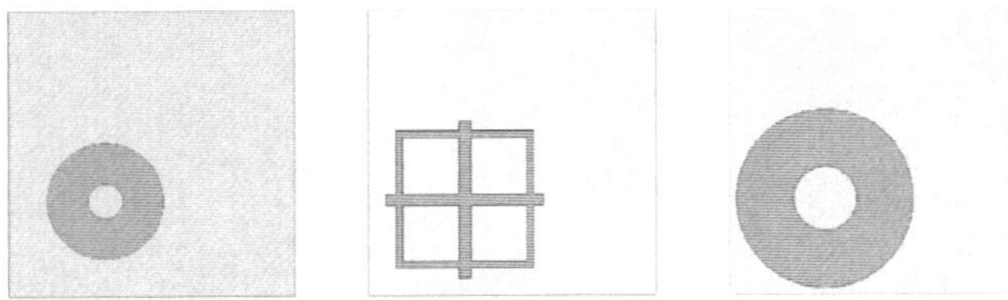

(a) (b) (c)

Figure 3. Wide Angle matching layer. (a) the top layer. (b)the middle layer. (c) the bottom layer.

The antenna elements proposed in this paper are designed and analyzed based on periodic boundary conditions of commercial simulation software HFSS. The final optimized size of the antenna is shown in table 1:

Table 1. Optimized value of the antenna

supz1	supz2	hr	ubz
1.016mm	0.508mm	2.1mm	0.254mm
mz	supx	supy	
3.048mm	7.3mm	7.3mm	

3. Simulation results and analysis

As can be seen from the comparison results shown in Fig. 4, before the addition of the dielectric matching layers, the lowest frequency point of the antenna is only about 5.65ghz, and the matching performance deteriorates from 7GHz to 8GHz. The VSWR is not ideal at 45°of H-plane. With the addition of multi-layer media patches, the low frequency point of E-plane is expanded by about 1GHz, and the wide-scanning Angle performance of antenna is improved at H-plane.

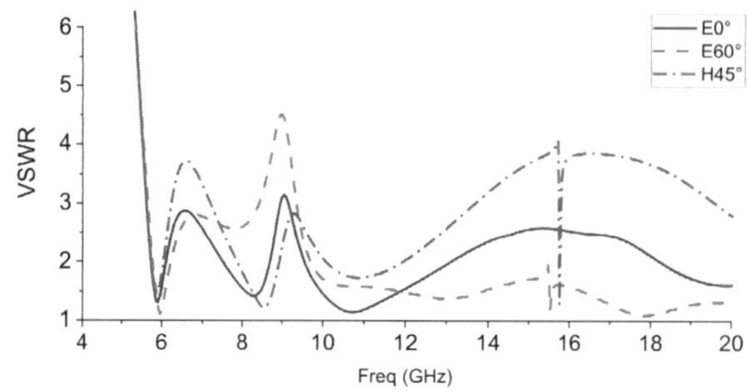

(a)

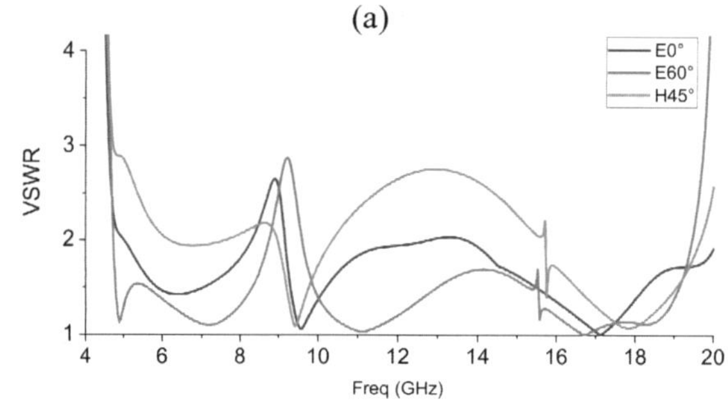

(b)

Figure 4. Comparison of simulated 4.11:1 PUMA array impedance performance for θ=0°, θ=45° and θ=60°, scans in the E-/H-planes. (a) Without wide angle matching layer. (b) With wide angle matching layer.

The VSWR of the improved antenna unit at different scanning angles in each plane is shown in Fig 5. The impedance bandwidth of VSWR<3 on the E-plane of the antenna element could reach up to 4.4:1, ranging from 4.61GHz to 20.44GHz. As the scanning Angle at E-plane reaches 60°, the VSWR worsens, but the impedance bandwidth still reaches 4.20:1, corresponding to a frequency range of 4.79GHz to 20.15GHz. When the antenna element scanning in the range from 0° to 60° at D-plane and 0° to 45° at H-plane, the VSWR can still be less than 3 in this band range, which achieves good standing wave property.

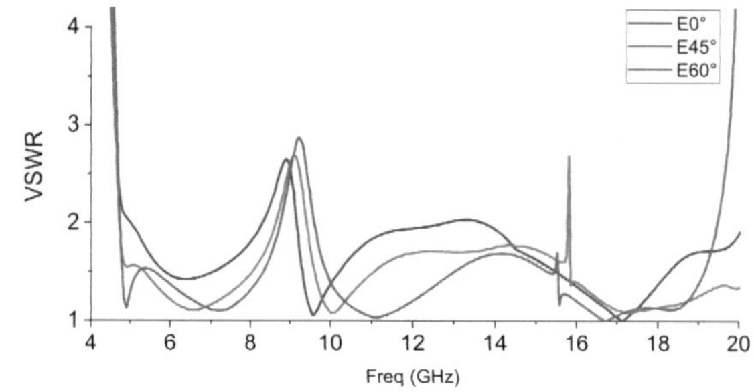

(a)

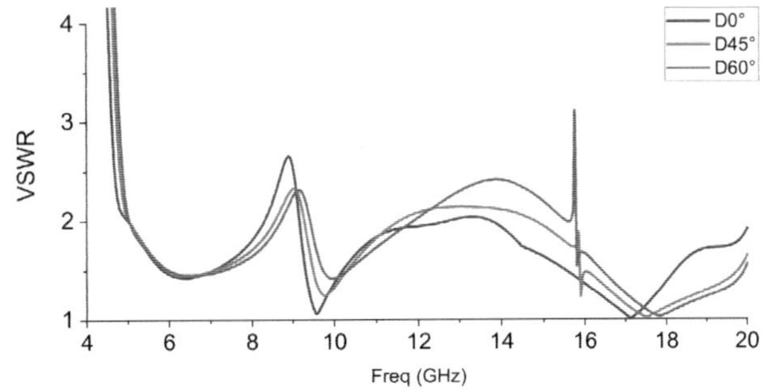

(b)

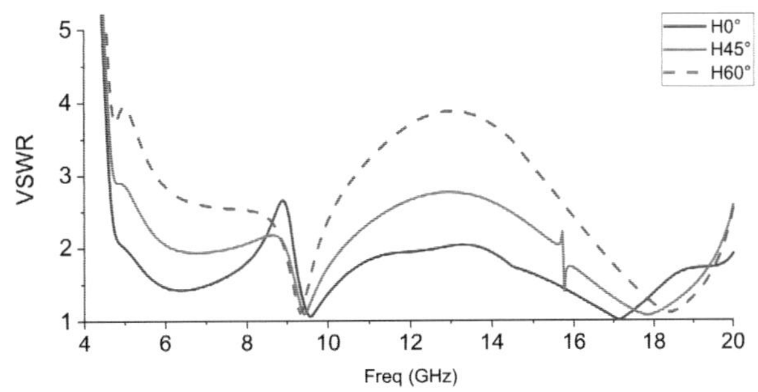

(c)

Figure 5. Measured and simulated VSWR versus frequency. (a)at E-plane. (b)at D-plane. (c)at H-plane.

Further, we measured the realized gain of the 16×16 scale antenna array. The simulation results shown in the following figure.

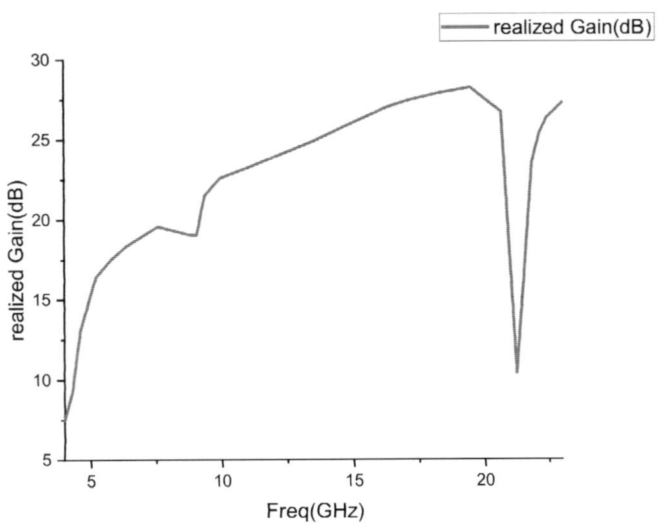

Figure 6. Simulated realized Gain of 16×16 array

4. Conclusion

In this paper, a dual-polarized antenna with ultra-wideband (UWB) and wide-scanning Angle is proposed. The ultra-wideband performance is realized by adopting tightly coupled array, and the wide scanning Angle performance is realized by adding wide angle matching layers. Under the periodic bound-ary condition, the impedance bandwidth (VSWR<3) of the designed ultra-wideband antenna array is 4.20:1 while scanning to 60°at E-plane and D-plane and to 45°at H-plane. The frequency range is from 4.79GHz to 20.15GHz.

Acknowledgment

The author would like to thank professor Cao for that he spends precious time offering helpful suggestions and supports.

References

[1] Munk, B. A., Janning, D. S., Marheka, R. J., McCann J. F and Schneider, S. W. (2010) "The design of wideband arrays of closely-spaced wire and slot elements," 2010 IEEE Antennas and Propagation Society International Symposium, pp. 1-4, doi: 10.1109/APS.2010.5561762.

[2] Holland, S. S and Vouvakis, M. N. (2010) "A 7–21GHz Planar Ultrawideband Modular Array," 2010 IEEE Antennas and Propagation Society International Symposium, pp. 1-4, doi: 10.1109/APS.2010.5561424.

[3] Wang H and Qu, S. W. (2021) "P-L band Planar Ultra-Wideband Modular Antenna (PUMA) Arrays," 2021 International Applied Computational Electromagnetics Society (ACES-China) Symposium, pp. 1-2, doi: 10.23919/ACES-China52398.2021.9581486.

[4] Duan, S., Zhang, C., Shen, J., Shao N and Zhang, X. (2020) "A Dual-Polarized Planar Ultra-wideband Modular Antenna (PUMA) Array With 9:1 Bandwidth," 2020 International Conference on Microwave and Millimeter Wave Technology (ICMMT), pp. 1-3, doi: 10.1109/ICMMT49418.2020.9387035.

[5] Magill E G , Wheeler H A . Wide-angle impedance matching of a planar array antenna by a dielectric sheet[C]// Antennas & Propagation Society International Symposium. IEEE, 1966.

[6] Munk B A . Finite Antenna Arrays and FSS[M]. 2005.

[7] Kindt, R., Mital, R., Logan, J., Lee M and Vouvakis, M. (2016) "A 6:1 bandwidth PUMA array at 7mm scale," 2016 IEEE International Symposium on Phased Array Systems and Technology (PAST), 2016, pp. 1-4, doi: 10.1109/ARRAY.2016.7832626.

[8] Logan, J. T., Kindt, R. W., Lee M. Y and Vouvakis, M. N. (2018) "A New Class of Planar Ultrawideband Modular Antenna Arrays With Improved Bandwidth," in IEEE Transactions on Antennas and Propagation, vol. 66, no. 2, pp. 692-701, doi: 10.1109/TAP.2017.2780878.

[9] Hill, R. A and Munk, B. A. (1996) "The effect of perturbating a frequency-selective surface and its relation to the design of a dual-band surface," in IEEE Transactions on Antennas and Propagation, vol. 44, no. 3, pp. 368-374, doi: 10.1109/8.486306.

[10] Pirhadi, A., Bahrami H and Nasri, J. (2012) "Wideband High Directive Aperture Coupled Microstrip Antenna Design by Using a FSS Superstrate Layer," in IEEE Transactions on Antennas and Propagation, vol. 60, no. 4, pp. 2101-2106, doi: 10.1109/TAP.2012.2186230.

Journal of Physics: Conference Series **2387** (2022) 012007 doi:10.1088/1742-6596/2387/1/012007

An Ultra-wideband, Low-profile tightly coupled array with inductive windows and split rings

Xin Quan[1], Zhenxin Cao*

State Key Laboratory of Millimeter Wave; Southeast University; Nanjing, China
[1]quanxin@seu.edu.cn

*caozx@seu.edu.cn

Abstract. A band-enhanced and low-profile tightly coupled array (TCA) with inductive windows (IWs) and split rings (SRs) is proposed in this paper. The IW cell is placed between the antenna plane and ground and consists of four parallel strips which introduces inductance to move the resonance frequency caused by the ground out of the operating band. Moreover, the SR is specially designed to thin the thick wide-angle impedance matching (WAIM) superstrates to reduce the TCA profile height. As a result, the proposed array has a 10:1 impedance bandwidth (0.6-6 GHz) with the active VSWR$<$3.0 and more than 90% radiation efficiency at broadside. The bandwidth reduces to 8.8:1 (0.7-6.2 GHz) when scanning to ±60° in E/D-plane and ±45° in H-plane, respectively. The profile height of the TCA with IWs and SRs is only 0.077λL with the distance of 0.5λH between the adjacent elements.

1. Introduction

With the development of wireless communication technology, a higher information data transmission rate is required. According to Shannon's theorem of wireless data transmission, the channel capacity of the communication system is directly proportional to the system bandwidth, which means that a wider bandwidth of the communication system will be required. As an indispensable part of the wireless systems [1], antennas can radiate and receive signal. A wideband antenna will save more space with the increasingly tense space demanding and a low-profile height will be maturely applied on specially platforms like aircrafts.

As early as 2003, Munk and his co-workers proposed a novel ultra-wideband antenna-tightly coupled array (TCA) and revealed its working mechanism [2]. Contrary to the traditional antenna design, the TCA is realized the mutual coupling between the adjacent elements, so that the antenna can work in traveling wave mode at low frequency, as a result the array has the characteristics of ultra wideband. For the bandwidth expansion of the TCAs, paper [3] proposed a "double balun" method based on Wheeler's equivalent waveguide theory [4]. The method is halving the distance of the array element along E-plane, which can effectively reduce the input impedance and make it easier to match to 50 Ω. A Marchand balun which is integrated with antenna is employed to feed the antenna in this design and broadens the bandwidth. The impedance bandwidth is 7.35:1 and the antenna can scan to ±45° in E/D/H plane. In order to further widen the bandwidth, a method of loading a resistance structure (resistant sheet or resistant FSS) between the antenna and the ground is proposed in [5-7]. After adding the resistance structure, the short-circuit effect of the ground on the input of the antenna at half wavelength [5] is mitigated, and then the bandwidth of the antenna can be doubled. The impedance bandwidth of this type of TCAs is more than 10:1, but due to the introduction of the loss materials, the broadside radiation efficiency is generally 76%. In addition, it can be realized by designing a thin electromagnetic-structure superstrate [8-9] to substitute the wide angle impedance matching layer (WAIM) which is a thick dielectric superstrate to thin the thickness of TCAs. [8] proposed a metasurface (MS) WAIM superstrate loaded over the TCA antenna plane. The profile height of the TCA is $0.088\lambda_L$ and the impedance bandwidth is 5.5:1. However, the design in [8] needs to hang up the MS-WAIM and keep a certain distance from the antenna surface, so the profile height cannot be further reduced. [9] designs a thin split ring (SR) over antenna plane of the TCA to reduce the profile height. The impedance bandwidth is 7.2:1 and 0.47 wavelength at the upper frequency.

To widen the impedance bandwidth without introducing insertion loss and reduce the profile height of the TCA. A ultra-wideband and low profile TCA with IW and SR is proposed in this paper. The following unique and attractive features will be demonstrated:

1) The IW is specially to mitigate the short-circuit resonance frequency when the distance between the antenna plane and ground is half of wavelength. The impedance bandwidth of the TCA is expanded to 10:1 after loading with the IW without reduce the radiation efficiency.

2) The designed SR-superstrate on the antenna plane reduce the profile height of the TCA to $0.077\lambda_L$ with the distance of $0.5\lambda_H$ between the adjacent elements (λ_H is the high operating frequency).

2. The design of the ultra-wideband and low profile TCA

2.1. The design of the IW

The working mechanism of the IW can be analyzed from the equivalent circuit mode (ECM) of the TCA in Fig.1. The short-circuit resonance frequency caused by the ground will be appeared at every half wavelength when IW unloaded. And the input impedance of the ground when looking into the ground can be calculated form:

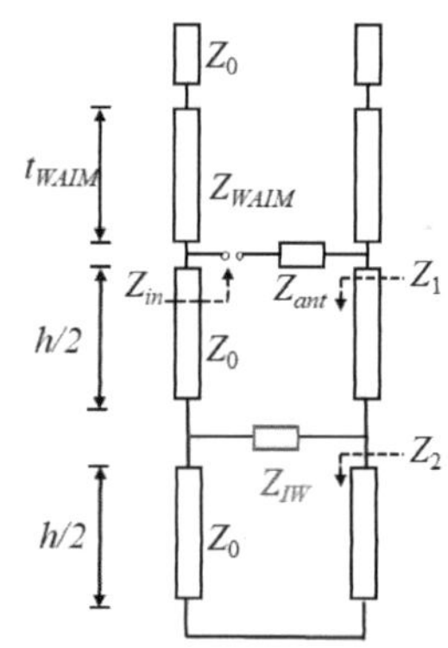

Figure 1. ECM of the TCA loading with IW.

$$Z_1 = jZ_0 \tan(\beta_0 h), \tag{1}$$

where Z_0 is the equal characteristic impedance of the transmission line between the ground and the antenna plane, h is the profile height from the ground to the antenna plane and β_0 is the transmission constant. The characteristic impedance Z_0 can be obtained by

$$Z_0 = \eta_0 \frac{d_E}{d_H} = \sqrt{\frac{\mu_0}{\varepsilon_0}} \frac{d_E}{d_H}. \tag{2}$$

When loading IW at h/2, the input impedance at the IW when looking into the ground is

$$Z_2 = Z_{IW} \parallel jZ_0 \tan(\beta_0 h/2), \tag{3}$$

where Z_{IW} is the equivalent impedance of the IW. Then Z_1 of will be adjusted to

$$Z_1' = Z_0 \frac{Z_2 + jZ_0 \tan(\beta_0 h/2)}{Z_0 + jZ_2 \tan(\beta_0 h/2)}. \tag{4}$$

It can be easily concluded from that after the introduction of IW, the short-circuit resonance generated by the ground can be adjusted if the IW structure is significantly designed.

To verify the above analysis, the electromagnetic simulation software HFSS is used for calculation. The basic unit of the TCA with IW is demonstrated in Fig. 2, and the E-plane distance between the element cells is half of the H-plane distance. The reference of the antenna input impedance is set as 100 Ω and the IW is placed at half of the distance between the antenna plane and the ground. The IW consists of four parallel strips along E-plane. And the four parallel strips are equivalent to four parallel inductors, and each inductance value can be obtained from [10]:

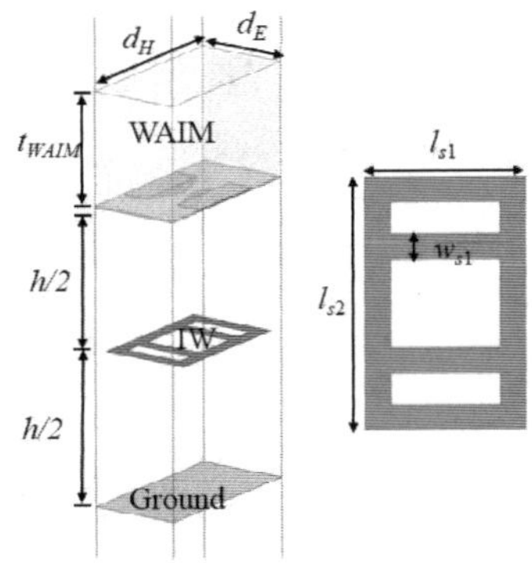

Figure 2. Unit cell of the TCA loading with IW.

$$L_{IW} = \frac{\mu_0}{2\pi} l_{s1} \ln(\frac{l_{s1}}{w_{s1}}) \tag{5}$$

Fig. 3 shows the relationship between the short-circuit resonance point and the number of parallel strips. And the number of the strips is 0, 2 and 4, respectively. It can be seen the resonance frequency moves higher when the number of the parallel lines increases, which means a lower inductance corresponding to a higher resonance frequency.

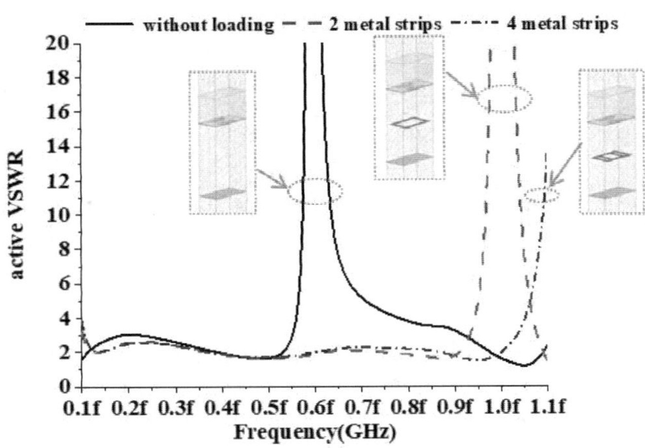

Figure 3. Simulated active VSWR when TCA loading with IW which consists of 0, 2, 4 strips.

2.2. The design of the SR

Paper [9][11] design a SR to reduce the profile height of the TCA. The SR offers extra reactive part to the TCA impedance matching which has the same function as WAIM. The SR has a thinner substarte compared with the traditional WAIM, which reduce the antenna profile height.

Three cases are simulated in Fig. 4. Case 1: the TCA with a traditional WAIM, the thickness of the WAIM is 16 mm. Case 2: the TCA with a thin traditional WAIM, the thickness of the WAIM is reduced to 0.254 mm (the thickness of the WAIM is far less than the operating wavelength). Case 3: the TCA with a SR superstrate, the substrate of the SR is also 0.254 mm. The superstrate in the three cases are all Rogers 5880 (ε_r=2.2). In Fig. 4, a thinner traditional WAIM corresponding to a worse active VSWR in the whole frequency band. After loading the SR, a lower active VSWR is achieved throughout the whole frequency band.

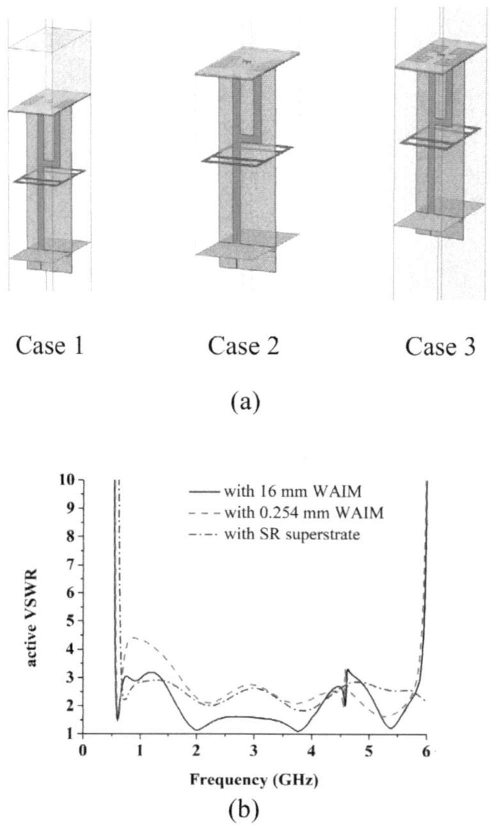

Figure 4. Schematic of the TCA with different superstrates and simulated active VSWRs of three cases. (a) Schematic of three cases; (b) simulated active VSWRs of three cases.

2.3. The design of the single-polarized TCA

According to the above analysis, the ultra-wideband and low-profile single-polarized TCA is designed in Fig. 5. Two cells are combined into one through the Wilkinson combiner and the input impedance is converted into 50 Ω. The radiation patches and parasitic patches are printed on a 0.254 mm Rogers 4350 (ε_r=3.66) substrate. The substrate of Marchand stripline balun is Rogers 4350, and two layers of the substrate of the balun is 1.016 mm. The IW is printed on a 0.508 mm thick PCB which is made up of Rogers 4350. The overall diagram of an 8×8 array is shown in Fig. 5(b) with the element distance 25 mm. The parameter values are summarized in Table 1.

Table 1. Parameters of the TCA with IW and SR (mm)

Parameters	Value	Parameters	Value	Parameters	Value
d_E, d_H	25	l_{b2}	10	w_{b2}	1
h	38	l_{c1}	7.5	w_{g1}	0.2
l_g	21	l_{p1}	3.45	w_{g2}	0.6
l_{b1}	14	w_{b1}	1.6	w_{l1}	0.15

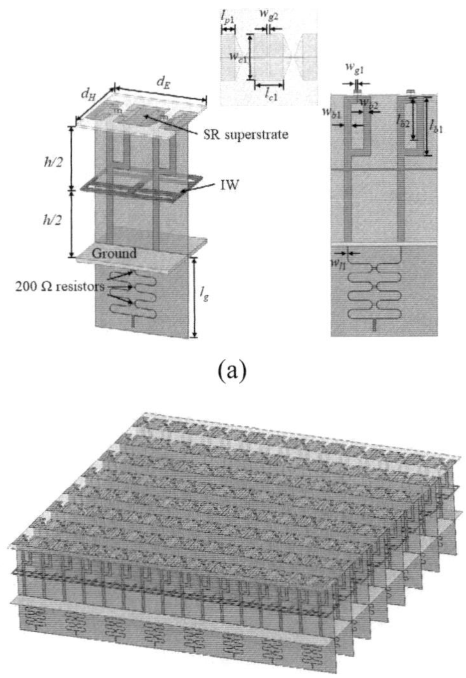

(a)

(b)

Figure 5. Geometry of the single-linear polarized TCA with IW and SR superstrate. (a) unit cell of the TCA; (b) an 8×8 single-polarized TCA.

Fig. 6 shows the active VSWR of the infinite TCA with IW and SR at broadside and when scanning in all plane. The results show that the active VSWR are less than 3.0 throughout the whole operating frequency when scanning to ±60° in E-plane and less than 3.5 throughout the whole operating frequency when scanning to ±60° in D-plane. The active VSWR is deteriorated to 4.2 when scanning to ±45° in H-plane. The scanning impedance matching is more difficult to achieved in H-plane than in E/D planes due to the increasing Floquet mode impedance [12].

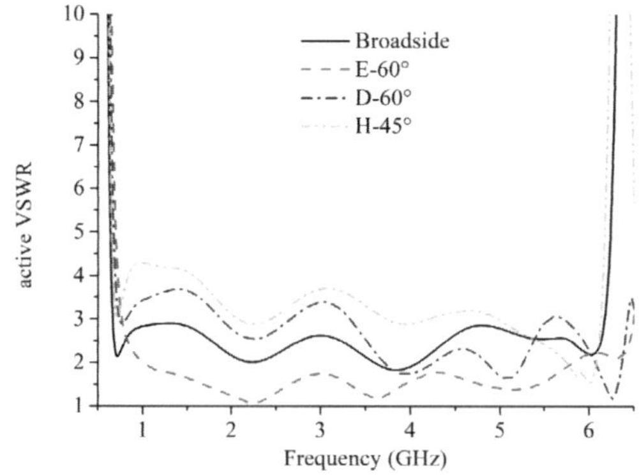

Figure 6. Simulated active VSWR of the proposed TCA with IW and SR superstrate.

Fig. 7 shows the broadside radiation performances of the 8×8 array. The ideal gain is also shown in Fig. 7. The ideal gain can be calculated by $4\pi A/\lambda^2$. The parameter A the antenna aperture area. It can be seen that the gain of the proposed array is near the ideal gain. The cross-polarization ratio of the array is below -20 dB throughout the whole operating frequency band. In addition, the radiation efficiency is more than 90%.

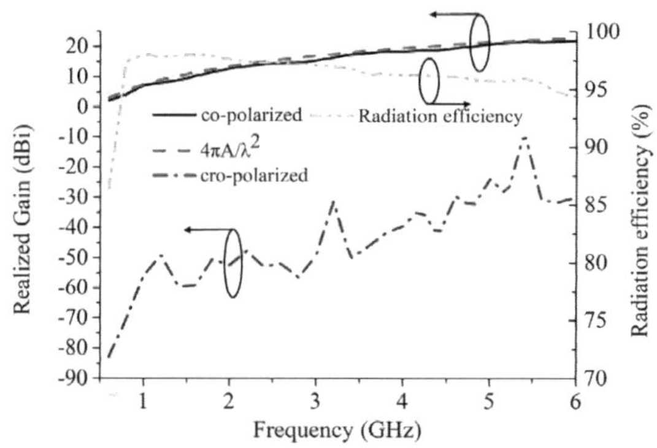

Figure 7. Simulated realized gain and radiation efficiency of the 8×8 single-polarized TCA with IW and SR superstrate.

3. Conclusion

This paper proposed an ultra-wideband and low-profile single-polarized TCA. The TCA loaded with IW achieves ultra-wideband impedance bandwidth for mitigating the short-circuit resonance frequency caused by the ground. The SR superstrate thins thickness of the traditional WAIM to reduce the profile height. The simulated results show that the TCA achieves a 10:1 (0.6-6 GHz) bandwidth for active VSWR less than 3 at broadside. The profile height of the TCA is only $0.077\lambda_L$ and has a high radiation efficiency of more than 90%. It is expected that the TCA with IW and SR superstrate could be widely used in communication and radars.

References

[1] Asplund, H, et al. (2020) Advanced Antenna Systems for 5G Network Deployments[M]. USA: Academic Press, pp. 639-676.

[2] Munk, B., et al. (2003) "A low-profile broadband phased array antenna," Proc. IEEE Antennas Propag. Soc. Int. Symp., vol.2, pp. 448-451.

[3] Doane, J. P., Sertel, K and Volakis, J. L. (2013) "A wideband, wide scanning tightly coupled dipole array with integrated balun (TCDA-IB)," IEEE Trans. Antennas Propag., vol. 61, no. 9, pp. 4538-4548.
[4] Bhattacharyya, A. K. (2006) "Floquet modal functions," in Phased Array Antennas: Floquet Analysis, Synthesis, BFNs and Active Array Systems, 1st ed. Hoboken, NJ, USA: Wiley-Interscience, 2006, ch. 3.
[5] Moulder, W. F., Sertel, K and Volakis, J. L. (2012) "Superstrate-enhanced ultrawideband tightly coupled array with resistive FSS," IEEE Trans Antennas Propag., vol. 60, no. 9, pp. 4166-4172.
[6] Moulder, W. F., Sertel K and Volakis, J. L. (2013) "Ultrawideband superstrate-enhanced substrate-loaded array with integrated feed," IEEE Trans Antennas Propag., vol. 61, no. 11, pp. 5802-5807.
[7] Papantonis, D. K and Volakis, J. L. (2016) "Dual-polarized tightly coupled array with substrate loading," IEEE Antennas Wireless Propag. Lett., vol. 15, pp. 325-328.
[8] Bah, A. O., Qin, P. Y., Ziolkowski, R. W., Guo Y. J and Bird, T. S. (2019) "A wideband low-profile tightly coupled antenna array with a very high figure of merit," IEEE Trans Antennas Propag., vol. 67, no. 4, pp. 2332-2343.
[9] Zhang, H., Yang, S., Xiao, S., Chen Y and Qu, S. (2019) " Low-profile, lightweight, ultra-wideband tightly coupled dipole arrays loaded with split rings," IEEE Trans. Antennas Propag., vol. 67, no. 6, pp. 4257-4262.
[10] Riviere, B., Jeuland, H and Bolioli, S. (2014) "New equivalent circuit model for a broadband optimization of dipole arrays," IEEE Antennas Wireless Propag. Lett., vol. 13, pp. 1300-1304.
[11] Zhang, H., Yang, S., Xiao, S., Chen, Y., Qu, S and Hu, J. (2019) "Ultrawideband phased antenna arrays based on tightly coupled open folded dipoles," IEEE Antennas Wireless Propag. Lett., vol. 18, no. 2, pp. 378-382.
[12] Wheeler, H. A. (1965) "Simple relations derived from a phased-array antenna made of an infinite current sheet," IEEE Trans. Antennas Propag., vol. AP-13, no. 4, pp. 506-514.

BSOT: Bandwidth-saving oblivious transfer protocol with confidential computing

Penglin Yang*, Huizheng Geng, Li Su, Li Lu, Tingting Yang

Department of Security Technology Research; China Mobile Research Institute; Beijing, China

*Email: yangpenglin@chinamobile.com

Abstract. Oblivious transfer is a private preserving algorithm for querying sensitive information without known by data owner. The disadvantage of this scheme is that the data owner needs to send amount of redundant data to puzzle the intended information, which will cause the waste of bandwidth. This paper introduces BSOT, a bandwidth-saving oblivious transfer protocol which uses query agent in confidential computing environment in the data source side to reduce bandwidth consumption. In this paper BSOT protocol is defined in CBOR format. Compared to the traditional OT method, this approach could push all the redundant data flow inside the data source instead of transferring by network, and could also restrict the private data from unnecessary leakage.

1. Introduction

Oblivious transfer (abbreviated as OT) protocol is a two-party computing protocol proposed by Rabin in 1981[1]. It is a basic protocol for PSI[2] and is widely used in secure multi-party computing. In OT protocol, the data owner sends a bunch of messages to the data Collector, but remains oblivious to message owner. This protocol is also known as the dazed transport protocol [3]. OT could be used in many scenarios like data center that carries mass data and has the obligation to provide certain private data to third party. And this third party wants to keep the query information secret from the data center. OT is encryption-based that consumes lots of time during processing. Some researchers improve this time complexity by introducing different encryption algorithms like [4] or modify the protocol sequence like [5][6]. While another main factor of efficiency is the transfer process of OT. In OT protocol, the data sender will send lots of encrypted redundant data to the data collector for obliviousness. The transmission of this mass data will cause a lot of bandwidth in network. This is the basic character of OT protocol which is hard to optimize only by algorithms.

Recent years, there is an emerging concept called confidential computing [7] could provide an appropriate solution to solve the OT bandwidth. Confidential computing is a computing hardware based technology that could provide integrity and confidentiality for runtime application and data. Computing units like SGX[8], TDX[9], SEV[10], CCA[11] and the newly issued NVIDIA[12] could provide confidential computing environment. The environment is protected by computing hardware architecture that even the privileged roles like kernel or hypervisor cannot access or temper the applications and data.

Confidential computing is originated from the concept of TEE. In fact, there already have researchers who combines OT with TEE like paper [13][14]. The major difference between TEE and confidential computing is the usage of remote attestation and trust execution environment provisioning. When using confidential computing, the app owner needs to provision the confidential computing environment with its application first. Then the app owner needs to confirm that the targeting confidential computing environment indeed has the ability of confidential and integrity protection. After verification, app owner will load its private data to the targeting confidential computing environment. The remote attestation procedure[15] in IETF could be used to provide the evidence of the target computing environment. The whole process of provisioning and remote attestation is referred to TEEP[16] in IETF.

Confidential computing's feature could be used in OT, and is more efficient than traditional algorithm improvement. In this paper, we design the bandwidth-saving oblivious transfer protocol using confidential computing and CBOR[17] coding format. Section 2 illustrates the architecture of BSOT, which includes remote attestation and trusted execution environment provisioning. Section 3 depicts the CDDL [18] described CBOR format in BSOT protocol. Section 4 makes a discussion about BSOT security analysis. In section 5 we make a performance analysis, in which the result shows major improvement of time efficiency. In the end, the conclusion of BSOT is generated.

2. Architecture of bost

2.1. Specific architecture

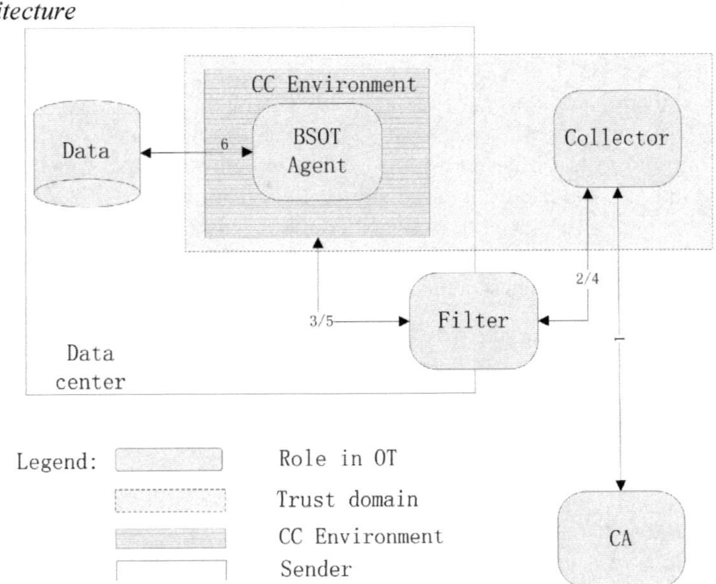

Figure 1. Architecture of BOST

As shown in figure 1 is the architecture of BSOT, three main components are Collector, Data Center and CA. The Collector is the data requester which wants to query private data without known by Data Center. The Data Center is the Data owner which has amount of private data that cannot perceive what data the Collector queries. Inside the Data Center have three components: Data, confidential computing environment and Filter. Data is the data set that contains private information. Confidential Computing (CC) Environment is hardware based that could provide confidentiality and integrity to application runs inside it. In BSOT architecture, this application is BSOT Agent which is in charge of querying private data by OT algorithm. BSOT Agent and dataset are in the same computing node. The third component in Data Center is the Filter, which is different to the concept of TEEP Broker in TEEP. Filter is not only used for relay information between Collector and BSOT Agent, but also in charge of filtering export information in case extra secret information may be leaked. CA in BSOT provides endorsement to confidential computing hardware, by which the Collector could trust the CC Environment.

The trust domain in BSOT is a virtual domain which includes Collector, CC Environment and BSOT Agent. Based on remote attestation, Collector, CC Environment and BSOT Agent could establish this virtualized trust relationship. Even though the BSOT Agent is inside the Data Center's computing node, the Data Center cannot access or tamper the information inside BSOT Agent.

OT algorithm aims to protect the privacy of Collector by collecting redundant private data. Even though this data is encrypted, the Data Center may still doesn't want this data to be leaked outside to avoid the potential credential stuffing and brute force attack. The trust domain could make sure query information cannot be exposed to Data Center. But the Data Center also needs to supervise the trust domain wouldn't export this encrypted redundant data outside. The Filter which is controlled by Data Center could perform this function.

Numbers inside the arrows in fugue1 shows the basic steps of how BSOT works. The explanation of these steps is shown below.

- Step 1, Collector request endorsement about CC Environment from CA. CA responses endorsement like X509 certificate to Collector.
- Step 2, Collector deploys private-free BSOT Agent to the Data Center during Filter.
- Step 3, Filter creates CC Environment in Data Center and install BSOT Agent in it.
- Step 4, Collector challenges the trustworthiness of CC Environment. CC Environment responses evidence about its identity, integrity and confidentiality features.
- Step 5, after verification, Collector establish security channel with BSOT Agent during Filter and then provisions privacy information like query details in BSOT Agent.
- Step 6, BSOT Agent process traditional OT procedure with dataset. This step should be processed inside the Data Center computing node, which could avoid network bandwidth and encrypted private data leakage caused by OT algorithm. And since the integrity and confidentiality provided by CC Environment, dataset cannot access or perceive what data is decoded in BSOT Agent. The Filter could only see the encrypted byte but cannot see the plaintext during the exportation.

2.2. Provision of BSOT agent

In this paper, we use SUIT[19] to provision the BSOT software to CC Environment. The implementation of BSOT Agent is based on the specification of Data Center. The SUIT architecture uses manifest and CBOR to express the specific software of BSOT Agent, which is more concise and precise than the raw binary package during the transmission.

2.3. Remote attestation to establish trust domain

Step 4 in BSOT architecture is the remote attestation action which is used for creating trust domain of CC Environment. Remote attestation is based on the trust anchor in confidential computing hardware, which is like SGX quote enclave[20], AMD SP[21], etc. The trust anchor cannot be compromised because of the protection of hardware, which is also the root of trust in BSOT. The trust anchor could provide a trusted measurement function to measure the CC Environment and make signature of the measurement result. The Collector will verify the remote attestation measurement result by the given endorsement of CA. In the industry implementation, the CA is an X509 PKIX which could provide certificate of confidential computing CPU. And the reference value could be generated by Collector or other third party authority.

2.4. CBOR encoding to fit filter

In our design, BSOT needs Filter to oversee the export data from BSOT Agent to prevent redundant encrypted data leakage. So the secret channel between BSOT Agent and Collector should use concise binary format to define every bit's definition. In this paper, we use CBOR as basic encoding format and COSE[22] as encryption format. . CBOR is as concise binary object representation which could represent data format in binary. The Filter will use the same format to extract payload to monitor the data flow.

3. CDDL based data format

The BSOT Data format is described by CDDL. CDDL is abbreviation of concise data definition language, which could represent CBOR by human readable language. Compared to other data format like JSON, CBOR is more efficient and short length. The following content in this section uses CDDL to express data format in BSOT. Need to notice that Step 1 in BSOT architecture could be an out of band method which does not have to be a protocol online. As a consequence, we use COAP [23] [24] as application layer protocol. COAP is more delicate than HTTP and can be used in Filter with better efficiency. For security concern, COAP usually uses DTLS [25] to build security channel.

The overall protocol steps are shown in figure 2 below. The Collector is the COAP Client, the BSOT Agent and Filter are two different COAP Servers. The overall CDDL could be found in https://github.com/PenglinYang/bsot.

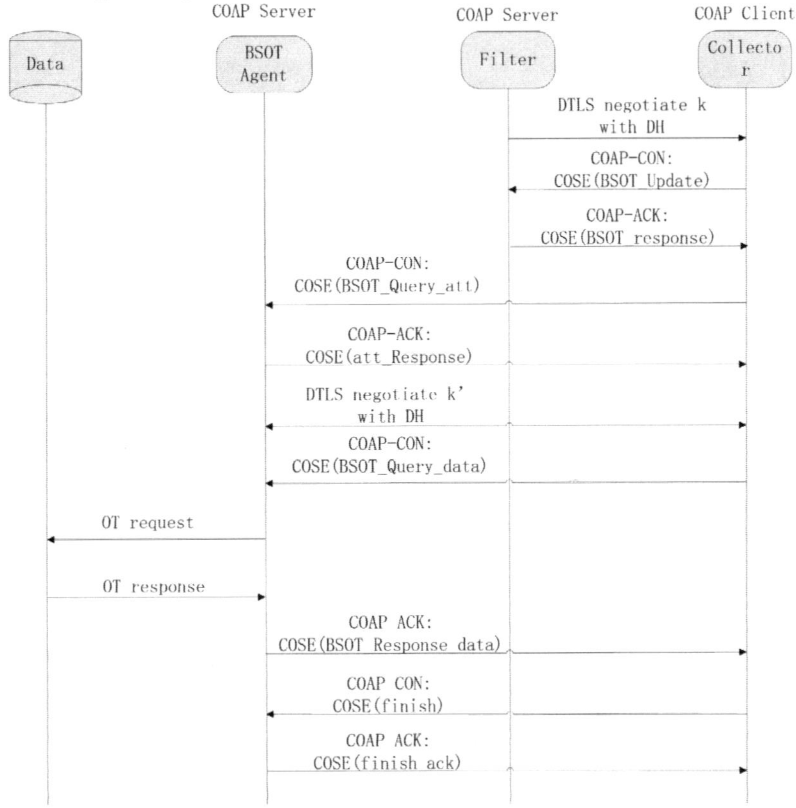

Figure 2. BSOT Protocol Steps

3.1. BSOT_update and response

BSOT Agent in charges of performing OT algorithm with data set. While in the original CC Environment there may have no BSOT Agnet on-premise. The BSOT_update message is used for deploying BSOT Agent in CC Environment as an opening step in BSOT protocol. BSOT Agent implementation and query mechanisms belong to the specific Collector and Data Center. For example, BSOT agent could uses memory cache to speed up the query time, or could uses other improved algorithm to replace the original OT algorithm. The following is the digest of BSOT_update message, which refers to the TEEP Update message and SUIT message.

```
BSOT_update = [
   manifest-list => [ + bstr .cbor SUIT_Envelope ],
   * $$update-extensions,
]
```

After the installment of BSOT Agent, the Filter will response a message to inform the Collector. The inform message also includes the network address of the COAP server in BSOT Agent. The CDDL digest is shown below.

```
bsot-response = [
msg => text .size (1..128),
bsot-agent-add => bstr .size(64)
   * $$teep-success-extensions,
]
```

3.2. BSOT query-att and response

This round uses EAT[26] to represent remote attestation evidence. EAT is a CBOR format data set which is used for representing remote attestation evidence. EAT is protected by signature based on COSE. The query and response message digest are shown below.

```
query-att = [
   challenge => bstr .size (8..512),
   versions => [ + version ],
   * $$query-request-extensions
]

att-response = [
   evidence => [+ bstr .cbor SUIT_Envelope ],
   * $$teep-option-extensions
 ]
```

3.3. BSOT query_data and response

After the query and response round for remote attestation, the DTLS needs to be used again for negotiating secret *k* between Collector and BSOT Agent. This key is used for encrypting query and response message of private information in Data Center. The information that the Filter needs to know is the requested item quantity and data length, by which the Filter could know how much information the Collector requests and the BSOT Agent sends. Any extra data will be cut by Filter in case the unnecessary leakage of private data.

```
query-data = [
   item-count => uint,
    item-key-list  => [ +bstr .size (8..512)],
   versions => [ + version ],
   * $$query-request-extensions
]

Data-response = [
   match-count => uint,
    item-list  => [ +bstr ],
   * $$query-request-extensions
]
```

3.4. COSE protected message

In BSOT, we use COSE to protect the integrity and confidentiality of BSOT messages. In our design, the key exchange in DTLS will be used for COSE encryption. First, DTLS establishes connection between Client and Server based on UDP. During this step, the ECHDE algorithm will be used to generate *k* as raw shared key. Then, the COSE Recipient architecture will be used to describe the parameters of ECDHE and KDF (Key Derivation Key), which will be used for generating CEK (Content Encryption Key) with the *k*. In the end, the COSE cyphertext architecture will be used to express the encryption content which is encrypted by the CEK. Both in the 3.1, 3.2 and 3.3 rounds use this process flow. While the difference is that 3.1 and 3.2 use the key generated between Fliter and Collector. And 3.3 uses the key generated between BSOT Agent and Collector. As a consequence, the Filter could monitor the data flow in COSE format but could not see the plaintext of it.

4. Security analysis

BSOT is a hardware and software integrated architecture in which its security is based on hardware trustworthiness and encryption algorithm security. From the perspective of Collector, it wants to make sure its query information is confidential and integrated. From the perspective of Data Center, it mainly concerns the potential unnecessary leakage of private data. The information exchange between Data Center and BSOT Agent is implemented by BSOT Agent, which is a software function based on specific Data Center architecture. The OT algorithm data flow could be demonstrated in Figure 3. In this figure, the value *Key* is the key parameter that the Collector wants to query. The *PK(N, e) SK(N, d)* is the asymmetric key pair parameters of RSA. Data Center generates key pairs and sends the PKs to the BSOT Agent. BSOT Agent chooses the PK_{key} and uses its e_{key} parameter to generate *v*. The Data Center then uses the *v* to generate m which will be used to create $m \oplus x$. The x is the map of *key* and its value. In the end, the BSOT Agent will resolve the x_{key} by XOR and send back to Collector.

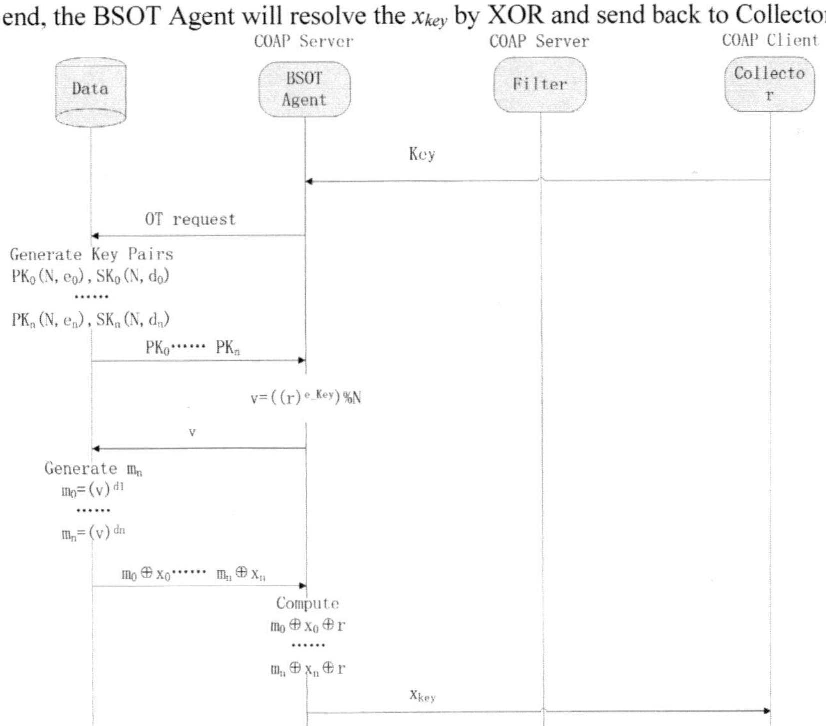

Figure 3. OT specification between Data Center and BSOT Agent

4.1. Security analysis of collector

In BSOT, security risks about Collector are determined by three elements, security of OT, security of confidential computing and security of COSE. We denote the security probability about these elements as P_{OT}, P_{CC}, and P_{COSE} respectively. The overall security probability about BSOT from the perspective of Collector would be denoted as $P_{BSOT}=P_{OT}*P_{CC}*P_{COSE}$.

P_{OT} is based on the algorithm of OT, in which its security is based on the difficulty of cracking *((r)e_key)%N* in figure 3. In this expression, *r* means a random chosen by Collector, and the *e_key* is the item in PK_{key} which mapped to the *Key* parameter. This is also the security principle of asymmetric encryption algorithm. We denote the security probability as P_{AE}. Thus $P_{OT} = P_{AE}$.

P_{CC} is based on the hardware design and remote attestation of CC Environment. We denote the probability of compromise of confidential computing hardware as P_{HW}. Right now P_{HW} cannot be defined as a deterministic expression because its trustworthiness is based on the endorsement of hardware manufacture. The remote attestation evidence in CC Environment is described by CDDL and encoded by COSE format. While the security principle is still based on basic encryption algorithms like signature and Hash functions. The security of Hash function could be denoted as P_H. The security of signature also based on the asymmetric encryption algorithm. Thus, the $P_{RA} = P_H*P_{AE}$. As a result, the $Pcc = P_{HW} * P_H * P_{AE}$.

P_{COSE} is the security possibility of security channel establishment, KDF and symmetric encryption of content, which denoted as P_{CE}, P_{KDF}, and P_{SE} respectively. The P_{CE} is based on the security of Diffie-Hellman algorithm, which could be represented as P_{DH}. Based on the definition of KDF in RFC8152, the implementation of KDF is based on Hash function like SHA256, which can be represented as P_H. Thus, the $P_{COSE} = P_{CE} * P_{KDF} * P_{SE} = P_{DH} * P_H * P_{SE}$.

As a summary, the security possibility of BSOT Collector could be represented as follow.

$$P_{BSOT} = P_{OT} * P_{CC} * P_{COSE} = P_{AE} * P_{HW} * P_H * P_{AE} * P_{DH} * P_H * P_{SE} = (2P_{AE})*(2P_H)*P_{DH}*P_{SE}*P_{HW} \quad (1)$$

The equation shows that BOST protocol relies on two kinds of security factors. The first factor is P_{HW}, which is based on the hardware design. The second factor is different encryption algorithms which have been proofed secure enough to be used in network. As long as the confidential computing hardware has no flaw in design and manufacturing, the BSOT protocol is secure enough to protect privacy of Collector.

4.2. Security analysis of data center

From the perspective of Data Center, the security concern is the potential unnecessary leakage of private data. There are two mechanisms to protect the private data. The first is the encryption of XOR. As shown in figure 3, the Data Center will export a XOR data set to BSOT agent which is represented as $m_n \oplus x_n$. Based on OT algorithm, the encrypted data set is practical impossible to decode. The second mechanism is the filter function executed by Filter. The Filter uses the COSE format to measure the actual transport message amount in case unnecessary leakage of private data.

5. Performance analyses

In this design, we use a simulation virtual machine environment to evaluate the performance of BSOT. The following table shows the parameter of this simulation platform.

Table 1. Parameter of simulation platform

Parameter	Value/Type
Confidential computing architecture	TrustZone
CPU basic frequency	2.6G HZ
Redundant data amount	1GB

Data unit size in Data Center	32Byte
Network bandwidth	100Kb/s
Network latency	5ms

In the original OT, time mainly spends in redundant data generating and transferring. In BSOT, time mainly spends in redundant data generating. The transferring of intended data is more efficient than original OT. The following figure shows the experiment result of both methods. The negotiation stage means the step 1-5 in section 2.1. The redundant generating stage means the OT generates the redundant data for obliviousness. The parse intended data stage means the XOR operation to generate intended data. The result transfer means the transfer between BSOT agent and data collector. As we expected, the BSOT method could save the transferring time in network. As a result, the BOST is much more efficient than the original OT method.

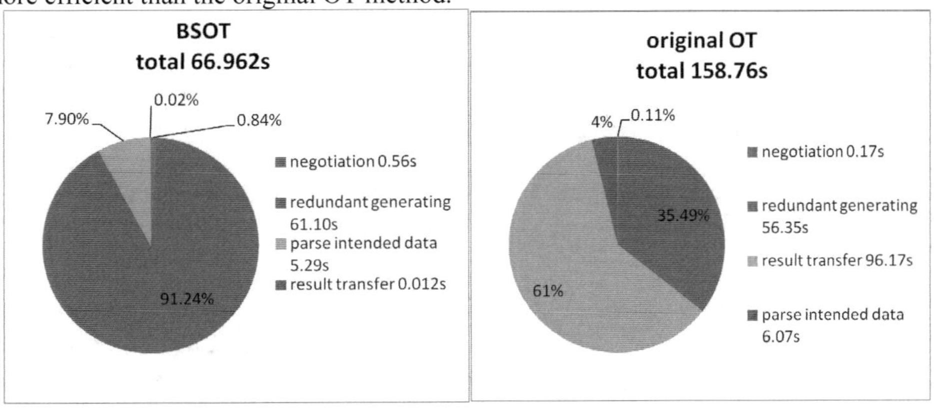

Figure 4. Performance Comparison between BOST and OT

6. Conclusion

In this paper, we introduced BSOT, which is an OT algorithm and confidential computing based query protocol. This protocol could restrict the transmission operation of original OT algorithm inside the computing node of data center and avoid the network bandwidth usage. In addition, we also use CBOR as basic protocol and use CDDL to describe the data format. As a summary, BSOT is more efficient than the original OT algorithm with the help of confidential computing and can be used in data center and other scenarios.

Reference

[1] Rabin, M. O. (2005) "How to exchange secrets with oblivious transfer," Cryptology ePrint Archive.

[2] Chase, M., Miao, P. (2020) "Private Set Intersection in the Internet Setting from Lightweight Oblivious PRF," Proc. Annual International Cryptology Conference (CRYPTO), Springer, pp. 34-63, doi: 10.1007/978-3-030-56784-2

[3] Wang, X., Li, Z. (2020) "Research on the security Oblivious Transfer protocol based on ECDDH," Proc. Journal of Physics: Conference Series, IOP Publishing, PP. 152.

[4] Chou, T., Orlandi, C. (2015) "The simplest protocol for oblivious transfer," Proc. International Conference on Cryptology and Information Security in Latin America (LatinCrypt 2015). Springer, PP. 40-58, doi: 10.1007/978-3-319-22174-8

[5] Peikert, C., Vaikuntanathan, V., Waters, B. (2008) "A framework for efficient and composable oblivious transfer," Proc. Annual international cryptology conference (Crypto 2008), Springer, pp. 554-571, doi: 10.1007/978-3-540-85174-5.

[6] Ishai, Y., Kilian, J., Nissim, K. (2003) "Extending oblivious transfers efficiently," Proc. Annual

International Cryptology Conference (Crypto 2003), Springer, pp. 145-161, doi: 10.1007/b11817

[7] Rashid, F. (2020) "The rise of confidential computing: Big tech companies are adopting a new security model to protect data while it's in use," IEEE Spectrum, vol. 57, pp. 8-9, doi: 10.1109/mspec.2020.9099920.

[8] Costan, V., Devadas, S. (2016) "Intel SGX explained," Cryptology ePrint Archive.

[9] Sardar, M., Musaev, S., Fetzer, C. (2021) "Demystifying attestation in intel trust domain extensions via formal verification," IEEE Access, vol. 9, pp. 83067-83079, doi:10.1109/ACCESS.2021.3087421 .

[10] SEV-SNP, A. M. D. (2020) "Strengthening VM isolation with integrity protection and more," White Paper.

[11] Mulligan, D., Petri, G., Spinale, N. (2021) "Confidential Computing—a brave new world," Proc. 2021 International Symposium on Secure and Private Execution Environment Design (SEED 2021). IEEE, pp. 132-138, doi: 10.1109/SEED51797.2021.

[12] Vaswani, K., Volos, S., Fournet, C., Diaz, A., Gordon, K., Vembu, B., Wilkinson, D. (2022) "Confidential Machine Learning within Graphcore IPUs," preprint arXiv:2205.09005.

[13] Yang, Y., Ding, X., Deng, R. (2008) "An efficient PIR construction using trusted hardware," Proc. International Conference on Information Security, Springer, pp. 64-79, doi: 10.1007/978-3-540-85886-7_5.

[14] Wang, S., Ding, X., Deng, R. (2006) "Private information retrieval using trusted hardware," Proc. European Symposium on Research in Computer Security, Springer, pp. 49-64, doi: 10.1007/11863908_4

[15] Birkholz, H., Thaler, D., Richardsom, M., Smith, N., Pan, W. (2022) "Remote Attestation Procedures Architecture," IETF, https://datatracker.ietf.org/doc/draft-ietf-rats-architecture/

[16] Pei, M., Tschofenig, H., Thaler, D., Wheeler, D. (2022) "Trusted Execution Environment Provisioning (TEEP) Architecture," IETF, https://datatracker.ietf.org/doc/draft-ietf-teep-architecture/

[17] Bormann, C., Hoffman, P. (2013) "Concise Binary Object Representation," IETF, https://www.rfc-editor.org/rfc/rfc7049.html

[18] Birkholz, H., Vigano, C., Bormann, C. (2019) "Concise Data Definetion Language(CDDL): A Notational Convention to Express Concise Binary Object Representation(CBOR) and JSON Data Structure," IETF, https://www.rfc-editor.org/rfc/rfc8610.html

[19] Moran, B., Tschofenig, H., Birkholz, H., Zandberg, K. (2021) "A Concise Binary Object Representation (CBOR)-based Serialization Format for the Software Updates for Internet of Things (SUIT) Manifest," IETF, https://datatracker.ietf.org/doc/draft-ietf-suit-manifest/

[20] Anati, I., Gueron, S., Johnson, S. (2013) "Innovative technology for CPU based attestation and sealing," Proc. Proceedings of the 2nd international workshop on hardware and architectural support for security and privacy. ACM.

[21] Kaplan, D., Powell, J., Woller, T. (2016) "AMD memory encryption," White paper.

[22] Schaad, J. (2017) "CBOR Object Signing and Encryption (COSE)," IETF, https://datatracker.ietf.org/doc/rfc8152.html

[23] Shelby, Z., Hartke, K., Bormann, C. (2018) "The Constrained Application Protocol,", IETF, https://datatracker.ietf.org/doc/rfc7252/

[24] Arvind, S., Narayanan, V. (2019) "An overview of security in CoAP: attack and analysis," Porc. 2019 5th international conference on advanced computing & communication systems (ICAconfidential computingS). IEEE, pp. 655-660

[25] Rescorla, E., Tschofenig, H., Modadugu, N. (2022) "The Datagram Transport Layer Security (DTLS) Protocol Version 1.3," IETF, https://datatracker.ietf.org/doc/rfc9147/

[26] Lundblade, L., Mandyam, G., O'Donoghue, J. (2022) "The Entity Attestation Token (EAT)," IETF, https://datatracker.ietf.org/doc/draft-ietf-rats-eat/

Multi-user detection for CPM spread system with turbo in VDES

Feiyu Wu[1]*, Qianwu Zhang[1, a], Qian Xiang[2, b], Huihui Yin[2,c], Xiyan Zhang[2,d]

[1] Shanghai University, Shanghai, China
[2] Shanghai Aerospace Electronic Technology Institute, Shanghai, China

*wfy2020@shu.edu.cn; [a]zhangqianwu@shu.edu.cn; [b]53811808@qq.com; [c]1840513481@qq.com; [d]zhangxiyan@sast804.com

Abstract. The multi-user detection technology uses the spread sequence, delay, amplitude and phase information of each user to jointly detect users, which can effectively suppress multiple access interference,It is one of the key technologies in satellite communication system to make full use of uplink spectrum resources and improve system performance and capacity. Continuous phase modulation spread signal can adapt to multi-user communication due to the random variation of spread sequence, which is gradually applied to satellite communication. QPSK-CPM spread modulation scheme is adopted in the random access channel of VHF data exchange system. Aiming at the multi-user detection problem of QPSK-CPM spread system, proposing a multi-user detection algorithm based on interference elimination joint Turbo iterative detection based on the characteristics of CPM spread sequence. Compared with the traditional serial interference cancellation algorithm, the algorithm can effectively eliminate the multiple access interference problem caused by user aliasing by using the characteristics of Turbo codes. The theoretical simulation shows that the algorithm has a good detection effect on multi-user aliased signals.

1. Introduction

VHF data exchange system (VDES) is a new generation of maritime data exchange system, which provides service requirements for global marine ship and satellite communication [1][2]. VHF data exchange (VDE) is a subsystem of VDES. VDE has a high bandwidth, and the information that ships can obtain is extremely enormous. Not only the location safety information can be transmitted through it, but also the port operation and freight information can be transmitted through it. VDE signals have 12 channels [3], where data confirmation signaling signals and random Access channels (RAC) use Link ID 20 physical layer frame format, The physical layer frame format of Link ID 20 is modulated by quadrature phase shift keying - continuous phase modulation spread (QPSK-CPM). Due to the multi-user competition during random access, The improvement of QPSK-CPM spread multi-user detection capability can effectively improve user random access capability, improve the overall performance of VDE communication, and lay a solid foundation for complex and multi-functional maritime VDE communication applications in the future.

For multi-user detection research, it was first applied to direct sequence spread system. The optimal multi-user detection algorithm proposed by Verdu S. has the best theoretical detection performance[4]. However, the algorithm has not been applied to the actual project due to the complexity of calculation. After that, the decorrelation detection algorithm[5] and linear minimum mean squared error (LMMSE)

[6][7]proposed by other scholars has not been applied to engineering because it involves the calculation of matrix. Until the idea of interference cancellation was introduced into the research of multi-user detection, the research of multi-user detection was not applied to practical projects. At present, the mainstream interference cancellation algorithms mainly include serial interference cancellation algorithm[8] and parallel interference cancellation algorithm[9]. Many subsequent algorithm studies are also based on the idea of interference cancellation.

For the study of CPM spread multi-user detection system, in [10][11] Cotttatellucci L and Muller R described the application of LMMSE in direct spread sequence system. The LMMSE method is used for multi-user detection, and the simulation results show that the decoding rate performance is very good. Later, Muller R used LMMSE method to detect multi-users in CPM spread system in [12], but the feasibility of using this method under ideal Gaussian white noise channel was only expounded , and theoretical analysis was only carried out, without simulation verification.

Turbo can realize reliable transmission close to Shannon theory, at the same time it is a combination of soft input/output decoding and iterative decoding, random and parallel cascade convolution code and other advanced technology, to make the most of the information contained in the signal [13]. as the basic idea of multi-user detection, which reduce the Multiple access interference, This is consistent with the decoding idea of Turbo. On this basis, this paper designs a QPSK-CPM spread joint Turbo iterative multi-user detection algorithm. The Turbo decoding based on the sliding window maximum A posteriori(SW-MAP) algorithm is cascaded with the multi-user detector, and the decoding results are feedback to the multi-user receiver for iteration. When the iteration is completed, Each decoder outputs the estimated value of each user's information sequence by hard decision.the algorithm can effectively remove multiple access interference. The structure of the algorithm is simple and can be applied to practical projects.

2. **System model**

Satellite component of the VDE (VDE-SAT) a slot length lasts 26.67ms, A frame of Link ID 20 has five slots, lasting 133.35ms, 263 symbol length. Frame contains training sequences, guard intervals, data areas and pilot sequences. The data area has a total of 96 bits, and the training sequences and pilots are inserted into the data area after Turbo coding. After that, QPSK-CPM spread modulation was performed. The specific uplink transmitter model is shown in Figure 1:

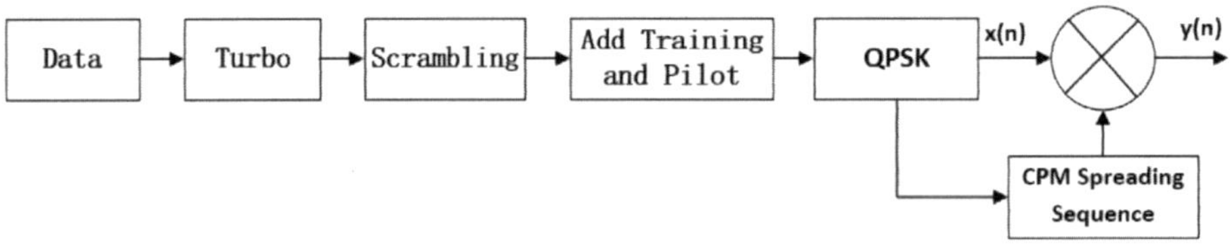

Figure 1. Transmitter Model

The QPSK-CPM spread communication system is shown in the Figure 2.

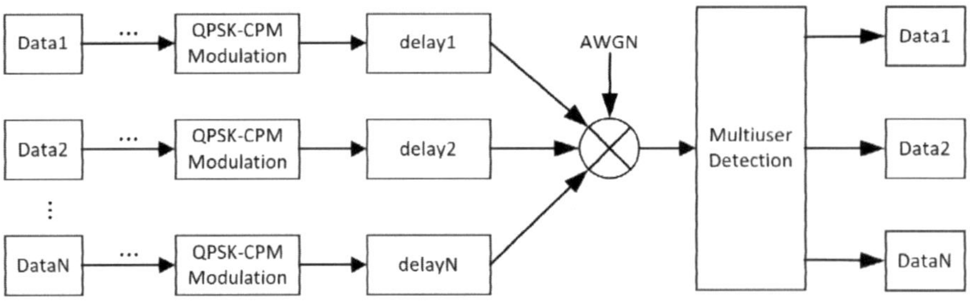

Figure 2. QPSK-CPM Spread Communication System

Similar to ground communication, the signals received by the multi-user receiver include random noise interference, multiple access interference (MAI), co-channel interference (CCI) and etc. After receiving signals are down-conversion and low-pass filtering, and sampled by an analog-to-digital converter, the receiving model can be expressed as :

$$r(t) = S(t) + n(t) = \sum_{k=1}^{K} A_k S_k(t - t_k) + n(t) \tag{1}$$

Where, let $A_k$is the amplitude of each user's signal, $t_k$is the delay time of user's signal, n(t)is additive Gaussian white noise, andS_k is the user's signal after CPM spread modulation. S_k can be expressed as:

$$S_k(t) = \sum_{k=1}^{K} x_{k,m} c_{k,m}(t - mT_s) \tag{2}$$

Where, $x_{k,m}$is the symbol of the user at the m time, $c_{k,m}$ is the CPM spread chip of the user at the m time, and T_s is the symbol cycle. CPM spread chip sequences is composed of CPA and CPE. A set of CPA and CPE spread sequences constitute CPM spread sequences, as {CPA, CPE}. Each group of CPA and CPE spread sequences share 4 groups respectively, so a group of CPM spread sequence has 16 combinations. Each constellation point symbol after QPSK mapping can be used to select the CPM spread sequence for modulation based on the differential calculation of the constellation points after QPSK mapping.

After CPM spread modulation for each user, the output modulation sequence y_k is given by:

$$y(k) = \begin{cases} x(n) \cdot cp_a\left(l_a, p_a(n)\right) \\ x(n) \cdot cp_e\left(l_e, p_e(n)\right) \end{cases} \tag{3}$$

where x(n) donates the QPSK modulated input signal of length BL symbols. Thus, $n \in [1, BL]$. Let SF donates spread factor, $SF = 16$. NS donates over sample, $\mathrm{NS} = 4, 8$ or 16. The first dimensional, representing sample time, table indexes l_a and l_e are given by:

$$l_a \in [SF \cdot NS \cdot n - SF \cdot NS - 1 : SF \cdot NS \cdot n] \tag{4}$$

$$l_e \in [SF \cdot NS \cdot n - SF \cdot NS - 1 : SF \cdot NS \cdot n] \tag{5}$$

The second dimensional table indexes, $p_a(n)$and $p_e(n)$, depend on $x(n)$ and are based on differential QPSK symbol quadrant computation. Given the applied Gray-coded QPSK bits-to-symbol mapping definition, the belonging quadrant is given by:

$$\mathrm{q}= \begin{cases} 0, \text{ for QPSK input bits equal to 11} \\ 1, \text{ for QPSK input bits equal to 01} \\ 2, \text{ for QPSK input bits equal to 00} \\ 3, \text{ for QPSK input bits equal to 10} \end{cases} \tag{6}$$

and the second dimensional table indexes

$$p_a(n) = \begin{cases} 0, & for\ n = 1 \\ \left(q(n) - q(n-1)\right) \% \, 4, & for\ n > 1 \end{cases} \tag{7}$$

and

$$p_e(n) = \begin{cases} \left(q(n+1) - q(n)\right) \% \, 4, & \text{for } n < BL \\ 0, & \text{for } n = BL \end{cases} \tag{8}$$

The multi-user detector cannot track which CPM sequence of the user is used to communicate a certain data symbol. Due to this lack of conditions, the detector needs to process the expected CPM spread sequence very similar for two reasons: 1) The difference in CPM spread sequences reduces the mean value of the medium waveform, thus reducing the energy of the received signal. 2) The greater the difference of CPM sequences of interfering users, the larger the subspace traversed by interference from this user[14]. The waveform differences that vary within the user's CPM spread sequence set are called guard intervals. Deploy the guard intervals to ensure they stay smooth during symbol transitions. The number of guard intervals required depends almost entirely on diffusion factors. Therefore, the relatives cost due to the guard intervals decreases as the expansion coefficient increases.A certain segment of the CPM spread sequence is shown in the Figure 3 . As can be seen from the figure, the left side is the CPA spread sequence, the right side is the CPE spread sequence, the CPA spread sequence has only a slight difference at the head, and the tail of the CPE spread sequence is slightly different. Multi-user detector detects multi-user information according to the difference of guard intervals.

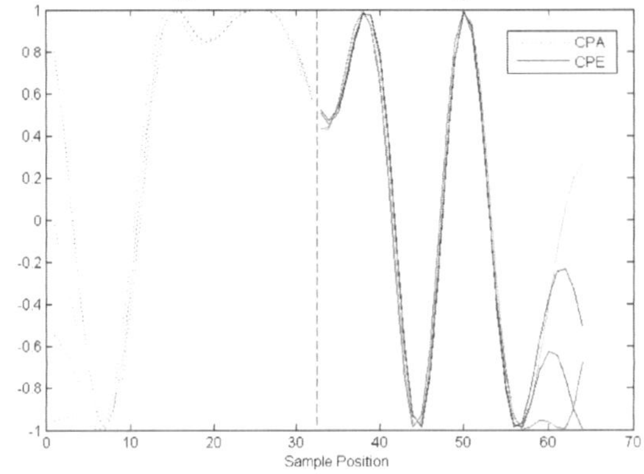

Figure 3. CPM spreading sequence

3. Multi-user detection

3.1 *serial interference cancellation (SIC)*

The basic idea of SIC is to reconstruct the signal of the user's decision result at the receiving end, and then subtract the reconstructed signal from the received signal and pass matched filtering to achieve the purpose of multiple access interference cancellation. The detector generally consists of multiple stages. The SIC takes the signal minus the estimated multiple access interference as the output of the next stage, and repeats this cycle to improve the detection accuracy. The structure of each stage is basically the same, which is convenient for engineering implementation. SIC first needs to sort the users according to the power of the users, and give priority to eliminating multiple access interference for users with higher power. For each user, only the interference caused by users with higher power is eliminated, and the interference caused by users with lower power is ignored. The specific detection process is as follows. Assuming that there are three users in the system, the SIC first sorts the users according to the signal power. The user with the highest power is User 1, the user with the next highest power is User 2, and the user with the lowest power is User 3. According to the power level, SIC first demodulates user 1, reconstructs and eliminates the information determined by demodulation user 1, and detects, reconstructs, and eliminates user 2; after eliminating the reconstruction information of user 1 and user 2, User 3 is detected. Finally, the original information of the three users is recovered.

3.2 *SIC with turbo iteration*

For complex QPSK-CPM spread multi-user systems, the combined Turbo detection and decoding can provide great coding gain, which is derived from the diversity effect of the interleaver. At the same time, multi-user detection through joint Turbo can effectively eliminate multiple access interference.The multi-user receiver designed is shown in Figure 4. Since the CPM spreading sequence is only different from the head and tail, the coherent soft output algorithm is used for de-spreading and demodulation. The demodulation data of these soft outputs are sent to the Turbo decoding module for decoding. The decoded data is divided into two channels, one for output and one for CPM spread modulation reconstruction. The reconstructed data is then sent to the coherent de-modulator to eliminate the user data so as to reduce the multiple access effect. That is to say, the previously detected user information is eliminated before demodulation of the next user. In this way, the system performance can be improved by eliminating the system through cascade iteration. The specific steps are as :

a) The coherent de-modulator performs coherent demodulation for the first intercepted user data. The user data is correlated with all CPA and CPE, calculated according to the maximum likelihood value q̇, and the demodulation soft information of the user is output according to the value. The output soft information of the demodulation *demod* is as:

$$\dot{q}(k,i) = max_i(\sum_{i=0}^{3} S_{k,\ i}c_{k,i}^{*}) \tag{9}$$

$$demod(k,i) = \sum_{BS*NS(i-1)}^{BS*NS*i} s_{k,\ i}c_{k,\ q(k,\ i)}^{*} \tag{10}$$

b) Before Turbo decoding, QPSK de-mapping is required for the information output by the de-modulator and effective data area is intercepted. the valid data is sent to the decoding module. Turbo decoding adopts soft decoding based on the SW-MAP algorithm, and the decoding output is a hard output, bit 0 or 1.

c) The Turbo decoding output is sent to the detection system in one way, and the CPM modulation reconstruction is carried out in the other way and sent to the coherent modem for elimination. The information after elimination of the modem is:

$$r(t) - \sum_{i=1}^{k} s_k(t - t_i) \tag{11}$$

In this way, the information of previously detected users is removed when the subsequent users are demodulated, which greatly improves the detection efficiency.

d) Iterate through the above steps until all users have been detected.

e) Turbo second iteration. All users' information output in the previous step is a subsystem and sent to the demodulator. Repeat the above steps again. In this step, all user reconstruction information except the target user is eliminated to reduce multiple access interference. The information after elimination before demodulation is as:

$$r(t) - \sum_{i \neq k}^{K} s_k(t - t_i) \tag{12}$$

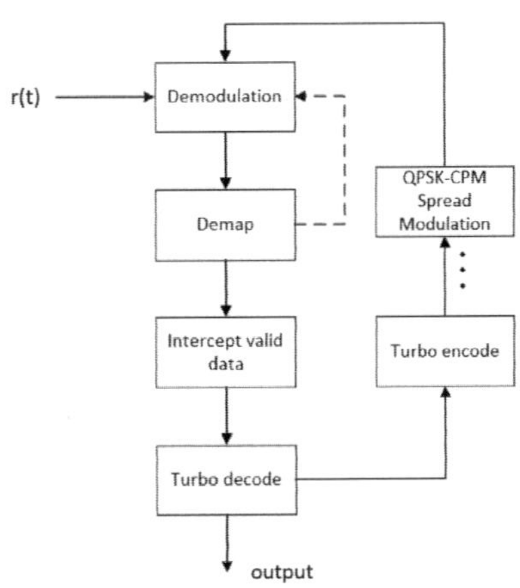

Figure 4. Multi-user Receiver

4. Simulation analysis

In order to verify the effectiveness of the proposed algorithm, the performance of the algorithm in the additive Gaussian white noise channel is simulated. When SIC is directly processed [8], the hard output of de-mapping module is a hard output in Figure 4, and the de-mapping data is directly transmitted to the demodulation module through QPSK-CPM spread modulation. As shown in Figure 5, the performance of direct SIC elimination and Turbo iterative elimination is simulated. Three users and five users were put into the system respectively, and the delay time difference of each user was 100 sample points. As can see from Figure 5, Turbo decoding has a lower frame error rate than SIC directly. This is because Turbo decoding iteration takes advantage of Turbo interweaving to restore the original information contained by users. In this way, the information of reconstructed error data can be reduced during iterative reconstruction , making the elimination cleaner and reducing multiple access interference.

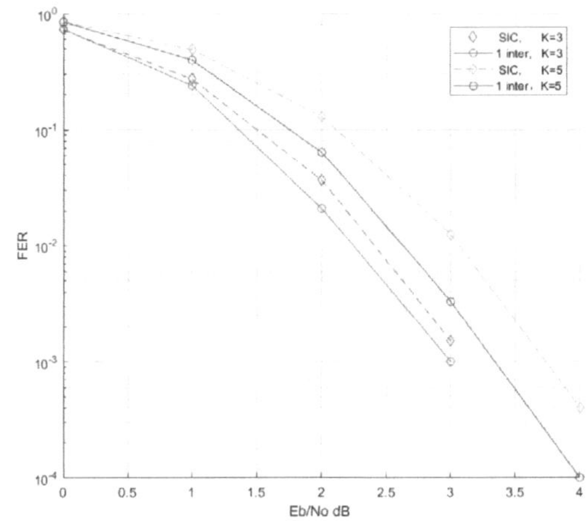

Figure 5. Comparison of SIC and Turbo iterative elimination algorithm

Figure 6 simulates the relationship between the average frame error rate and the signal-to-noise ratio of multi-user detection when the number of iterations of interference cancellation is different. This receiver system has 3 users and 5 users respectively, and the delay time difference of each user is 100 sample points. It can be seen from the figure that when there is three user in the system, the frame error rate performance of the second iteration is slightly better than that of the first iteration. When the number of users increases to 5, the second iteration detection is better than the first iteration. performance is better. This is because the user information eliminated in the second iteration is the interference information of all users except the demodulated user information, which causes less interference to CPM de-spreading and demodulation.At a frame error rate of 10^{-2}, the second iteration outperforms the first iteration by about 0.6 dB.

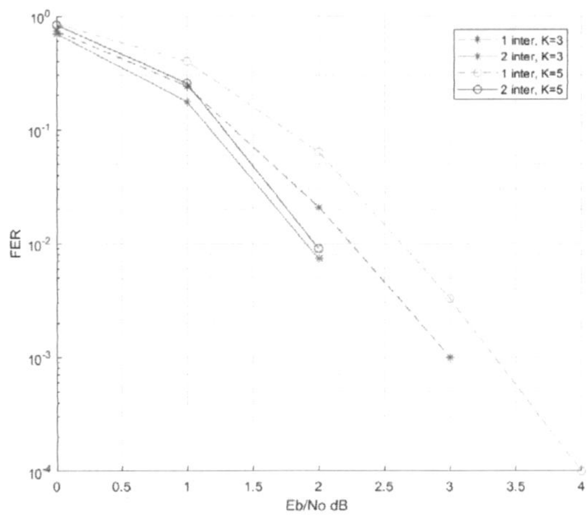

Figure 6. Comparison of second-iteration with first-iteration algorithm

Figure 7 simulates the relationship between the frame error rate curve and the signal-to-noise ratio when different users are aliased under different signal-to-noise ratios when the second iterative elimination algorithm is simulated. It can be seen from the figure that when the number of users is low, $K<=5$. the second iteration under low signal-to-noise ratio is not sensitive to the number of users' aliasing. When the number of access users is relatively large, $K>=7$, and the frame error rate increases with increase with the increase of the number. When the bit error rate is 10^{-2}, the performance decreases by about 0.2 dB for every two additional users.

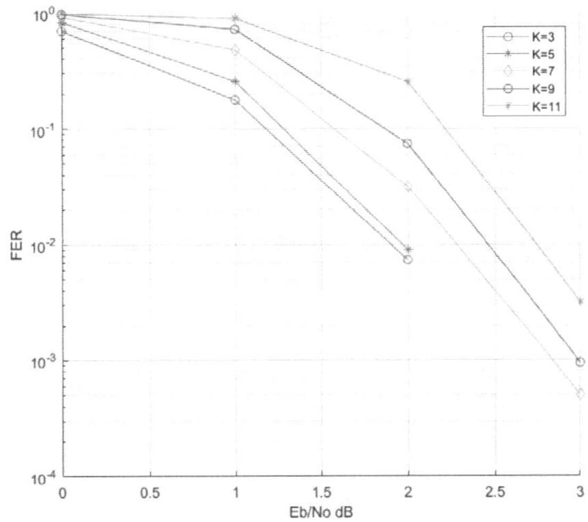

Figure 7. Comparison of the number of different users in the second iteration algorithm

5. Conclusion

This paper introduces the modulation method of the QPSK-CPM spread system in VDES, and designs a multi-user detection method for the QPSK-CPM spread system. Using the second Turbo iteration cancellation algorithm can effectively eliminate multiple access interference and improve user aliasing. The algorithm is simulated and verified. The results show that the detection algorithm has better frame error rate performance than the traditional SIC algorithm, which can meet the needs of multi-user for real-time communication and ensure the needs of ships operating in the ocean.

In the follow-up research, it is necessary to study the influence of Doppler frequency offset on the multi-user detection performance. At the same time, the acquisition of multi-user signals can also be studied in depth, and the multi-user signal acquisition algorithm is taken as the research direction. Combined multi-user acquisition algorithm and multi-user detection algorithm to analyze the performance of the entire receiver system.

References

[1] ITU-R, (2015) "Technical characteristics for a VHF data exchange system in the VHF maritime mobile band," Rec. ITU-R M.2092.

[2] Safar, J., (2014) " VHF data exchange system channel sounding campaign," Report ITU-R M.2317-0.

[3] Lazaro, F., Raulefs, R., Wang, W. (2018) "VHF Data Exchange System (VDES): an enabling technology for maritime communications," CEAS Space Journal, DOI:10.1007/s12567-018-0214-8.

[4] Verdu, S. (1986) "Verdu.Minimum probability of error for asynchronous Gaussian multiple access channels," IEEE Transactions on Information Theory,vol.32, no.1, pp. 85-96.

[5] Shao, X., Yuan, J. and Shao, Y.(2007) "Error performance analysis of linear zero forcing and MMSE precoders for MIMO broadcast channels," IET Communications, vol.5,no.1,pp.1067-1074.

[6] Ping, L., Zhengyuan, X.(2008) "Blind MMSE-Constrained Multiuser Detection, " IEEE Transactions on Vehicular Technology, vol.57, no.1, pp. 608-615.

[7] Hui, C., Shing, C.C. (2007) " Blind Linear MMSE Receivers for MC-CDMA Systems," IEEE Transactions on Circuits and Systems-I:Regular Paper,vol.54, no.2, pp. 367-376.

[8] Kohno, R. (1999) "Combination of an adaptive array antenna and a canceller of interference direct sequence spread spectrum multiple access system," IEEE Journal on Selected Areas in Communications, vlo.8,no.1, pp.675-682.

[9] Chan, S.C., Chu,Y.J. (2015) "Performance of NLMS-based Parallel Interference Cancellation (PIC) For Up-Link CDMA Systems," Journal of Signal Processing Systems,vlo.78, no.2, pp.155-162.

[10] Cottatellucci, L., Muller, R. R and Debbah, M. (2010) "Asynchronous CDMA Systems With Random Spreading—Part I: Fundamental Limits," IEEE Transactions on Information Theory, vol. 56, no. 4, pp. 1477-1497, doi: 10.1109/TIT.2010.2040890.

[11] Cottatellucci, L., Muller, R. R and Debbah, M. (2010) "Asynchronous CDMA Systems With Random Spreading—Part II: Design Criteria," IEEE Transactions on Information Theory, vol. 56, no. 4, pp. 1498-1520, doi: 10.1109/TIT.2010.2040898.

[12] Muller, R. R. (2011) "Multi-user detection for continuous phase CDMA," 2011 IEEE-APS Topical Conference on Antennas and Propagation in Wireless Communications, pp. 591-594, doi: 10.1109/APWC.2011.6046787.

[13] Maunder, R. G., "A Fully-Parallel Turbo Decoding Algorithm," IEEE Transactions on Communications, vol. 63, no. 8, pp. 2762-2775,doi: 10.1109/TCOMM.2015.2450208.

[14] Haghighat, P. and Ghrayeb, A. (2007) "A Hybrid Serial-Parallel Interference Cancellation Scheme for WCDMA Systems," 2007 Canadian Conference on Electrical and Computer Engineering, pp. 621-624, doi: 10.1109/CCECE.2007.160.

Journal of Physics: Conference Series **2387** (2022) 012010 doi:10.1088/1742-6596/2387/1/012010

Optical arrangement method of lighting units in switchgear based on partial discharge luminosity distribution characteristics

Jin Miao[1,*], Junfeng Wu[1], Wei Liu[1], Fen Huang[1], Yufeng Wu[1]

[1] State Grid Wuxi Power Supply, Wuxi, Jiang Su, China

*1526314454@qq.com

Abstract. Partial discharge optical measurement has the characteristics of high sensitivity and strong anti-interference, but its detection effect is greatly affected by the deployment position of the sensor. Therefore, based on the radiation characteristics and transmission characteristics of optical signals, the deployment position of sensors in equipment is optimized. In this paper, an optimal layout scheme of switch cabinet lighting unit based on the distribution characteristics of partial discharge photometric is proposed. Firstly, the optical simulation model of switchgear cable room is designed, and the discharge light source is set at the typical fault location. According to the optical simulation results, the position sensor with high response intensity is selected for different discharge light sources. In order to prove the effectiveness of the layout scheme, the discharge optical detection experiment of switch cabinet was carried out. The results show that the sensor can detect discharge effectively in the selected installation position.

1. Introduction

Switchgear is widely used in power system for power generation, transmission, distribution and power conversion control and protection. Switchgear is one of the important equipment in power grid. However, potential insulation defects may occur in the switchgear during long-term operation, which may cause partial discharge in the switchgear and even discharge faults such as short-circuit arc[1]. Therefore, partial discharge detection is an important means of switchgear operation and maintenance. At present, the commonly used methods for partial discharge detection of switchgear include ultra-high frequency method[2], transient ground voltage method[3], and ultrasonic method[4]. The transient earth voltage method is installed outside the equipment, which is susceptible to external corona interference in practical work and has the disadvantage of being difficult to quantify. The ultrasonic detection method is simple in operation and mature in positioning algorithm, but it has low sensitivity and large signal attenuation in field use. Ultra-high frequency detection method is widely used in the field of partial discharge on-line detection due to its strong corona resistance. However, its detection sensitivity is low in practical application, and the detection frequency band is not consistent with the corona discharge frequency band of switch cabinet [5-7]. Partial discharge optical measurement is a new detection method for measuring and analyzing discharge optical signals. It can realize the qualitative, quantitative analysis and reliable detection of discharge faults, and has the advantages of strong anti-interference and high sensitivity. It is widely used in the field of partial discharge detection. The optical measurement methods used in the field include arc monitoring method[8], fluorescence fiber method[9] and ultraviolet pulse method[10]. However, arc detection cannot effectively detect discharge in the early stage of fault development, and its detection results are difficult to provide effective reference for maintenance and operation. Fluorescent fiber is not easy to be arranged in the field and cannot work for a long time. The shortcomings of UV pulse method are high cost and single detection analysis criterion.

The new generation of micro-weak light sensor represented by silicon photomultiplier tube (SiPM) has the advantages of strong anti-electromagnetic interference ability, small size and high quantum efficiency[11], which provides a new carrier for the detection of partial discharge of insulation in the equipment. Ren et al. proposed a practical partial discharge weak light detection scheme based on the SiPM sensor[12], which realized the discharge detection and analysis at the single photon level. However, when the weak light sensor is installed inside the power equipment, the complex refraction and reflection environment will lead to serious attenuation of optical signal, and the illumination distribution on the reserved layout surface of the sensor is extremely uneven, so that the detection performance of the device for discharge cannot be effectively guaranteed. Therefore, selecting the installation position of the light signal strength to meet the lower limit of the sensor response, so as to ensure the effective acquisition of the sensor light signal, is the primary prerequisite for the application of optical measurement method in practical switching equipment.

In view of the above research status, this paper puts forward the optimization layout method of indoor lighting unit of switchgear cable to solve the problem of difficult application of optical measurement method in equipment. In this paper, the optical simulation model of the switch cabinet cable chamber is established, and the light distribution data of the switch cabinet when the discharge occurs at the typical fault location are obtained by using the ray tracing algorithm. Then the installation position of the sensor is determined based on the photometric distribution characteristics of the inner wall of the switch cabinet and the detection performance of the sensor. Finally, the discharge optical signal is measured in the switch cabinet to verify the rationality of the deployment scheme. This study provides a reference for the effective application of optical measurement in switchgear cable room.

2. Optical simulation model of cable chamber

The main places of the actual switchgear failure may be the bus room, cable room and handcar room. Partial discharge occurs frequently in the cable chamber, but its internal structure is complex, and serious refraction and reflection will lead to attenuation of optical signals, which is difficult to detect. Therefore, it is of great significance to select reasonable lighting unit layout in this room. In this chapter, the optical signal propagation characteristics of the cable chamber during discharge are simulated, and the illumination distribution image on the reserved layout surface of the lighting unit is obtained.

2.1. Design of cable chamber model

In this paper, the optical path simulation of TracePro software is used to optimize the position of the internal discharge sensor in the cable room. TracePro simulation software is an optical analysis simulation software based on Monte Carlo method and solid model. At each surface and interface of solid model, each light emitted by light source strictly follows the laws of absorption, reflection, refraction, diffraction and scattering.

Referring to the actual typical distribution switch cabinet KYN28 (as shown in Fig. 1), in order to ensure the internal insulation margin of the switch cabinet, the surface of the shell, wall bushing and pillar insulator is usually smooth without excessive burrs. In order to be closer to the actual situation, the surface materials of the cable chamber model are all smooth metal materials, and the parameters of the ABg model are A = 0.0025, B = 10 − 5, g = 2.5.

The actual switch cabinet has complex internal structure, many internal components and disorderly installation position. If the actual switch cabinet is directly simulated, the model is complex and too detailed, which not only greatly improves the difficulty of modeling and is difficult to carry out experimental analysis. Therefore, without affecting the accuracy of the simulation results, the internal structure of the switch cabinet is simplified as follows:

a) There are tiny slits and observation windows in the cable cabinet. Due to the obvious difference between the uniform light and the fault light, the influence of the light transmitted by the slit on the test is not considered.

b) According to the arcing specification of switch cabinet, the bottom of switch cabinet is fixed during the test. Therefore, in the simulation process, a fixed constraint is set at the bottom.

According to the physical structure of KYN28, the optical simulation model of cable room is established in TracePro software by 1:1. The model and size parameters are shown in Fig. 2.

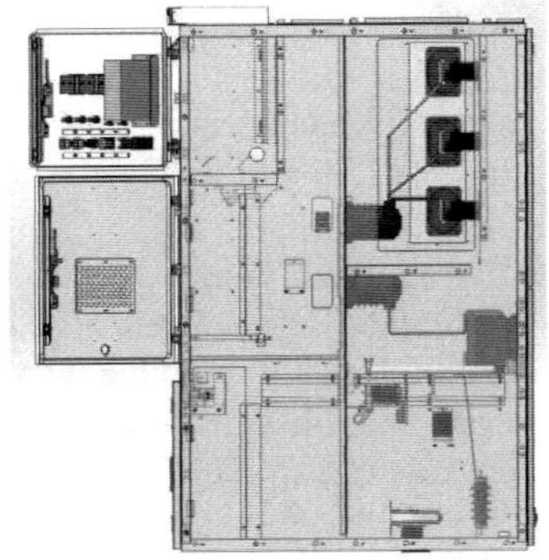

Figure 1. Switchgear KYN28 physical model

Figure 2. Cable room simulation model

2.2. Simulation settings

There are many precision components in the cable room. Components are prone to failure under long-term high load operation, and the probability of live display failure is high. The main reasons for the failure of the arrester are sealing failure[13], porcelain sleeve pollution and valve aging caused by long-term operation. Insulation faults occur in insulators under the conditions of conductor phase or pillar porcelain flashover[14]. Contact box often occurs contact melting, poor contact and other issues[15]. In summary, the arrester, pillar insulator and contact box in the cable room are the locations where faults often occur.

Therefore, for these nine discharge positions, the discharge light sources are set respectively, and the above ABg model is used to simulate the discharge photorefractive reflection inside the cable chamber. The internal equipment of the cable room is complex, the light blocking and refraction will be more serious, and the light attenuation of the discharge power supply radiation will be more serious. The discharge attenuation of the upper part of the cable chamber near the bus chamber and the upper wall at some positions is obvious, so it is not included in the consideration.

The illumination distribution inside the equipment is simulated when discharge occurs in the cable chamber. Partial discharge light source can be regarded as point light source, in the simulation set the number of light is 1000000, light source wavelength is 380-450nm.

On the sensor reserved layout plane, taking the left inner wall of the cable room as an example, the illumination image of this element array is derived by simulation, and the proportion of illumination on each element of the image to total discharge light energy is analyzed. The sensor layout range that can achieve reliable discharge detection can be obtained.

3. Sensor position optimization based on photometric results

3.1. Simulation result

The faults of the cable room of the switch cabinet at different power supply positions are simulated, and the illuminance distributions of the lower, left, right, front and rear inner walls are counted. The results of the right wall are shown in Fig. 3.

It can be seen from the simulation results that the optical signal radiated by the partial discharge power has been greatly attenuated when it reaches the cable wall[16]. Figure 4 shows that the irradiance in most areas is less than 1W/m2, and there is a large irradiance only in areas close to the power supply. It can be found from the figure that the radiation flux, radiation intensity, radiation brightness, position and range of high light intensity are significantly different due to the different positions of the discharge power. The location of the high illumination region depends on the position of the arranged light source and the internal structure of the cable chamber. The size of the high illumination region is positively correlated with the discharge intensity.

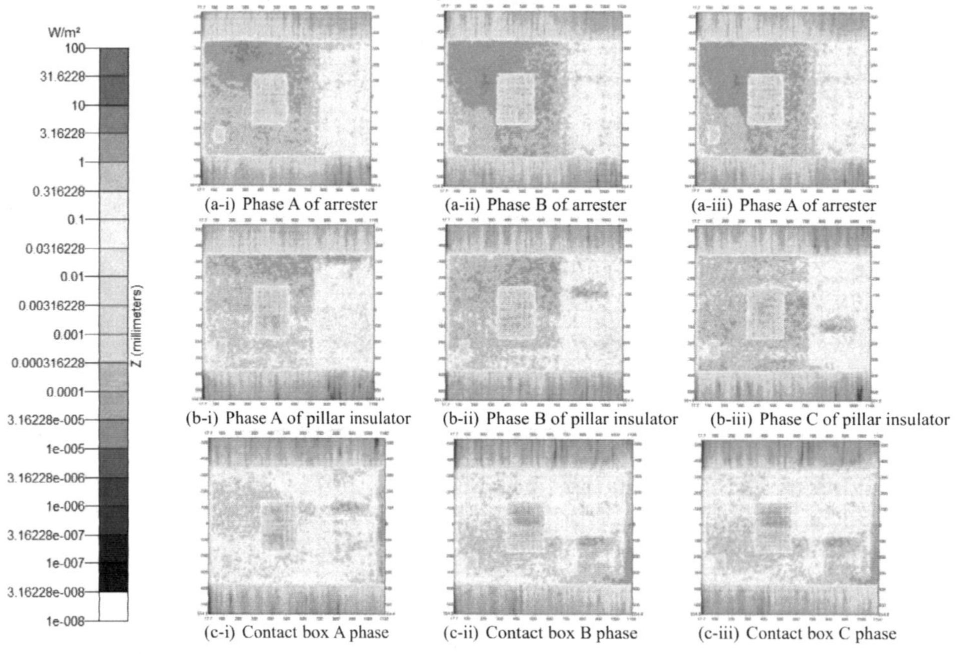

Figure 3. The irradiance of the right wall of the cable room at each discharge position

3.2. Installation location selection principle

The selection of sensor installation location in switch cabinet cable room follows the following four principles:

1) The safety principle is to ensure the safety margin of the live parts in the sensor and the shielded wire and the switchgear cabinet.

2) Due to the normal heating of the cabinet or the working of the heater in the cabinet, the installation position of the photoelectric sensor and the shielding wire needs to avoid the high temperature area in the cabinet to prolong the life of the sensor.
3) The refraction and reflection of light in the switch cabinet will consume the energy and light intensity of light, so the luminous flux incident to the selected position after refraction and reflection needs to reach a certain threshold to ensure that the discharge at each position can respond.
4) The fewer sensors put into a switch cabinet, the better. Therefore, it is necessary to ensure that only one sensor can be installed and sufficient light intensity can be incident into the sensor for the above nine discharge positions.
According to the above principles, the layout positions of the inner walls of each surface of the switch cabinet cable room are screened, and the screening results are shown in Fig. 4. Firstly, according to the safety principle and long-term working principle, the vicinity of the grounding switch installation hole, arrester and glass sealing plate is listed as the prohibited area, as shown in the gray area in Fig. 4.
Secondly, the sensor layout area is selected according to the detection performance principle. The proportion of illumination on each element in the illumination distribution map to the total light energy of the discharge is analyzed. Compared with the calibration results of the integral sphere, the sensor layout range that meets the requirements of discharge detection can be obtained. The detection range is shown in the red area in Fig. 4. It can be seen that the detection performance of the sensor on the left and lower inner walls is poor. In order to select a more economical sensor layout scheme, the sensor will not be arranged on this surface. The final installation range of the sensor is the red area of the front, right and rear walls of the cable room in Fig. 4.

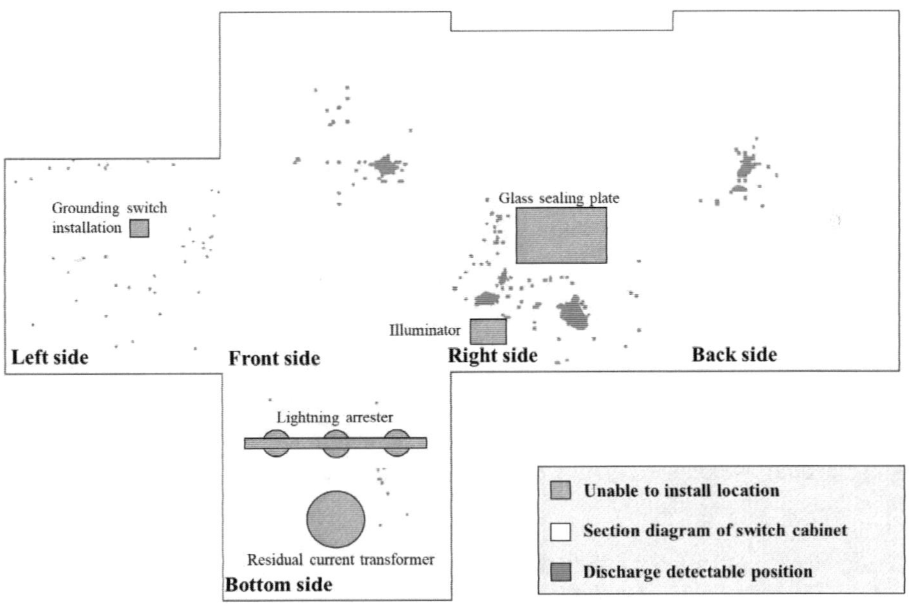

Figure 4. Selected sensor location on each wall of the cable compartment for each discharge location

4. Conclusion

This paper presents an optimal layout method of switchgear lighting unit based on the distribution characteristics of partial discharge photometric. Firstly, the optical simulation model of the switch cabinet cable chamber is built, and the irradiance map of the typical fault location on the reserved layout plane of the sensor is obtained by simulation. On this basis, the deployment scheme of the sensor is optimized, and the effect of the scheme is verified by experiments. The main conclusions of this paper are as follows:

1) In the simulation experiment, the optical signal radiated by the partial discharge power has been greatly attenuated when it reaches the cable wall. As shown in Fig. 4, the irradiance in most regions is lower than 1W/m2, and there is a large irradiance only in the region close to the discharge power position.

2) The simulation results show that the intensity and range of irradiance are significantly different due to different power supply. The installation position of the sensor is selected as shown in the red region in Fig. 4.

Acknowledgment

The work was supported by the Science and Technology Program of the State Grid Corporation of China, whose name is Research on integrated early warning technology of partial discharge and arc light for medium voltage switchgear based on spectrum (J2021059).

References

[1] Gui, Y. G., Zhang, X. X., Zhang, Y., Qiu, Y. J., and Chen, L. C., (2016) "Study on the characteristic decomposition components of air-insulated switchgear cabinet under partial discharge", AIP Advances 6, 075106.

[2] Judd, M. D., Farish, O and Hampton, B. F. (1996) "The excitation of UHF signals by partial discharges in GIS," in IEEE Transactions on Dielectrics and Electrical Insulation, vol. 3, no. 2, pp. 213-228.

[3] Ren, M., Zhou, J., Yang, S., Zhuang, T., Dong, M and Albarracín, R., (2018) "Optical Partial Discharge Diagnosis in SF6 Gas-Insulated System With SiPM-Based Sensor Array," in IEEE Sensors Journal, vol. 18, no. 13, pp. 5532-5540.

[4] Lundgaard, L. E. (1992) "Partial discharge. XIII. Acoustic partial discharge detection-fundamental considerations," in IEEE Electrical Insulation Magazine, vol. 8, no. 4, pp. 25-31.

[5] IEC International Standard 60270. (2000) High Voltage Test Techniques - Partial Discharge Measurements. International Electrotechnical Commission (IEC), Geneva, Switzerland, 3rd edition.

[6] Jiang J., et al. (2020) "Characteristics of the propagation of partial discharge ultrasonic signals on a transformer wall based on Sagnac interference." Plasma Science and Technology, 22 (02):12-20.

[7] Zheng, Q., Luo, L., Song, H., Sheng, G. and Jiang, X. (2020), Intelligent learning approach for UHF partial discharge localisation in air-insulated substations. High Volt., 5: 583-590.

[8] Peng, Z., et al., (2020) "Partial Discharge Monitoring and Arc Protection Based on Multispectral Detection in High Voltage Switchgear," 2020 Asia-Pacific Conference on Image Processing, Electronics and Computers (IPEC), pp. 344-348.

[9] Xia, C. et al., (2022) "Multispectral Optical Partial Discharge Detection, Recognition, and Assessment," in IEEE Transactions on Instrumentation and Measurement, vol. 71, pp. 1-11, Art no. 7002911.

[10] Ma, D., Qiao, X., Jin, H., Jin, L and Wang, H. (2018) "Discrimination of Insulator Discharge Phenomena Based on Characteristics of UV Pulse," 2018 28th International Symposium on Discharges and Electrical Insulation in Vacuum (ISDEIV), pp. 127-130.

[11] Eckert, P., Schultz-Coulon, H. C., Shen, W., Stamen, R., Tadday, A. (2010) Characterisation studies of silicon photomultipliers, Nuclear Instruments and Methods in Physics Research Section A: Accelerators, Spectrometers, Detectors and Associated Equipment, Volume 620, Issues 2–3, Pages 217-226,

[12] Ren, M., Wang, S. Y., Zhou, J. R., et al., (2018) Multispectral detection of partial discharge in SF6 gas with silicon photomultiplier-based sensor array, Sensors and Actuators A: Physical, Volume 283, Pages 113-122.

[13] Chang, H., Gu, F., Chen, H and Kuo, C. (2011) "Partial discharge measurement and pattern recognition in gas insulated switchgear," 2011 7th Asia-Pacific International Conference on Lightning, pp. 459-462.

[14] Biswas, A., Khan, M. E., Islam, M. R. (2013) "Enhancement of power system capacity with existing switchgear using superconducting fault current limiter," 2013 International Conference on Informatics, Electronics and Vision (ICIEV), pp. 1-5.

[15] Zhao, L., Wu, Y., Huang, X., Hong, G., Ren, J., Ning, W., Wang, L., Sun, T. and Yang, S. (2022), Research on the Temperature Rise Characteristics of Medium-Voltage Switchgear under Different Operation Conditions. IEEJ Trans Elec Electron Eng, 17: 654-664.

[16] Xia, Y. et al., (2014) "Research on arrangement of Ultra High Frequency sensors based on GIS partial discharge," 2014 ICHVE International Conference on High Voltage Engineering and Application, pp. 1-4.

Research on transformer condition evaluation based on oil chromatography differentiation threshold

Pei Yang[1,*], Junfeng Qiao[1], Gang Wang[1]

[1] State Grid Smart Grid Research Institute CO.,LTD, Nanjing 210003, China & State Grid Key Laboratory of Information & Network Security,Nanjing,210003,China.

*yangpei@geiri.sgcc.com.cn; 404648438@qq.com

Abstract. As one of the important power transmission and transformation equipment, the reliability of transformer is directly related to the safe operation of power system. At present, the on-line monitoring technology based on dissolved gas analysis in oil is widely valued by researchers all over the world because it can continuously monitor the operation status of transformers and timely and effectively find the early faults of transformers. In this paper, the on-site oil chromatography data are differentiated and classified according to the voltage level of the transformer. The curve fitting method is used for statistical analysis and to build a distribution model. The cumulative probability is associated with the defect rate and failure rate of the equipment, and the corresponding threshold value is obtained by using the inverse cumulative distribution function as the attention value and warning value. By using this method, the attention values of the volume fraction of dissolved gas in oil and the gas production rate of dissolved gas in the existing national standards are improved, and the corresponding warning values are calculated to make the warning mechanism more flexible and reliable.

1. Introduction

In terms of operation inspection, for a long time, China's transformer maintenance strategy[1] mainly adopts regular maintenance based on time. Although regular maintenance can generally find defects in the equipment during maintenance, which plays an important role in ensuring the safe and economic operation of the equipment, the defects of "excessive maintenance" and "insufficient maintenance" in regular maintenance not only cause blind maintenance of some transformers, resulting in a large waste of human and material resources, but also increase the probability of new hidden dangers and reduce the reliability of power supply. In the current situation of power enterprises moving towards the market, it is an inevitable trend for power enterprises to replace the traditional regular maintenance system with the mode of condition based maintenance. Condition based maintenance is based on the current actual working condition of the equipment[2], and judges the status of the equipment through advanced condition monitoring means, reliable evaluation means and life prediction means. The judgment made by condition based maintenance on the location, severity and development trend of the fault can identify the early symptoms of the fault, and carry out maintenance before the equipment performance drops to a certain degree or the fault is about to occur according to the analysis and diagnosis results.

2. Difference classification of original data and fault information statistics

The volume fraction data, account information and defect information of dissolved gas in oil collected by all sensors in a regional power grid in one year were collected in this study.

Take each oil chromatography data[3] as the characteristic value, and supplement the manufacturer and number of the sensor collecting the chromatography data, as well as the manufacturer, wiring mode, power supply company, voltage grade and other information of the transformer installed with the sensor. Thus, a unified data information base can be formed.

In this paper, the oil chromatography data collected in this paper are mainly from transformers with three kinds of voltage levels, they are 35 KV, 110 KV and the other one is 220 KV. Considering that voltage level is one of the most basic attribute characteristics of transformers, this study takes the voltage level of transformers as an example to carry out differential classification on the collected oil chromatography data, and make statistics on the defect rate and fault rate of transformers under each classification. The results are shown in Table 1.

Table 1. Differential classification and defect fault statistical results.

Voltage class /KV	Number	Data volume	Defect rate/%	Failure rate/%
35	8	1 844	2.82	0.05
110	173	51 236	4.54	0.18
220	220	61 246	1.79	0.9
Total	401	114 326		

3. Distribution of dissolved gas data in oil

Taking the volume fraction of H2 dissolved in 35kV transformer oil and the volume fraction of total dissolved hydrocarbon in 220kV transformer oil as examples, the frequency histogram and its fitting curve were drawn, the curve is as shown in Figure 1. According to the frequency histogram and fitting curve in the figure, the distribution law of dissolved gas volume fraction in oil conforms to Weibull distribution[4].

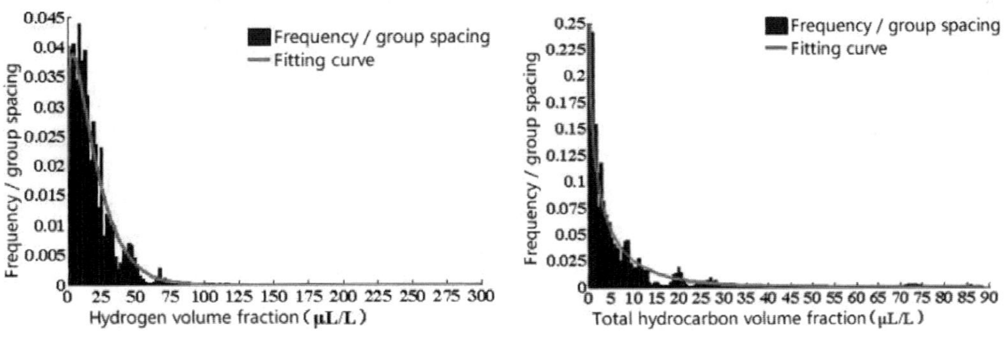

Figure 1.(a)Frequency histogram and fitting curve of H2; (b)Frequency histogram and fitting curve of total hydrocarbon.

To make a correct judgment on the severity of the fault only based on the absolute value of the dissolved gas volume fraction in the oil is difficult, because the fault often starts with a low-energy latent fault. It could lead to a major accident[5]. Therefore, it is necessary to consider the development trend of the fault, that is, the gas production rate at the fault point. The gas production rate is directly related to the energy consumed by the fault, the fault location, the temperature of the fault point, etc.

The gas content values for 10 consecutive days are used for linear fitting, and the slope of the fitted curve is taken as the gas production rate. For the example data set: {64102,40,26,62,52,54,96,64132}, the straight line y=4.274 can be obtained after fitting*X+46.3, with the slope of 4.274 taken as the gas production rate. The scatter diagram of original data and the curve obtained by fitting are shown in Figure 2.

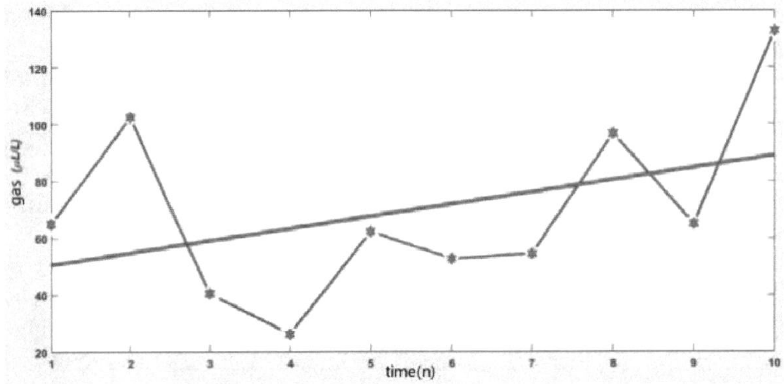

Figure 2. Scatter plot of original data and curve obtained by fitting.

The gas production rate of each gas in each classification is calculated by the above method. Taking the gas production rate of dissolved C2H2 in 110KV transformer oil and the gas production rate of dissolved C2H6 in 220KV transformer oil as examples, the frequency histogram and its fitting curve are drawn. According to the frequency histogram and fitting curve in the figure, the distribution law of gas production rate of dissolved gas in oil is similar to that of gas volume fraction, which conforms to Weibull distribution.

4. Construct intelligent diagnostic analysis model

The dissolved characteristic gases in the oil of the online monitoring system mainly include H2, CH4, C2H6, C2H4, C2H2 and other characteristic gases, the fault case table is as Table 2.

Table 2. The fault case table.

NO.	H2	CH4	C2H4	C2H6	C2H2	State
1	27	18	6.3	5.3	0	normal
2	27	71	86	44	0	partial discharge
3	27.15	26.51	2.81	1.88	0	normal
4	27.44	35.01	29.4	5.74	2.36	high temperature overheating
5	27.79	1.57	0.8	0.11	1.43	low energy discharge
6	28	90	87	48	0	medium overheat
7	28.35	76.92	98.12	29.97	0.51	high temperature overheating
8	29	17	1.6	5.2	0	normal
9	30	7.4	1.8	8.5	19	low energy discharge
10	30	62	460	60	3.4	high temperature overheating
11	32	13	8.9	2.9	0	normal
12	32.4	5.5	12.6	1.4	13.2	high energy discharge
13	xx	xx	xx	xx	xx	normal

The model building process includes the following steps:
(1) A total of 680 cases of transformer faults based on characteristic gases dissolved in transformer oil with clear conclusions collected from published literature, including the type, value and status of transformer equipment dissolved in transformer oil, as transformer faults Case Library.
Among them, the characteristic gases dissolved in the transformer oil include five gases such as H2, CH4, C2H6, C2H4, and C2H2. The data types of the transformer equipment status include 9 categories, which are normal, partial discharge, low-energy discharge, high-energy discharge, low temperature overheating, and medium temperature overheating. , high temperature overheating, high energy discharge and overheating, low energy discharge and overheating.
(2) Using the transformer fault case database with clear conclusions collected in step 1, use random sampling to use 70% of the fault cases as the training data set of the transformer fault diagnosis and analysis model, and use 30% of the fault cases as the test data. set to verify the accuracy of the transformer fault diagnosis model.
(3) The construction of the transformer fault diagnosis analysis model can be completed with the assistance of the IBM SPSS Modeler modeling tool. Using the C5.0 classification decision tree algorithm provided by the modeling tool, 70% of the transformer fault cases randomly sampled are used as the training set. The C5.0 classification algorithm performs unsupervised machine learning, and uses 30% of the transformer fault cases randomly sampled as the test set to verify the accuracy and rationality of the analysis model.
By setting the relevant parameters of the C5.0 classification decision tree algorithm, the fault diagnosis data mining flow based on the dissolved gas in the transformer oil is constructed, and the transformer fault diagnosis analysis model based on the dissolved gas in the oil is formed.
In the process of model building, based on the C5.0 unsupervised machine learning algorithm, the fault discrimination rule set and fault discrimination decision tree for transformer equipment based on dissolved gas in oil are formed, and the analysis model is used to identify the fault of transformer equipment based on dissolved gas in oil. in accordance with.
(4) Using the post-pruning algorithm of the C5.0 classification decision tree[6], according to the error estimation of the test sample set and the pruning standard of the C5.0 algorithm, by setting parameters such as the percentage of model pruning severity and the minimum number of records for each sub-branch, the basis for The decision tree formed by the training sample set is pruned to prevent the problem of reducing the predictive analysis results caused by over-fitting of the analysis model.
(5) Repeat steps 3 and 4, and use the C5.0 algorithm to optimize the analysis model, and finally form a transformer fault diagnosis analysis model based on dissolved gas in oil.
(6) Evaluate the accuracy, rationality and applicability of the model prediction and analysis results through the model analysis function provided by the modeling tool. According to the model evaluation results, continue to step 4 until the accuracy of the model prediction analysis meets the business requirements.
(7) Collect the characteristic values of dissolved gases in transformer equipment oil that exceed the alarm value by the state monitoring system, including five gases such as H2, CH4, C2H6, C2H4, and C2H2, and input the fault diagnosis and analysis model to output the fault type of the transformer.

5. Field verification

Select the online monitoring data of a transformer in a certain period, calculate the gas production rate, and draw the change trend of gas volume fraction and gas production rate, as shown in the figure.
It can be seen from Figure 3(a) and (b) that the volume fraction of ethylene exceeded the warning value and its gas production rate exceeded the warning value around March 15, and the volume fraction of ethylene exceeded the warning value and its gas production rate exceeded the warning value around May 25. Therefore, the monitoring system will give an alarm about ethylene.

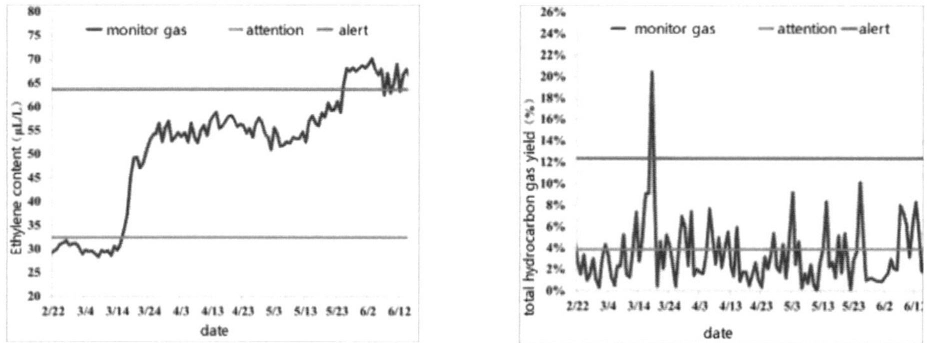

Figure 3.(a)Variation trend of ethylene volume fraction; (b)Variation trend of ethylene gas production rate

It can also be seen from Figure 4(a) and (b) that the volume fraction of total hydrocarbon exceeds the attention value and the gas production rate exceeds the warning value on March 15, and the volume fraction of total hydrocarbon exceeds the warning value and the gas production rate also exceeds the warning value on May 25, so the warning system will give an alarm about total hydrocarbon.

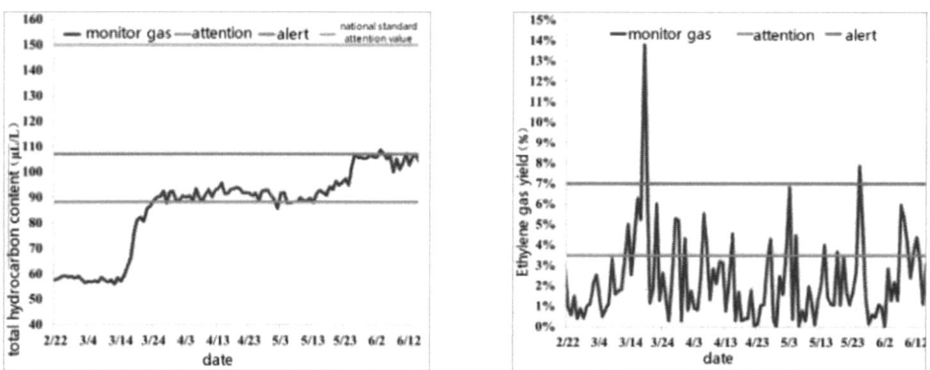

Figure 4.(a)Change trend of total hydrocarbon; (b)Variation trend of total hydrocarbon gas production rate

In addition, it can be seen from Figure 5(c) that if the 150 μL/L given in the national standard is set as the attention value, the monitoring system will not give a prompt on March 15 and May 25, that is, there is a missing report.

To sum up, applying the attention value and warning value calculated in this study to the early warning system, the early warning results are consistent with the actual results, and more accurate than the traditional attention value.

In terms of fault judgment, among the 60 samples, there are 10 partial discharge, low-energy discharge, high-energy discharge, low-temperature overheating, medium-temperature overheating, and high-temperature overheating. It is higher than the three-ratio method and David's triangle[7] method. At the same time, it can solve some faults caused by incomplete fault codes and missing codes in the traditional three-ratio method[8]. When the faults work together, the ratio combination corresponding to the fault cannot be found and the diagnosis cannot be carried out. At the same time, the fault diagnosis rate is improved. The result is shown as Table 3.

Table 3. Fault Model Validation.

NO.	State	sample	Decision Tree Classification Algorithm		three ratio method[8]		David's triangle	
			Positive judgment	Recognition rate	Positive judgment	Recognition rate	Positive judgment	Recognition rate
1	partial discharge	10	9	90.00%	5	50.00%	1	10.00%
2	low energy discharge	10	7	70.00%	2	20.00%	6	60.00%
3	high energy discharge	10	7	70.00%	2	20.00%	6	60.00%
4	low temperature overheating	10	7	70.00%	7	70.00%	5	50.00%
5	medium overheat	10	9	90.00%	8	80.00%	4	40.00%
6	high temperature overheating	10	9	90.00%	8	80.00%	10	100.00%
7	total	60	48	80.00%	32	53.33%	32	53.33%

Acknowledgments

This work was financially supported by the science and technology project to State Grid Jiangsu Electric Power Co.,Ltd. Research Institute "Research on data-driven panoramic visualization and operation optimization of distribution network(SGGR0000JSJS2100729)".

References

[1] La, Y., Wang, H. B., Chen, Z. D. (2013) Research and application of transformer condition based maintenance technology [j] Transformer,50 (8): 5.

[2] Zhang, Y. Y., Liao, R. J., Yang, L. J., et al. (2014) Optimization of power transformer maintenance scheme considering reliability and economy [j] Journal of electrotechnics (11): 10.

[3] Liu, D. L. (2008) Analysis and treatment of abnormal chromatogram data of transformer oil [j] Transformer, 45 (3): 3.

[4] Zheng, R. (2015) Parameter estimation of three parameter Weibull distribution and its application in reliability analysis [j] Vibration and shock, 34 (5): 4.

[5] Zhang, Y. J., Feng, J., Li, D. Y., et al. (2021) New optimization strategy of transformer fault symptom based on oil color spectrum data [j] Power grid technology, 45 (8): 8.

[6] Chen, W. J., Ma, G. Q., Chen, T. J., et al. (2021) Classification method, device and electronic equipment based on classification decision tree model: CN112749749A[P].

[7] Yuan, J. S., Qin, Y., He, Y. J. (2015) Basic Trust Assignment Function for Fault Diagnosis Of Transformer based on David Triangle Method [J]. Electrical Measurement & Instrumentation, 52(1):6.

[8] Yang, T. F., Liu, P., Li, J. L., et al. (2007) FCM combined with IEC three-ratio method to diagnose transformer faults [J]. High Voltage Technology, 33(008):66-70.

Full scale experimental and numerical simulation of outside transformer fire

Yashan Hu[1,2], Bo Yan[1,2], Yan Li[1,2], Zheng Huang[1,2], Decheng Wang[1,2], Jinghui Sun[1,2]

[1]Jiangsu Branch of State Grid Economic and Technological Research Institute Co., Ltd. Jiangsu, China
[2]State Grid Jiangsu Electric Power Design Consulting Co., Ltd. Jiangsu, China

*Email: 1003472829@js.sgcc.com

Abstract. To study the real scale transformer fire and fire control process, this paper uses FDS to simulate the transformer fire combustion process and analyzes the data of transformer fire temperature and flow field. Through data analysis, this paper establishes a fire visualization system based on the unity3d engine and uses modeling software to complete the Substation Scene Modeling, the visualization process of fire particles and fire water jet in the whole process of transformer fire occurrence and fighting is realized. The research results are of great significance for substation fire construction and emergency rescue.

1. The introduction

The power industry supports the national economy and occupies an indispensable position in people's life. Once an accident occurs, it will have a serious impact on people's production and life. To ensure the safe operation of the power system and the People's Daily electricity consumption, the state has issued a series of power safety regulations and fire protection regulations, requiring comprehensive attention to the substation fire protection level. However, according to the current on-site accident data of substation fire, the supporting data of fire extinguishing research cannot be obtained, and the development stage of transformer fire cannot be accurately understood. Therefore, people need to study the occurrence and fire extinguishing process of a transformer fire and apply the research results to the construction of substation fire safety. To maximize the awareness of fire and enhance the substation's fire response-ability.

Because of the high cost of the real transformer in the fire experiment, the scholars adopt the method of computer simulation to simplify the transformer model experiment. Through computer simulation, we can easily and quickly predict the transformer fire development process and its impact on the surrounding equipment. In addition, we can also get the temperature and smoke diffusion law when the fire happens. The fire reach mainly focuses on fire dynamics, including smoke movement, flame, radiation, etc. Scholars study fire processes by building models. For example, NIST [1] used FDS software to reproduce the fire of the oil pool (tank) on the ground, providing a numerical research method for subsequent simulation research. Zhuang Lei [2] et al. used FDS fire simulation software to conduct a numerical simulation of diesel tank combustion and studied the thermal radiation of oil tank fire under different wind speed conditions, analyzed its intensity and spatial distribution, and studied the change of radiation intensity with L/D. Li Jia et al. [3], based on virtual reality technology, use

Torque engine to build 3D scenes and fire models to realize the visual simulation of traditional fire drills. Wang Xingbo [4] et al. used Delta3D to model the tank farm and completed the performance of fireworks special effects and human activities. Wang Xiaoping et al. [5] set up a fire simulation drill system and proposed that OpenCV's and Visual C++ could be used to realize firefighting simulation. In 2009, Yin Jingang [6] established a fire spread model to simulate transformer fire in substation and completed visualization and real-time data display of fire spread and fire extinguishing process by using particle system and collision detection. With the development of The Times, more and more fire simulation visualization methods are updated. Chang Zhuang et al. [7] established fire model and explosion model by using cellular automata theory to achieve special fire effects in Vega, reproduce the spread of fire, and achieve dynamic control of firefighting. In 2010, Ren Hongxiang [8] selected LBM (Lattice Boltzmann model) discrete method and combined with texture technology to simulate the flame burning scene and make the flame more realistic. In 2011, Qi Hongyu [9] built a GPU-based particle system and used the body light source model to establish flame light source to achieve real lighting effects. Random number control and turbulence equation were added to the calculation to improve the reality of simulation. Lu Yang et al. [10] simulated the fire scene of subway station based on OSG (Open Scene Graph) virtual reality technology and realized the dynamic control of the scene by virtualizing the 3D fire scene after receiving external input fire information. Zhang Yong et al. [11] realized the simulation modeling of buildings based on the 3d terrain visualization technology of Terra Vista and Global Mapper software and the simulation modeling technology of static objects based on Creator software. In Vega, particle system module was used to the fire process, smoke diffusion and explosion are visualized, and the visual effect is relatively simple. Balci M et al. [12] adopted the spring-particle model and added the characteristics of fuel's own attributes and the influence of environmental combustible into the modeling process. Based on the fire dynamics, the single flame model was merged into the overall fire diffusion model to realize the three-dimensional diffusion visualization process. To predict the damage caused by wildfires, Roger V. Hoang et al. [13] established a V-Fire wildfire simulation and visualization system, using Huygens ellipse equation to calculate the fire spread rate by constructing the spread equation, generating data virtual environment from the real world related to topology and vegetation, and updating particle positions in real time to realize the fire spread. Nima Nikfetrat et al. [14] proposed PCA based on the technical and statistical analysis of images to analyze the tracks of various fire samples by visualizing their main features and synthesize realistic flames by combining fire samples recorded at various high frame rates to edit the given fire animation sequence.

Through the analysis of the current situation of visual stimulation of fire, we can find that scholars have carried out an extensive simulation of fire scenes based on various techniques by numerical simulation. The simulation results can provide many fire data which is difficult to obtain in practice and is of great significance to fire prevention. However, there is little research on the visual simulation technology of Transformer Fire. In this paper, the visualization system of transformer fire based on Unity3D engine is established utilizing numerical simulation, and the visualization process of fire particles and fire water is realized in the whole process of transformer firefighting and extinguishing. It provides theoretical and technical help for substation fire prevention and control.

2. Numerical simulation theory and model

2.1. *Basic theory of FDS*

FDS uses numerical solution to solve Navier-Stokes equations, which include continuity equation, energy conservation equation, momentum conservation equation and component conservation equation, as follows:

Continuity equation:

$$\frac{\partial \rho}{\partial t}+\nabla\cdot(\rho u)=0 \qquad (1)$$

Momentum conservation equation:

$$\frac{\partial}{\partial t}(\rho u)+\nabla\cdot\rho uu+\nabla \mathrm{p}=pg+f_b+\nabla\tau \quad (2)$$

Energy conservation equation:

$$\frac{\partial(\rho h)}{\partial t}+\nabla\cdot(\rho hu)+\frac{d\rho}{dt}=\dot{q}'''+\nabla\cdot((k\nabla T)+\sum_i\nabla(h_i\rho hD_i\nabla Y_i-\dot{q}_r''))+\varepsilon \quad (3)$$

Component equation:

$$\frac{\partial(\rho Y_i)}{\partial t}+\nabla(\rho Y_i\vec{u})=\nabla\cdot(\rho D_i\nabla Y_i)+\dot{W}_i \quad (4)$$

Where: ρ is density, kg/m3; U is the velocity vector, m/s; ∇ is a Hamiltonian operator; f_b is the external force vector acting on the fluid, kg/ (s2·m); T is time, s; H is specific enthalpy, KJ; τ is the viscous force tensor, kg/ (s2·m); P is pressure, Pa; W is vorticity, s-1; G is the acceleration of gravity, m/s; $\vec{q}_\tau$ is heat radiation flux, KW/m2; T is temperature, K; K is thermal conductivity, W/(m2·k); Yi is the component mass fraction; $\dot{W}_i$ is the component formation rate; $\dot{q}'''$ is the heat generation rate per unit volume, kW/m3; $\dot{q}_r''$ is radiant heat flux, W/m2; Di is component diffusion coefficient, m2/s; I is the ith component.

2.2. *Model establishment*

Transformer fire is mainly caused by the initial root burst of the insulator, a partial burst of the oil tank, and an integral burst of the oil tank. Transformer fire is also a special combustion phenomenon, which follows the basic laws of the combustion process. The transformer fire model is mainly composed of fuel components (transformer oil and the combustible gas generated from the pyrolysis of transformer oil), the spatial arrangement of the oil pan, and the ignition mode. Referring to the actual scene of the substation and the basic operation principle of FDS simulation software, the transformer is placed in the x-Y plane center at the bottom of the workshop, with a length of 4 m × 4 m × 5 m and a STEEL boundary type on the surface. The grid size is set as 0.25×0.25×0.25m, the grid is set as open type, and the bottom surface is set as an adiabatic boundary. The grid independence and model validity verification refer to the research on the influence of fire source power on indoor transformer fire combustion characteristics carried out by Chen Peng et al. [15]. The arrangement of transformer and thermocouple is shown in Figure 1. A group of thermocouples is arranged in the horizontal and vertical directions of the central axis of the transformer with a spacing of 0.2m, and four groups of thermocouples are arranged in the same way with a spacing of 0.5m above the group. The simulation model of transformer fire is shown in Figure 1:

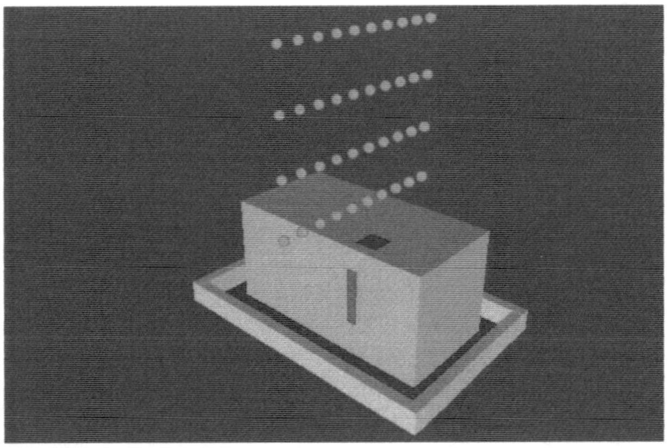

Figure 1. Transformer and thermocouple layout

3. Visual system design

3.1. System software

The software used for system development is the virtual reality engine Unity3D. Unity3D is a multi-platform comprehensive development tool developed by Unity Technologies that can create architecture, real-time 3D animation, and other content. In this paper, 3DMax is used to conduct 3D modeling of substations and various kinds of equipment, and the completed model is exported to FBX format files. Then, import the FBX file into Unity3D resource to load the scene, which is divided into three steps: first, construct the scene, such as lighting, skybox, baking, etc.; Use UGUI to make the user interface, including the main navigation bar in the system and the choice of requirements; Finally, the realization of each functional module of the system. Finally, it is distributed as an exe executable, and the 3DMax version used in this article is 2018.

3.2. Overall system design

This paper introduces virtual reality technology to develop a substation fire visualization system and realizes the purpose of creating transformer fire, combustion, and fire extinguishing scene simulation in a real substation environment. The overall design of the substation fire visualization system is to the first model through 3D Max, and then reproduce the substation fire occurrence process based on the virtual reality engine Unity3D and the data of fire characteristic parameters in numerical simulation to complete the visualization process of fire occurrence, diffusion, and firefighting. The development technical route of the system is shown in Figure 2.

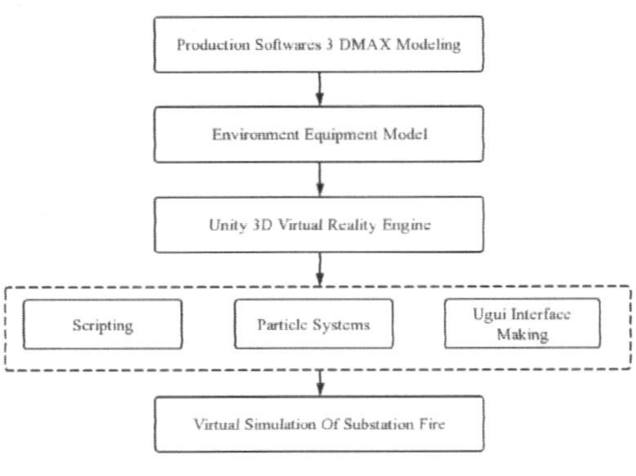

Figure 2. Virtual scene architecture diagram of the visualization system

The formation of the scene model is the foundation of the whole virtual reality system. The higher the quality of the model, the more realistic it is, and the more immersive the virtual scene will be. However, due to the limitation of computer performance, we will simplify as far as possible in combination with the actual situation, and create models of characters, scene environment, and fire equipment and facilities.

The models in the visualization system of substation fire mainly include transformers, high-voltage circuit breakers, disconnecting switches, busbars, arresters, capacitors, reactors, relay protection devices, automatic devices, measurement, and control devices (current transformers, voltage transformers), metering devices, etc., which are all made by 3DMax 2018. 3DMax is widely recognized and used in the media industry, game design industry, and industrial modeling field. It can realize system development by generating FBX format files and importing them into the Unity3D engine.

According to the data and photos of some substations and the site civil construction data, including size and shape, simulation modeling is carried out in proportion to approximate the realistic presentation as far as possible. Then modeling of key parts is carried out according to the visual requirements. The 3D modeling effect of a certain substation is shown in Figure 3.

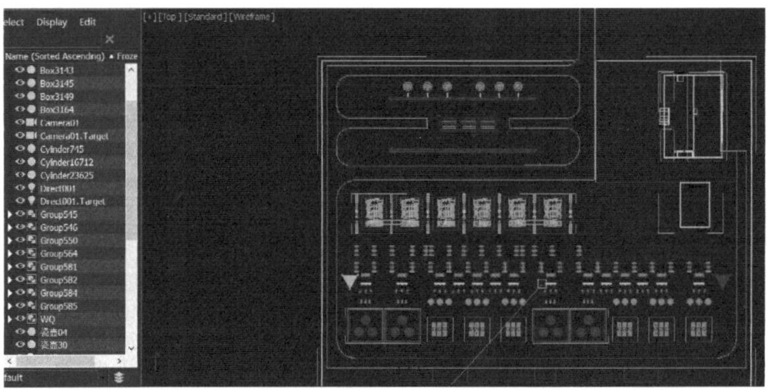

Figure 3. Substation 3dmax modeling

After completing the substation modeling in 3DMax, export the FBX format file and drag it to Unity3D to edit it on the interface. Figure 4 shows the overall scene of the substation. Add other resources, particle systems, various components, scripts, and UI interfaces. After the system starts running, the perspective will be initialized near the transformer, and the character can walk around and interact with the scene.

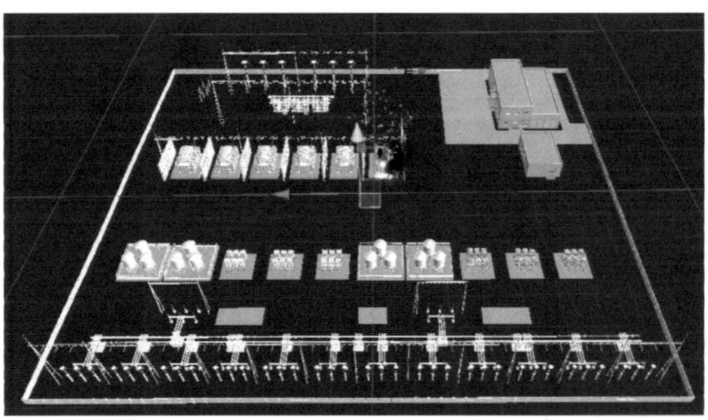

Figure 4. Substation 3d modeling simulation effect diagram

4. Results and discussion

4.1. Analysis of numerical simulation results

4.1.1 Flow field analysis

The flow field morphology affects the heat transfer, mass transfer, and chemical reaction of fire and smoke spreading behavior. The flow field distribution during fire combustion is shown in FIG. 5. In the early fire combustion heat release a large amount, heating the air besides, hot air rising form buoyancy-driven, flame upward vertical movement, in the whole section, in the middle of the maximum speed, smaller air velocity on both sides of the center, this is due to the buoyancy-driven in the process of combustion, to strengthen the horizontal and vertical airflow. As the combustion progresses, the vertical wind speed remains stable.

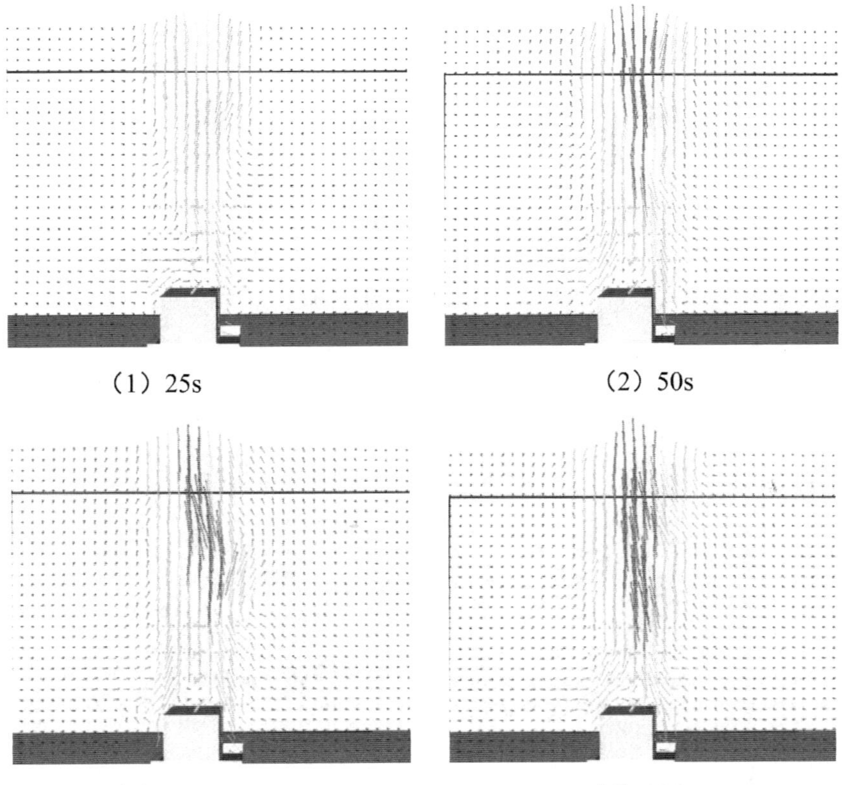
(1) 25s (2) 50s
(3) 100s (4) 200s
(a) Flow field in x-Z section

2387 (2022) 012012 doi:10.1088/1742-6596/2387/1/012012

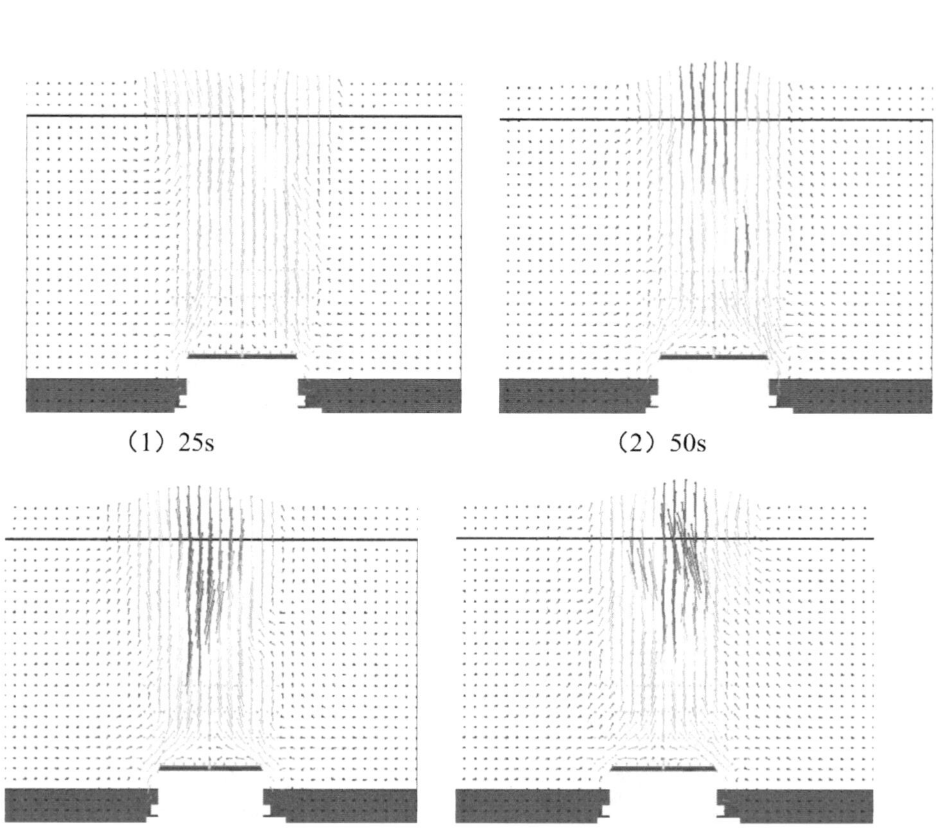

(1) 25s (2) 50s

(3) 100s (4) 200s

(b)Flow field in y-Z section

Figure 5. Flow field distribution

4.1.2 Temperature analysis

Figure 6 shows the temperature profile distribution of the x-Z section in the combustion process of transformer oil. The flame form of the upper oil pool is unstable. From the 50s, high temperature covers the whole transformer. At 100s, there is a buoyancy plume area and discontinuous temperature gradient phenomenon, indicating that flame and flue gas pulsate behavior, temperature gradually rises, and the temperature gradient stratification phenomenon occurs.

FIG. 7 Temperature distribution rule of an x-Y section of the transformer. In the fire development stage, the oil pool flame below the transformer forms a typical bipeak structure. After 25s, the temperature gradually rises, and the heated air flows float along the outer wall of the transformer. In the process of rising, the fire plume absorbs a large amount of air around it and expands the flame by increasing the combustion reaction rate of unburned matter. The high-temperature area continues to expand, and the thermal radiation area also further expands.

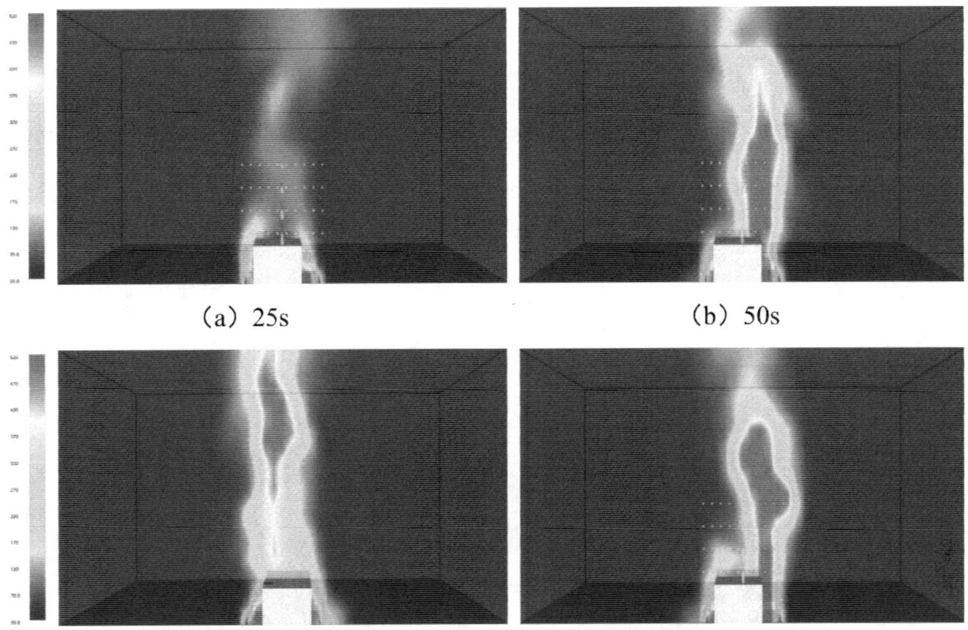

(a) 25s (b) 50s

(c) 100s (d) 300s

Figure 6. X-Z section temperature

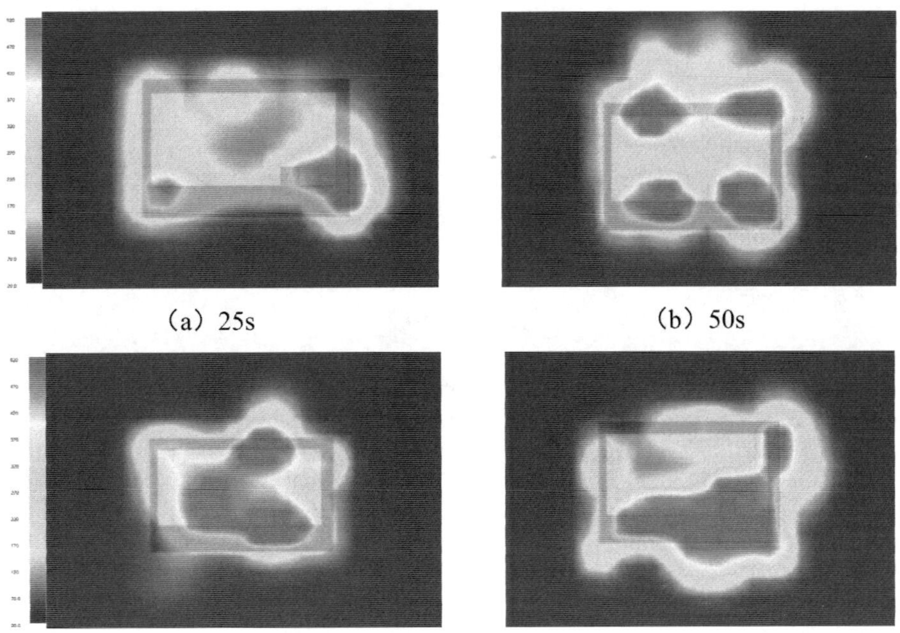

(a) 25s (b) 50s

(c) 100s (d) 300s

Figure 7. X-Y section temperature

4.1.3 Heat flow analysis

FIG. 8 shows the distribution of radiant heat flux in the process of transformer oil combustion. The distance between radiant heat flux and the front of the transformer increases and decreases, and the radiant heat flux reaches the peak value starting from the 50s. In the early stage of the fire, the radiant heat flux began to increase gradually with the increase of entrainment. In the development stage of the fire, the radiation intensity of the transformer remained unchanged until the end of combustion. After 25s, the temperature rises gradually, and the heated airflow floats along the outer wall of the transformer.

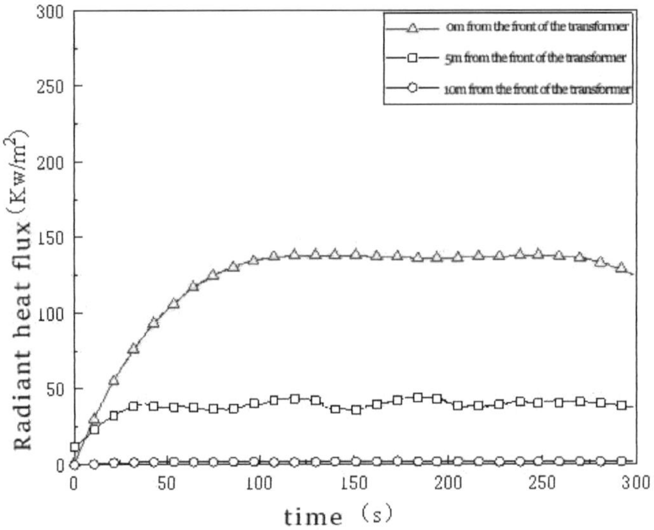

Figure 8. X-Y section temperature

This section discussed the results of transformer fire simulation analysis, summarizes the development of transformer fire temperature, the flow field and heat flux distribution, simulated data-intensive applications next substation 3 d scene construction, and the development of the visualization system provides important help, make the model used to simulate the transformer substation fire process is closer to the actual situation.

4.2. System construction

4.2.1 Flame simulation

Flame the particle's initial position is determined by the transformer fire location, its particles generated by substation fire development rule, the first is casing is broken, and in the casing appear upward jet flame, after spraying spreading oil combustion, flow form of fire, and gathered to form oil pool fire down, finally, external three-dimensional fire and casing gas-liquid mixing combustion of pillar, The combined combustion phenomenon of oil pool fire, transformer body boiling over, and jet fire at casing is formed. In addition, according to the development of the fire, smoke particles are synchronously set in the system to simulate the upward movement of smoke in the fire. Figure 9 shows the fire progress in the substation at different stages of fire simulation. As can be seen from the figure, there is less smoke in the early stage, and then the smoke gradually spreads to the upper part of the transformer, which is close to the smoke-spreading process obtained in the experiment.

(a) Initial flame and smoke effects of fire simulation

(b) Fire simulation effects of mid-stage flames and smoke

(c) Late flame and smoke effects of fire simulation

Figure 9. Renderings of different stages of fire simulation

4.2.2 Simulation of water mist

In this paper, a variety of parameters and materials have been tried, and better water mist simulation results have been obtained. The specific steps are as follows:

(1) Create a new particle system, edit the particle generator properties through the component, set the particle system.

(2) Set the particle's initial state properties, such as start time, delay time, and life cycle, to initialize the setting.
(3) Adjust the particle rendering, set the particle speed vector, shape, color change with time, particle size, material, and so on, until approaching the actual situation.
Because the fire extinguishing effect is to be realized, the C# script program is written to add a Collision detection function interface to realize the extinction of flame particles. After completing the above Settings, water mist particles realize the basic effect. The particle simulation effect is shown in figure 10.

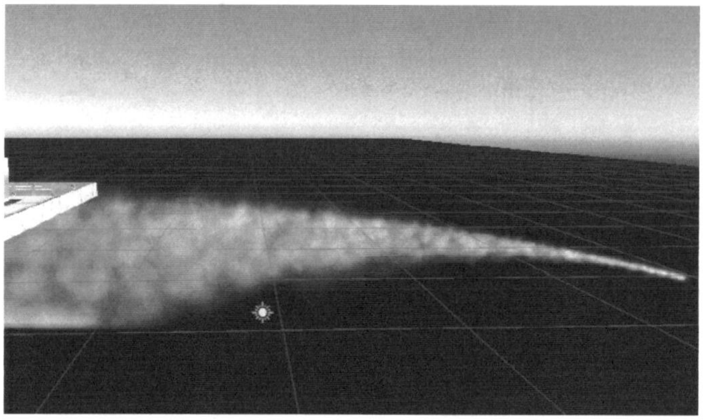

Figure 10. Simulation effect of water mist particles

4.2.3 Simulation of thermal radiation
Based on the shader of Unity3D, this paper writes a script program to display circles of different sizes in the substation in real-time during the development of fire, to achieve the purpose of displaying thermal radiation radius (damage/damage distance threshold). The effect of code operation is shown in Figure 11 below, which is a circular aperture. The color of the aperture and the size of the aperture position can be modified in the particle system to achieve different effects, and the changes over time can be realized by supplementary codes. Run after visualization system, fire radiation scope as shown in figure 12, white representatives have caused severe burns the distance threshold R3, yellow representatives distance threshold R2 caused by mild burns, blue representatives will cause discomfort distance threshold R1, the range can be the modification of code based on the experimentally measured data, and the development process and fire.

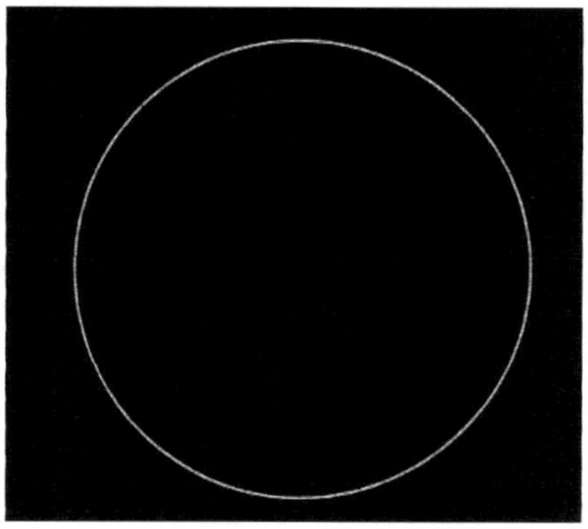

Figure 11. Aperture algorithm demo

Figure 12. Schematic diagram of heat radiation

4.3. System operation

The Unity3D used in this paper runs in the Windows10 system. The visualization system supports the generation of program packages suitable for Windows, Android, and IOS platforms, and supports the import of virtual reality hardware devices. In this paper, the scene construction of a substation and the visualization of the fire development process are completed, and the Windows application program is generated and tested in the Windows environment. The program can run on ordinary computers.

Start the desktop applications, can enter the program start screen, choose virtual substation scene back Angle appeared in the transformer substation, the actual running effect as shown in figure 13, using the keyboard mouse to interact, can be realized through the WASD key direction of movement, using drag the mouse to, implement panoramic roaming in the scene. The visualization of the spread process of the fire is more in line with the actual situation, and the system runs smoothly, the simulation degree of Flame and water mist in the scene needs further improvement. When selecting the casing burst fire diffusion model, as shown in Figure 13, simulate loading a substation fire scenario.

In the scenario, combustion development stages and phenomena are as follows:

In FIG. 12 (a), the combustible gas is explored, the casing breaks, and the burning transformer oil is sprayed out, causing spray combustion of transformer oil. Later, the transformer oil is emitted, and the transformer oil keeps spreading and burning on the rheological body. In figure 12 (b), is given priority to medium flow, pipe burst, transformer oil from rupture overflow, along with the lower casing wall transformer for lower flowing, in the process of flow, due to the oil temperature has reached a certain temperature, and below the flames to the outflow of transformer oil for heating and lighting, wall flows on the transformer oil in combustion, pour out many transformers oil casing. The spreading flow fire on the top of the rheological exchange, the transformer oil collects in the oil collecting pit, forming the oil pool fire, the side wall flows the fire, the oil collecting pit oil pool fire three combustion forms; In FIG. 12 (c), the casing flange fire, top surface fire, side wall flow fire, oil pool fire in oil collecting pit, and various forms of combined combustion form a multi-scene three-dimensional fire, including the platform fire at the bottom of the casing, the side-wall fire at the bottom of the casing, and the oil pool fire at the bottom of the transformer. In Figure 12 (d), the fire extinguishing system starts to work to realize the fire extinguishing process.

(a) (b) (c) (d)

Figure 13. Transformer fire scene

5. Conclusion

Based on Unity3D, this paper carries out simulation research on substation fire and draws the following conclusions:

(1) Using the numerical simulation method, the temperature, flow field, and heat flow distribution data in the process of transformer fire are analyzed to provide data support for the establishment of a visual system.

(2) based on the Unity particle system and C# programming, simulation of water mist particles and transformer fire spread and development, the establishment of the fire thermal radius and radiation damage distance threshold aperture display, integrated transformer fire dynamic development, visual radiant heat damage, and other functions.

(3) Based on virtual reality technology, the data simulated by FDS is applied to the visualization process of substation fire. Modeling software 3dsMax is used to complete the main scene modeling of the substation, and the fire visualization system is established by using Unity3D engine to simulate the occurrence of a transformer fire and the process of water mist extinguishing. The visual process of fire particles and fire water injection is realized based on relevant motion law, and the virtual simulation of substation fire is realized.

The research results of this paper can provide a real fire diffusion scene, strengthen the cognition of the fire development process, and provide theoretical and technical support for the fire control and control of the substation.

Funding
Science and Technology Project of State Grid Jiangsu Electric Power Design and Consulting Co., Ltd. ——Key Technology Research on WebGIS-based Substation Fire Detection（JE202201）

References

[1] Cooper, L. Y., (1995) Combined buoyancy and pressure-driven flow through a shallow, horizontal, circular vent[J]. Journal of Heat Transfer, 117(3):659-667.
[2] Zhuang, L., Chen, G. Q., Sun, Z. Y., et al. (2008) Journal of Safety and Environment, 04:110-114.
[3] Li, J., Wang, Y., Wang, Y., (2013) Research on collaborative fire emergency drill platform based on VRML [J]. Computer measurement and control, 21 (12).
[4] Wang, X. B., Yuan, D. Y., Jin, X. S., (2014) Research on simulation training system for individual soldier in firefighting and rescue operation [J]. Journal of system simulation, 26 (6).
[5] Wang, X. P., Qi, Y., Zhang, Y., (2001) Research on DIS system framework of firefighting simulation drill [J]. China safety science journal, 11 (6).
[6] Yin, J. G., (2009) Dalian: Dalian Maritime University. (In Chinese)
[7] Chang, Z., Qiu. J. S., Zhang, X. S., (2009) Research on architecture of Warship Virtual Fire Visualization System Based on Virtual reality Technology [J]. China Ship Research, 4(3):56-61.
[8] Ren, H. X., Jin, Y. C., Yin, Y., (2010) Journal of Wuhan University of Technology (Transportation Science and Engineering), 34(1):19-22.
[9] Qi, H. Y., (2011) Research on key Technology of Flame Special Effect in augmented reality System [D]. Beijing: Beijing Institute of Technology.
[10] Lu, Y., Qian, X. J., (2012) Urban Rail Transit Research, 15(1):62-65. (In Chinese)
[11] Zhang, Y., (2013) Research on Emergency Management System based on Virtual reality [D]. Shenyang: Shenyang Aerospace University.
[12] Balci, M., Foroosh, H., (2006) Real-time 3D fire simulation using a spring-mass model[C]// Multi-Media Modelling Conference Proceedings, 2006, International. IEEE, 8.
[13] Hoang, R. V., Sgambati, M. R., Brown, T. J., et al. (2010) Graphics for Serious Games: VFire: Immersive wildfire simulation and visualization[J]. Computers & Graphics, 34(6):655-664.
[14] Nikfetrat, N., Lee, W. S., (2013) Fire pattern analysis and synthesis using EigenFires and motion transitions [J]. Computer Animation & Virtual Worlds, 24 (3-4): 225-235.
[15] Chen, G., Jing, W., Wang, Z. G., Xu, L., Chen, P., (2019) China Safety Science and Technology, 15(7):186-192. (In Chinese)

Journal of Physics: Conference Series **2387** (2022) 012013 doi:10.1088/1742-6596/2387/1/012013

Optimization and research of coil based on marine heavy oil electromagnetic heater

Ziqiang Yang[1], Cunxi Zhang[2,*], Rui Wang[2]

[1] School of Ocean Engineering Equipment, Zhejiang Ocean University, Zhoushan 316022, China;

[2] School of Shipping and Maritime Engineering, Zhejiang Ocean University, Zhoushan 316022, China.

*Email:zhangcx@zjou.edu.cn

Abstract. Based on the theory of electromagnetic field, the Marine heavy oil electromagnetic induction heater uses the principle of electromagnetic induction to realize the energy conversion and transfer, so as to heat the heavy oil to the inlet temperature to achieve the purpose of heating.By means of electromagnetic induction, eddy currents are generated on the tubing wall under the action of a magnetic field. The wall of the heated oil pipe conducts heat energy to the heavy oil in the pipe. In this heating mode, the coil is directly around the oil supply unit pipeline of the ship for heating, and the heating effect depends on the heating electromagnetic field.Therefore, to study the electromagnetic field and its influencing factors of heavy oil electromagnetic induction heating is the premise of designing heavy oil electromagnetic induction heater.In this paper, the magnetic induction intensity generated by the electromagnetic induction heater with different pitch is analyzed theoretically, and simulated and optimized by COMSOL Multiphysics software.

1. Introduction

At present, the boiler steam heating method is commonly used on ships to heat heavy oil, but this heating method has the disadvantages of slow heating speed and easy failure.With the development of electromagnetic induction heating technology, some scholars have proposed to apply electromagnetic induction heating technology to ship heavy oil heating.In this paper, the influence of pitch of the ship heavy oil electromagnetic induction heating coil on the heating effect is studied, and the analysis method of the cylindrical spiral coil is divided into two types: tight winding and thin winding.According to the Bio-Saver theorem[1]. Numerical solutions are used to solve and analyze the magnetic field of electromagnetic coils with different pitches.For the theoretical analysis and then through the COMSOL Multiphysics software, the magnetic induction intensity generated by the electromagnetic coil with different pitches is distinguished, and the heating effect of the electromagnetic induction coil is densely wound and loose.

2. Study of magnetic field strength of electromagnetic induction coils with different pitches

2.1. Calculation of the magnetic induction intensity of the densely wound cylindrical spiral coil

When the pitch is 0, the spiral becomes a circular line,its magnetic field distribution is the distribution of circular linear electromagnetic fields, with only magnetic fields in the z-axis direction and no magnetic field components in the x,y-axis direction [2]. Therefore, when analyzing densely wound cylindrical spirals, only the magnetic field in the z-axis direction is analyzed under idealized conditions.

Schematic diagram of the calculation of establishing a loose cylindrical spiral in a spatial cartesian coordinate system as shown in Figure 1, the coil equation is:

$$\begin{cases} x' = a\cos\theta \\ y' = a\sin\theta \\ z' = \dfrac{h\theta}{2\pi} \end{cases} \tag{1}$$

Where a is the radius of the helical coil section; h is the pitch; θ is the helix angle.

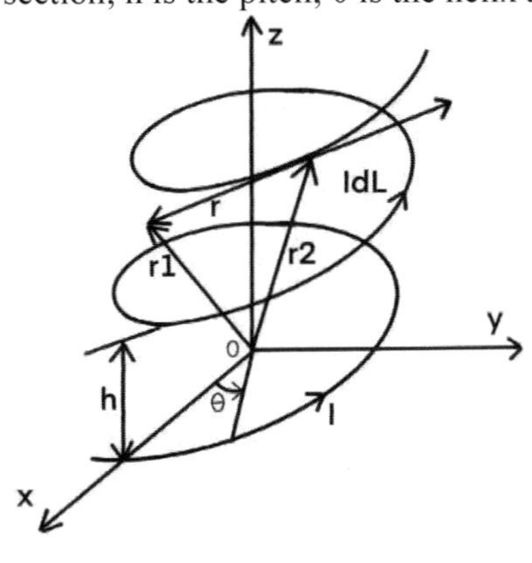

Figure 1. Schematic diagram of a cylindrical spiral coil

Take any bit of P(x,y,z) in space,in the spiral coil, the current element *Idl* is taken,then the vector r of the P point relative to the current element can be expressed as [3]:

$$r = r_1 - r_2 = (x - a\cos\theta)e_x + (y - a\sin\theta)e_y + (z - \frac{h\theta}{2\pi})e_z \tag{2}$$

According to equation (1-2), the membrane of the vector r can be calculated as $r = \sqrt{(x-a\cos\theta)^2 + (y-a\sin\theta)^2 + (z-\frac{h\theta}{2\pi})^2}$ .According to the spiral equation,the axis component form of the coil microele dl is $dl = -a\sin\theta d\theta e_x + a\cos\theta d\theta e_y + \frac{h}{2\pi}d\theta e_z$ .From this, it can be obtained that the vector product of the coil micron dl and the vector r is:

$$\begin{aligned} dl \times r = & (az\cos\theta - a\frac{h}{2\pi}\theta\cos\theta - \frac{h}{2\pi}y + a\frac{h}{2\pi}\sin\theta)d\theta e_x \\ & + (\frac{h}{2\pi}x - a\frac{h}{2\pi}\cos\theta + az\sin\theta - a\frac{h}{2\pi}\theta\sin\theta)d\theta e_y \\ & + a(a - y\sin\theta - x\cos\theta)d\theta e_z \end{aligned} \tag{3}$$

According to the Bio-Saval theorem , the magnetic induction intensity generated by current unit Idl at P at this point is [4]:

$$dB = \frac{\mu_0 Idl \times r}{4\pi r^3} \tag{4}$$

Since there is no magnetic field component on the x,y axis, only the coordinate axis (z-axis) component of the magnetic induction intensity generated by the current element Idl at P is calculated, and the relevant parameters are brought to Equation (1-4) to obtain:

$$dB_z = \frac{\mu_0 Ia(a - y\sin\theta - x\cos\theta)}{4\pi\left((x-a\cos\theta)^2 + (y-a\sin\theta)^2 + (z-\frac{h\theta}{2\pi})^2\right)^{\frac{3}{2}}} d\theta e_z \tag{5}$$

As can be seen from the spiral equation, $\theta = 2\pi z'/h$ .At this time, the calculation formula for the past coordinate axis component of the magnetic induction point generated by the spiral coil at the P point is:

$$B_z = \frac{\mu_0 Ia}{4\pi}\int_0^{2\pi z'/h} \frac{a - y\sin\theta - x\cos\theta}{\left((x-a\cos\theta)^2 + (y-a\sin\theta)^2 + (z-\frac{h\theta}{2\pi})^2\right)^{\frac{3}{2}}} d\theta \tag{6}$$

Therefore, the approximate magnitude of the total magnetic induction intensity of the P-point is:

$$B = B_z \tag{7}$$

P is any point in space,so the coordinate axis component size and the total magnetic induction intensity of any point in space can be calculated by equation (1-4)-(1-7).

Take the P-point coordinate as (0,0,z), and the bring-in formula (1-4)-(1-7) can be obtained to calculate the magnetic induction intensity on the z-axis.

$$B_z = \frac{\mu_0 I}{4\pi}\int_0^{2\pi z'/h} \frac{a^2}{\left(a^2 + (z-\frac{h\theta}{2\pi})^2\right)^{\frac{3}{2}}} d\theta \tag{8}$$

Here, take $\mu_0 = 4\pi \times 10^{-7}$ H/m, $I = 1A$ .When h takes a very small value, that is, h = 1, and h takes 10 and 20 respectively, the characteristics of the magnetic induction intensity along the radial change are shown in Figures 2 and 4 below.

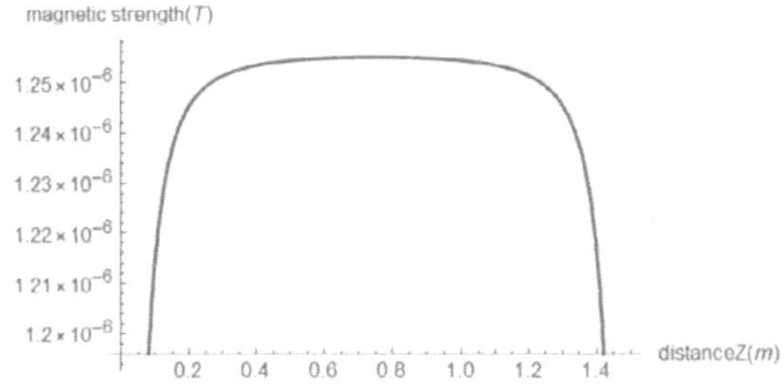

Figure 2. When h=1, the magnetic induction intensity changes along the radial characteristics of the plot

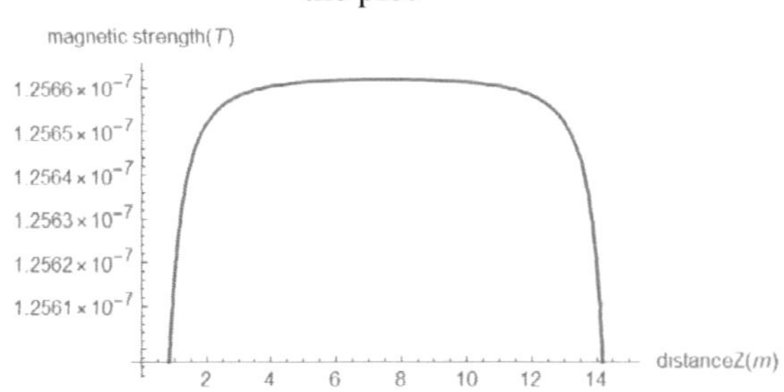

Figure 3. When h=10, the magnetic induction intensity changes along the radial characteristics of the plot

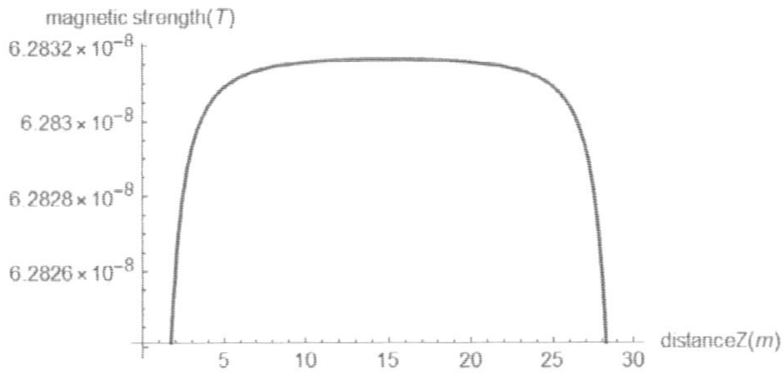

Figure 4. When h=20, the magnetic induction intensity changes along the radial characteristics of the plot

As can be seen from the above figure, the smaller the value of h, the greater the value of the maximum magnetic induction intensity; when the value of h is larger, the value of the maximum magnetic induction intensity is smaller.

2.2. Calculation of magnetic induction intensity around cylindrical spiral coils

When the pitch is not 0, the magnetic field of the spiral current can be approximated using the magnetic field of the circular line current [2]. In the Rectangular Coordinate System, take any point P(x,0) as the observation point on the abscissa coordinate,and establish a schematic diagram of the calculation of magnetic induction intensity in the plane of the circular spiral coil, as shown in Figure 5 below.

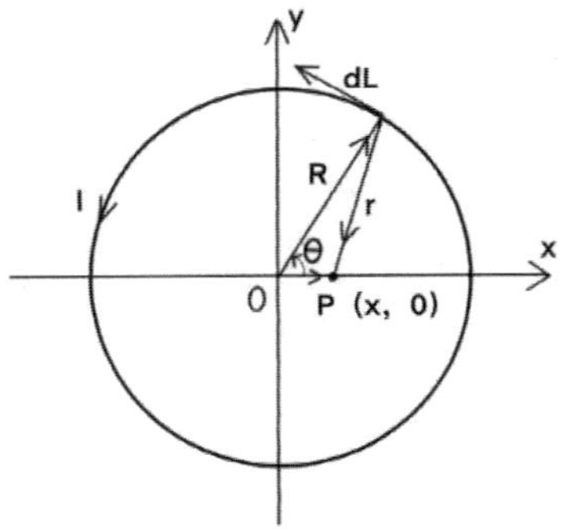

Figure 5. Schematic diagram of calculation of magnetic induction intensity in the plane of circular spiral coil

As can be seen from Figure 3, $OA = x_A i + y_A j$; $OP = xi$; $AP = r = (x - x_A)i - y_A j$ .It is known that the radius of the coil is R,and the constant current value in the coil is I,so $x_A = R\cos\theta$; $y_A = R\sin\theta$ .And because $dl = dx_A i + dy_A j$,know:

$$\begin{aligned}|dl \times r| &= (-R\sin\theta d\theta i + R\cos\theta d\theta j) \times \left[(x - R\cos\theta)i - R\sin\theta j\right] \\ &= R(R - x\cos\theta)d\theta\end{aligned} \tag{9}$$

According to the Bio-Saver theorem,the magnetic induction intensity generated by a unit current element is calculated at this time as [5]:

$$dB = \frac{\mu_0 I dl \times r}{4\pi r^3} = \frac{\mu_0 IR(R - x\cos\theta)}{4\pi(R^2 - 2Rx\cos\theta + x^2)^{\frac{3}{2}}} d\theta \tag{10}$$

Integrating the above equation shows that the magnetic induction intensity value generated by the coil is:

$$B=\frac{\mu_0 IR}{4\pi}\int_0^{2\pi}\frac{(R-x\cos\theta)}{(R^2-2Rx\cos\theta+x^2)^{\frac{3}{2}}}d\theta \tag{11}$$

Here, take $\mu_0 = 4\pi\times10^{-7}$ H/m, $I=1A$， $R=5.6\times10^{-6}$ Ω,At this time, the magnetic induction intensity changes along the radial characteristics as shown in Figure 6 below.

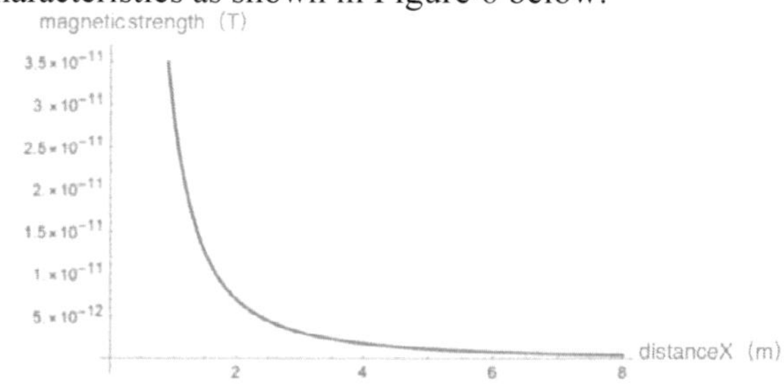

Figure 6. Characteristics of magnetic induction intensity changing radially

From the above figure, it can be seen that when x is closer to the center of the circle, the greater the value of the magnetic induction intensity, and the distribution of magnetic field lines at the center of the circle is also denser; The farther x is from the center of the circle, the smaller the value of the magnetic induction intensity at this time, and the sparser the distribution of magnetic field lines.

2.3. Calculation of magnetic induction intensity in the coil plane under alternating current

When an alternating current with a maximum value of I0 and a sinusoidal change is passed into the coil, the magnetic induction intensity is calculated as follows:

$$B(x，t)=\frac{\mu_0 RI_0\sin(\omega t+\varphi)}{4\pi}\int_0^{2\pi}\frac{(R-x\cos\theta)}{(R^2-2Rx\cos\theta+x^2)^{\frac{3}{2}}}d\theta \tag{12}$$

Obviously, the magnetic induction intensity is not just a function of x, but a two-dimensional function of x and t, and its value will vary from time to space. If the time period of the alternating current is T, the average magnetic induction intensity in one period is:

$$\vec{B}(x,t)=\frac{2}{T}\int_0^{\frac{T}{2}}\left(\frac{\mu_0 RI_0\sin(\omega t+\varphi)}{4\pi}\int_0^{2\pi}\frac{(R-x\cos\theta)}{(R^2-2Rx\cos\theta+x^2)^{\frac{3}{2}}}d\theta\right)dt \tag{13}$$

For a certain instantaneous time-varying current,the current value is constant, and the change trend of the magnetic induction intensity generated by it along the radial direction is the same as the change trend of the magnetic induction intensity generated under the constant current along the radial direction.

3. Optimized simulation of the effect of electromagnetic induction coils with different pitches on magnetic induction strength

In order to verify whether the relationship between magnetic induction intensity and h is inversely proportional, through the COMSOL Multiphysics software, modeling is based on a single-turn coil, as shown in Figure 7 is the actual coil.

Figure 7. Actual single-turn coil

After simplifying the actual coils, coordinate modeling is built on top of THE COMSOL Multiphysics software. This is shown in Figure 8 and Figure 9.

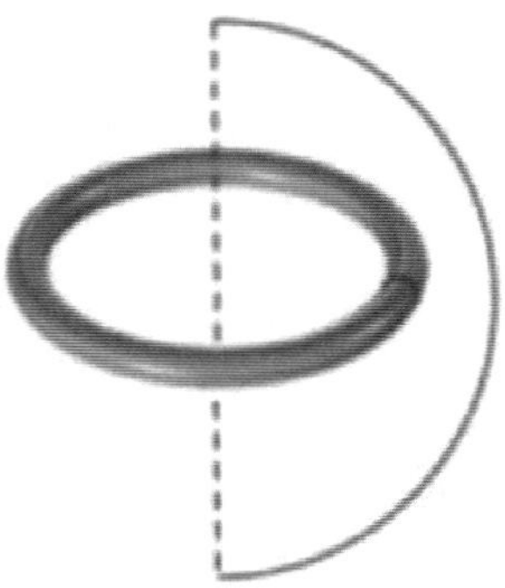

Figure 8. Simplified coil

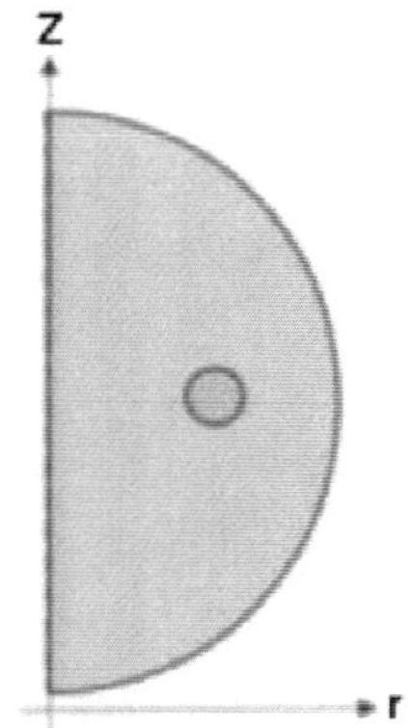

Figure 9. Coordinate construction

The power expression for pipe vortex heating is:

$$P = \lambda f^2 B_m V \tag{14}$$

Equation (2-1), f is the resistance coefficient of the iron core material; B_m is the maximum magnetic induction strength in the core; V is the effective volume of the iron core material. The size of B_m is mainly determined by the magnitude flux φ size, by Ohm's law $\Phi = \dfrac{NI}{R_m}$ [6]. Therefore, control f =1KHz,Number of coil turns N=9 and I=1A,Divide the pitch into pitches h=0 与 h≠0 Two cases.For the case of h≠0,in this experiment, h=10 and h=20 are taken to verify, and the effect can be seen. The influence of the electromagnetic coil with different pitches on the magnetic induction strength is shown in figures (10-12) below.

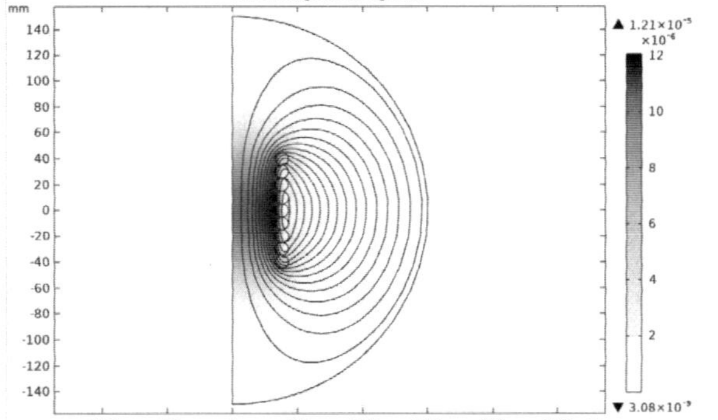

Figure 10. Magnetic induction strength diagram of the spiral coil at h=0

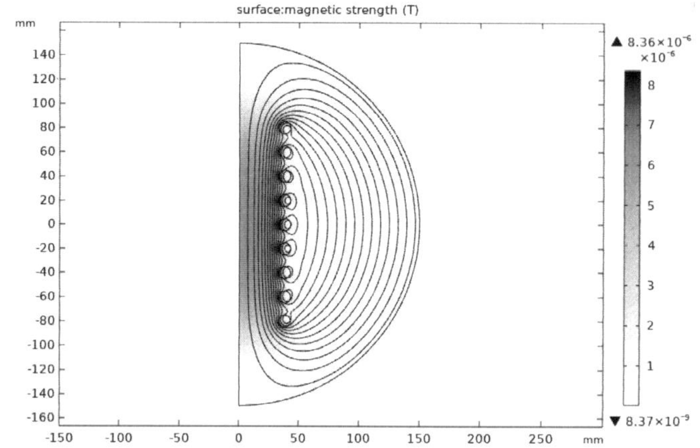

Figure 11. The magnetic induction intensity of the spiral coil at h=10

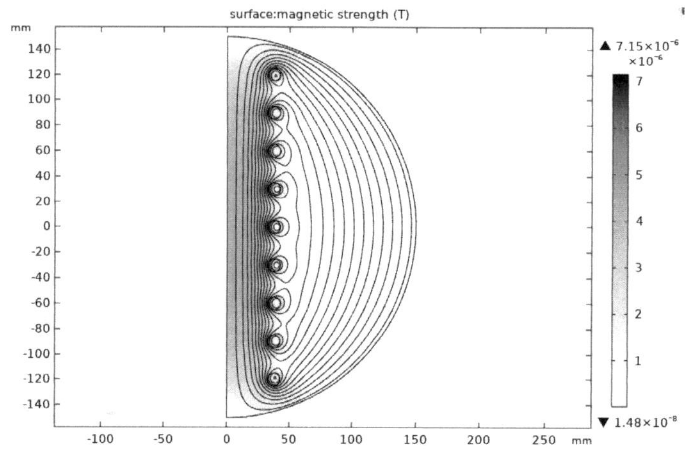

Figure 12. h=20 magnetic induction strength diagram of the spiral coil

It can be seen from the figure that in the case of the same frequency, the number of coil turns and the current size, when h = 0, the maximum magnetic induction intensity of the spiral coil is1.21×10-5;when h = 10, the magnetic induction intensity of the spiral coil is 8.36×10-6 at the maximum;when h = 20, the magnetic induction intensity of the spiral coil is 7.15×10-6 at the maximum.It can be seen that when other conditions are unchanged, when the spiral coil pitch is 0, the greater the magnetic induction intensity, and the heating effect of the electromagnetic coil is the best at this time. The larger the pitch, the smaller the magnetic induction intensity, and the worse the heating effect of the electromagnetic coil.

4. Conclusion

According to the Bio-Savard theorem, the spiral coils with different pitches are analyzed, and the characteristics of the magnetic induction intensity in the plane of the circular spiral coil are drawn along the radial change.It can be concluded that when a constant current is passed into the spiral coil,when the value of h is larger, the value of the magnetic induction intensity is smaller, and in the plane of the coil, the closer to the center of the circle, the smaller the magnetic field strength, and the closer to the coil the greater the magnetic field strength.And whether the pitch is 0, its magnetic field distribution can be equivalent to the distribution of circular linear electromagnetic fields,only the magnetic field in the z-axis direction, and no magnetic field component in the x,y-axis direction.Through the COMSOL Multiphysics analysis software,the magnetic induction intensity generated by the electromagnetic coil with different pitches is optimized for simulation, and it can be concluded that the smaller the pitch,the greater the magnetic induction intensity,the better the heating effect of the electromagnetic coil, and the simulation results are basically consistent with the theoretical results.According to the experiment, on the actual oil tanker electromagnetic heating coil,it is necessary to use the dense winding method to wind the electromagnetic coil as much as possible,and the electromagnetic induction heating method of the oil tanker heavy oil heating method has a better practical reference value.

References:

[1] Yu, S. X., Zhang, S. J. (1999) Simple Formula for Uniform Analysis of Helmholtz Coil, Straight Solenoid and Circular Current Magnetic Field[J]. University Physics, (8):1-3.
[2] Feng, W. J. (2001) Calculation and analysis of magnetic field on the axis of helix current[N].Journal of Gansu University of Technology, 12(4).
[3] Luo, H., Cai, M. (2006) Distribution of Circular Current Full-Space Magnetic Induction Intensity B[J]. Journal of Shenyang Institute of Aeronautical Industry, 23(1):79-80.
[4] Zhang, X. H. (2006) Distribution of Circular Current Magnetic Induction Lines and Function Expression of Magnetic Induction Intensity[J]. University Physics, 25(1):32-37.
[5] Jiang, J. Q. (2015) Spatial distribution of magnetic field of cylindrical solenoids with finite length constant current[J]. Journal of Guangdong Second Normal University, 35(3):53-57.
[6] Luo, W. Z., Shen, J., Zhao, H. Y., Zhang, F. S. (2013) Study on magnetic field distribution in different heating elements under electromagnetic induction heating conditions[J].Thermal Processing Technology, 42(22):95-98.

Development of low power and high voltage RF signal generator

Man Wang, Tianhong Zhou*, Jie Shen

Wuhan Business University, Wuhan, China
The first author's e-mail: wangman@wbu.edu.cn

*corresponding author's e-mail: 20150594@wbu.edu.cn

Abstract. Ion confinement technology can be widely used in quantum computation and precision measurement. In ion trapping system, ions are stably trapped in ion trap by the action of external forces.Multi-ion confinement requires the joint action of electric field and magnetic field to make the trapped ions overcome the interaction force between ions and be stably confined in the same specific space. Therefore, the stability of the electric force and the magnetic force determines the stability of ion confinement . Although the electric field and magnetic field provided by traditional commercial instruments is stable, it is not conducive to the miniaturization of the device. According to the previous experiment of ion confinement, we have designed a radio frequency(RF) signal generator that can provide continuous low power and stable high voltage RF electric field for multi-ion confinement system to guarantee the stability of trapped ions in quadrupole linear ion trap, also to improve the tapped ions density.

1. Introduction

There are semiconductor integrated circuits inside the classical computer to form the computing unit, which can only record 0 and 1 states, and the efficiency of data processing is limited to a certain extent. With the continuous development of computer science and technology, people have derived computing science into the microscopic world, and quantum computing has developed accordingly. In 1995, J.I. White and P.Zoller first proposed the study of quantum computation by particle mass analyzer[1]. Particle mass analyzer is a quadrupole type trap device based on alternating electric field, which can make the target particles exist stably in ultra-high vacuum environment without boundary interference. The linear trap of particle mass analyzer uses the combination of electrostatic field and radio frequency field to make charged ions move back and forth like simple harmonic oscillator at the mechanical equilibrium point and realize ion confinement. When the ion group is stably imprisoned in a certain range of space, the interaction force between ions needs to be overcome[2]. Therefore, electric and magnetic fields are needed to achieve the stable confinement of ions[3]. It is found that the stability and density of ion confinement are closely related to the electromagnetic field[4]. Because different ions have different requirements for the electric field, the electric field frequency of some ion confinement needs to reach more than several MHz and the amplitude is as high as several thousand volts, which is difficult for general equipment to meet such requirements[2]. It is necessary to design a low-power high-voltage RF signal generator.

In this paper, we have formulated the multiion confinement mode and designed a miniaturized radio frequency circuit to control the quadrupole linear ion trap radio frequency field. The research on the stability of radio frequency field was firstly carried out[5]. Then the radio frequency circuit is designed according to the experimental results[2]. The miniaturized radio frequency circuit not only has excellent stability, but also has small volume and low power consumption.

2. Ions imprison stability condition

The linear quadrupole rf ion trap is a kind of mass filter, which trapped target ions by electric and magnetic force. Here, the quadrupole ion trap is taken as the research object.The linear quadrupole rf ion trap is made up of four axial poles, and two end-cap. DC voltage is applied between two end-caps to provide axial electrostatic force, while rf voltage is applied between two asymmetric axial poles to provide a alternating electric field[2].

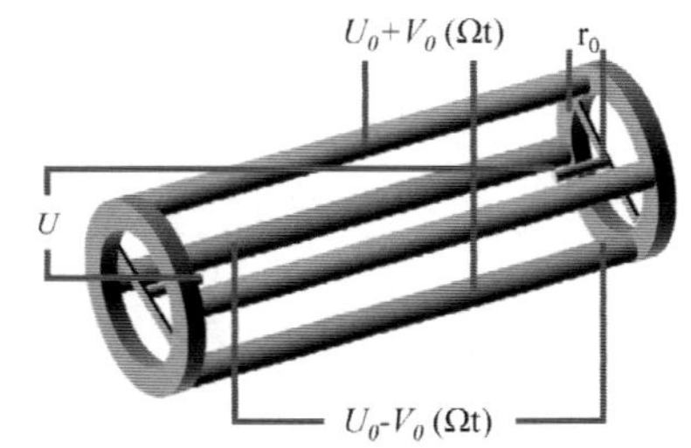

Figure 1. The model of the quadrupole linear ion trap.

U is the DC voltage which applied between two end-caps.U_0-V_0 (Ωt) is the rf voltage which applied between one group of the asymmetric axial poles, U_0+V_0 (Ωt) is applied between another. U_0 is a DC bias voltage which is always set 0V. V_0 is the peak-to-peak voltage of the rf voltage and Ω is the angular frequency of rf signal.
According to the Mathieu equation, the potential of any point in linear quadrupole rf ion trap could be described as[6]:

$$\Phi=U_0+V_0\cos(\Omega t)\frac{(x^2-y^2)}{2r_0} \tag{1}$$

Where r_0 is the distance from the center of the linear quadrupole rf ion trap to the pole[7].
Because of Newton's second law of motion:

$$F=ma=-e\nabla\Phi \tag{2}$$

Where e is the charge of the trapped ions.
We can solve the Mathieu equation by giving the parameters:

$$a=a_x=-a_y=\frac{4eU_0}{mr_0^2\Omega^2} \tag{3}$$

$$q=q_x=-q_y=\frac{4eV_0}{mr_0^2\Omega^2} \tag{4}$$

As is known that, the stability of Mathieu equation's solution is determined by the value of a, q. When $|a| << |q| <1$, ions could trapped in the linear quadrupole rf ion trap steadily. So, rf voltage applied between rod electrodes could affect ion imprison in the trap directly[8].

3. The system design

The traditional RF source uses a commercial signal generator to output a small RF signal. Then the small RF signal voltage is boosted through the boost coil,and finally connected to the pole of a linear trap[9]. Because the commercial signal generator has large volume and high operating power, it is not conducive to miniaturization. At the same time, because the signal generator output signal power is small, and the coil boost capacity is limited, the RF signal with higher voltage amplitude cannot be obtained[10]. We propose a design of a high stability low power RF source over kilovolts. It is a combination of digital circuit and analog circuit. The circuit can output radio frequency signal with high amplitude and stability. The volume power consumption is also much lower than that of traditional RF sources. Its structure diagram is shown in Figure 2. The RF circuit consists of two parts, the digital module and the simulation module. The digital module ensures the stability of the signal, and simulation module ensures the integrity of the high voltage RF signal output. The voltage amplitude of the output signal is adjustable within a few kilovolts. The whole circuit has an excellent performance of small volume and low power consumption, which can help to miniaturize the ion trapping system.

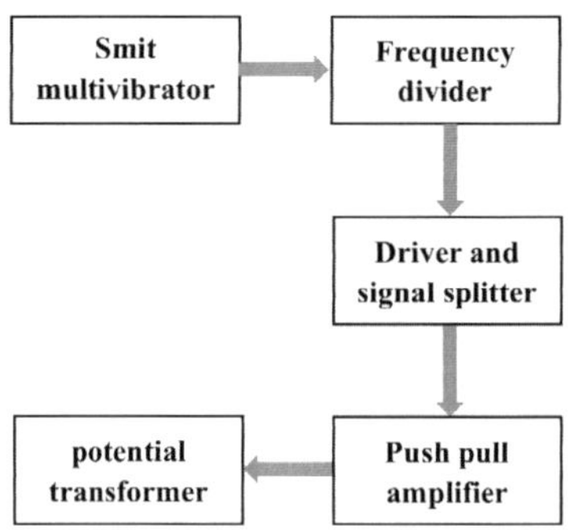

Figure 2. Block diagram of miniaturized rf circuit.

Firstly, schmidt flip-flop can generate continuous anti-wave signal by resonance, the frequency divider , the driver and the signal splitter are the digital modules of miniaturized RF circuit. The power amplifier and the voltage booster are the simulation module. The two modules combine to implement the RF signal output. The circuit diagram of the design are shown in figure 3 and figure 4.

Figure 3. The digital part of the rf circuit.

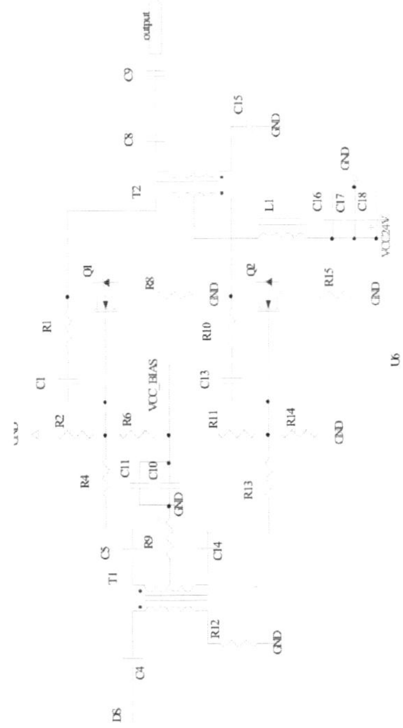

Figure 4. The simulation part of the rf circuit.

The miniaturized RF circuit is constructed by integrated circuits. The square wave is generated by schmidt flip-flop and converted into a sinusoidal signal by a power push-pull amplifier, so the signal has extremely high stability and anti-interference ability. The use of the high efficiency booster core can reduce the volume and power consumption of RF field controller. The most important feature of this circuit is that no input signal is required. The Schmidt flip-flop with closed loop connection can output continuous square wave signal. The signal passes through the frequency divider and becomes a two-way frequency signal to provide a backup channel for signal transmission, and improve drive capability by the drive. The signal splitter converts the signal into two symmetric signals and outputs to the power push-pull amplifier. Then through the booster, the signal voltage range up to thousands of volts.

4. Testing and analysis

The signal voltage peak value of the miniaturized radio frequency circuit is as high as several kilovolts, so the method of partial voltage test is used for circuit testing. Figure 5 shows the result of our test. A tenth of the total voltage is about 180V, the frequency of the RF circuit is about 1.4MHz.

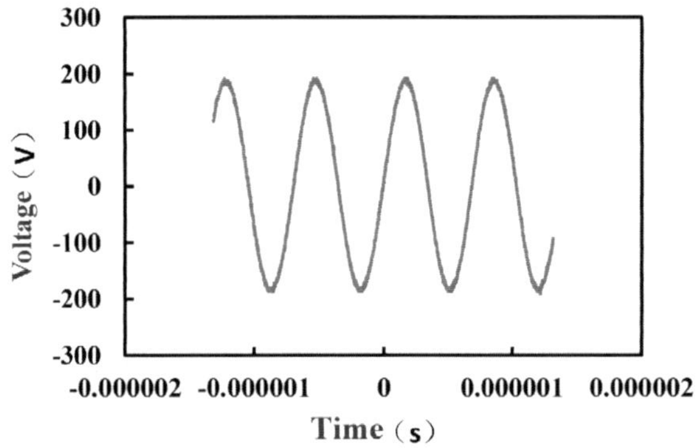

Figure 5. One tenth of the output voltage

In order to verify whether the design circuit can meet the requirements of the experiment, we tested the amplitude stability and frequency stability of the circuit.

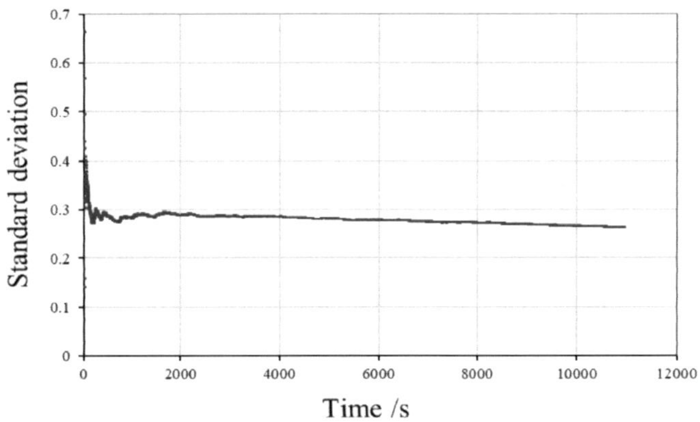

Figure 6. The amplitude stability of 10000 sec.

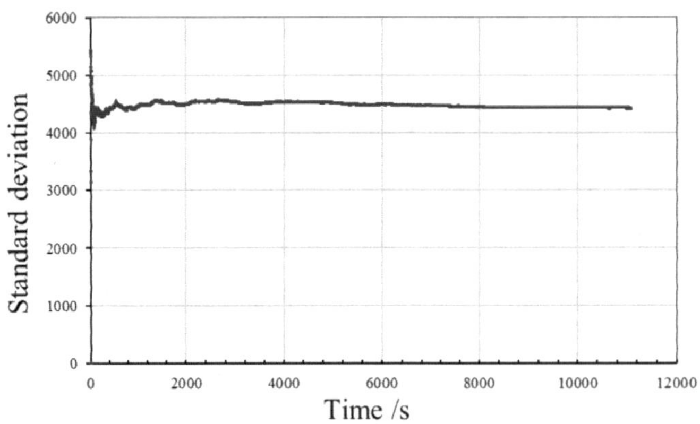

Figure 7. The frequency stability of 10000 sec.

It can be seen from Fig.6 and Fig.7, the amplitude and frequency of the output signal tend to be stable when the RF source runs for a long time. After calculation, the stability of amplitude and frequency are 0.267% and 0.443%, respectively. RF signal output has an excellent amplitude stability and frequency stability. However, through the ions confinement experiment, the RF signal can have a significant effect on the imprison ions in the case of large RF voltage change. Moreover, when a and q are satisfied with the stability condition of Mathieu equation, the higher the RF voltage is, the more it help for ions confinement. Therefore, the design of the RF circuit conforms to the experimental requirements and is far superior to commercial instruments in terms of volume and power consumption.

5. Summary

The whole work is centered on the development of miniaturized radio frequency circuit. Through the experiment, we found that under the stability condition of Mathieu equation, the higher the RF voltage, the greater the number of ions in captivity, the smaller the center frequency and the center spectrum width. what's more the number of ions increased degree is more significant than center frequency decreases. It is noteworthy that RF field parameters in the ion quadrupole trap was confirmed by the experiment[2], then the miniaturized RF circuit was designed. The voltage amplitude stability and the signal frequency stability of the circuit have been tested. It can be seen from the test that the performance index of RF circuit is superior that it can meet the requirements of the experiment completely and can be adjusted in a wide range, which can contribute to multiple ions confinement. The volume of the circuit is only $0.06*0.06*0.04m^3$, and the power is only about 3.5W, which is more than ten times lower than the volume and power consumption of commercial instruments used in the past.

Acknowledgment

This work was supported by the Guiding Project of the Scientific Research Plan of the Hubei Provincial Department of Education (No. B2021272), the Knowledge Innovation Program of Wuhan-Shuguang Project 2022010801020429, and Doctoral Fund of Wuhan Business University (No. 2019KB004).

References

[1] Jiang, J. L., Zhu, L., et al. (2018) Multiband bonding/anti-boding interaction and selective electric field confinement in the complementary metamaterial[J]. Optics letters, 43(24): 6065-6068.

[2] Wang, M., Liu, H., et al. (2019) Radial confinement potential of quadrupole linear ion trap based on trapped 199Hg+ ion, Chinese Journal of Quanturn Electronics, 36(1): 28-33.

[3] Mathur, R., O'Connor, P. B. (2006) Design and implementation of a high power rf oscillator on a printed circuit board for multipole ion guides[J]. Review of Scientific Instruments, 77(11):531.

[4] Bardoczi, L., Carter, T. A., Haye, R. L., et al. (2017) Impact of Magnetic Island - Turbulence Multi-Scale Interaction on Plasma Confinement and Magnetic Island Stability[J]. APS Meeting Abstracts.

[5] Li, D. Q., Zhao, X., Liu, S. B., Liu, M. L., Ding, R. X., Liang, Y. H., Zhu, Z. M. (2022) Radio frequency analog-to-digital converters: Systems and circuits review[J]. Microelectronics Journal, 119, 105331.

[6] Gao, B., Zhang, F., Zhou, P., et al. (2017) A Frequency-tunable Two-tone RF Signal Generator by Polarization Multiplexed Optoelectronic Oscillator[J]. IEEE Microwave and Wireless Components Letters, PP(2):1-3.

[7] Romashov, V. V., Doktorov, A. N., Yakimenko, K. A., et al. (2020) Application high-speed digital-to-analog converters for direct digital synthesis of high-frequency radio signals[J]. Journal of Physics: Conference Series, 1632:012023.

[8] Johnson, J. H., Gilbert, J. A. (2018) Radiofrequency amplifier impedance optimization:, US10105172B2[P].
[9] Chiba, D., Matsukura, F., Ohno, H. (2006) Electric-field control of ferromagnetism in (Ga,Mn)As[J]. Applied Physics Letters, 89(16):951.
[10] Lottermoser, T., Lonkai, T., Amann, U., et al. (2004) Magnetic phase control by an electric field.[J]. Nature, 35(6999):541-544.

Journal of Physics: Conference Series **2387** (2022) 012015 doi:10.1088/1742-6596/2387/1/012015

Research and design of capacitance plug-in assembly manipulator

Jiangtao Qin [1,*], Jirong Wang[1,3,4], Yihao Xiao [1], Jun Li [2]

[1] College of Mechanical and Electrical Engineering, Qingdao University, Qingdao 266071, China;
[2] College of Computer Science and Technology, Qingdao University, Qingdao 266071, China;
[3] Weihai Innovation Research Institute of Qingdao University, Weihai 264200, China.
[4] wangjirong43@163.com

*E-mail: qinjiangtao12345@163.com

Abstract. In order to improve the efficiency of PCB (Printed Circuit Board) capacitor plug-in, reduce labor input and improve the accuracy of plug-in, a capacitor plug-in assembly manipulator was designed. The manipulator was mainly composed of X-axis module, Y-axis module and Z-axis module. The mechanical gripper of Z-axis module could adjust the distance between two mechanical grippers to meet the needs of different capacitance sizes. By adopting double z-axis design, the work efficiency of rectangular coordinate manipulator was improved. The addition of the X-ray axis improved the synchronization of the movement at both ends of the X and enhanced the rigidity of the structure. Siemens PLC S7-200 motion library function controlled manipulator grasp, positioning and plug-in. The efficiency of the plug-in unit was 2.5 s/board, and the manipulator could complete capacitor plug-in efficiently and accurately. The whole system has the advantages of high control accuracy, simple operation, stable operation and low noise, and has important practical significance for enterprises.

1. Introduction

PCB is an important electronic component, the carrier of electrical connection of electronic components, and the parent bearing various capacitors. The PCB board used in this paper is shown in Figure 1. The automatic capacitor plug-in equipment is an electromechanical integration equipment that automatically and standard inserts regular electronic components into the conductive through hole of PCB board [1-2]. Nowadays, there are two kinds of capacitor plug-ins on the market: manual operation and automatic mechanical operation. The production mode of manual plug-ins is inefficient and the product quality is not easy to guarantee. Today, due to the shortage of labor and the development of science and technology, manual plug-ins are far from meeting the needs of mass production. Therefore, the design and research of plug-in machine is gradually attracting the attention of enterprises [3-4]. There are two main forms of plug-in work by machine: SCARA manipulator and cartesian coordinate manipulator [5-6]. Based on the advantages of simple structure, high speed and high precision of cartesian coordinate manipulator [7], In this paper, we designed a cartesian coordinate manipulator with capacitor plug-in through structural modeling, PLC control and the actual production requirements of two different capacitor components (safety capacitor and electrolytic capacitor) [8-9].

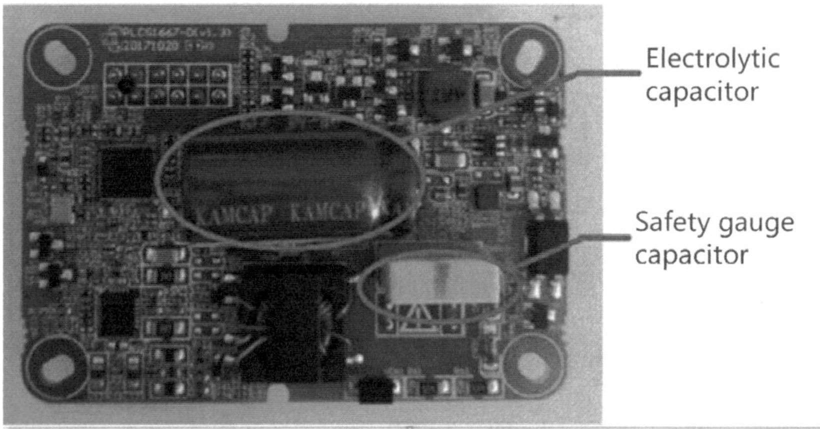

Figure 1. Printed Circuit Board

2. Machine structure and working principle

2.1. Overall structure

The cartesian coordinate manipulator is mainly composed of 1 electrical cabinet, 2 bracket, 3 Y-axis module, 4. claw and 5.Z-axis module. The manipulator adopted cartesian coordinate structure, which had the advantages of simple structure and high efficiency. Its structural form is shown in Figure 2. The manipulator is located on the bracket of Q235 steel material of the electrical cabinet. The bracket used hollow design to increase its stability and reduce its quality. The Y axis of the manipulator adopted linear module and linear slide rail on both sides respectively. In order to ensure the synchronicity of both sides of the X axis when moving forward and backward, to design a relatively complex z-axis module counterweight and to prevent the vibration and the rollover when the Z-axis module moves, a synchronous optical axis is added [10-11]. Z axis module used THK ball screw spline, could simultaneously achieve Z axis direction of linear motion and rotation around the Z axis, this was the biggest characteristic of the manipulator. The cartesian coordinate manipulator of the plug-in combined the SCARA manipulator and the cartesian coordinate manipulator together, which not only reduced the design space, but also simplified the mechanical structure. At the same time, the design of the dual manipulator effectively improved the efficiency of the manipulator plug-in [12-13].

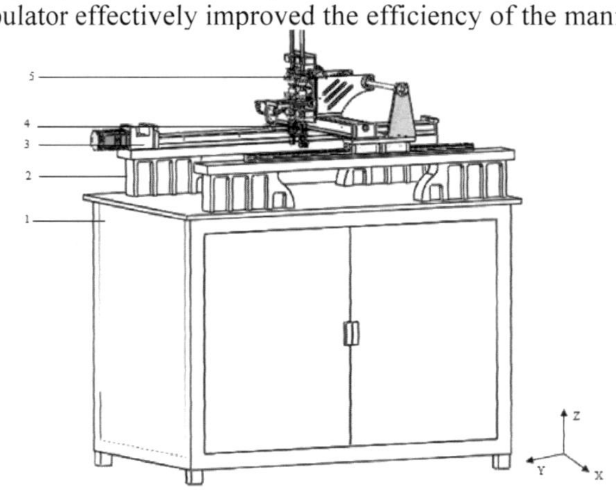

1. Electric cabinet; 2.Bracket; 3.Y-axis module; 4. Claw; 5.Z-axis modul

Figure 2. Structural Chart of Plug-in Cartesian Coordinate Manipulator

2.2. Plan analysis of z-axis motion coordination relationship

The plug-in rectangular coordinate manipulator to complete a plug-in process is mainly divided into four steps: small claw clamping, visual compensation, plug-in, reset. Each PCB board needs two capacitors: electrolytic capacitor and gauge capacitor. The X and Y modules in all schemes have the same motion, but the difference lies in the design of the motion relation of the Z-axis modules.

(1) The module action flow chart of scheme 1 is shown in Figure 3. The Z-axis module has one small claw. It takes 1.81s to complete a capacitor plug-in and 3.62s to complete a PCB small board plug-in.

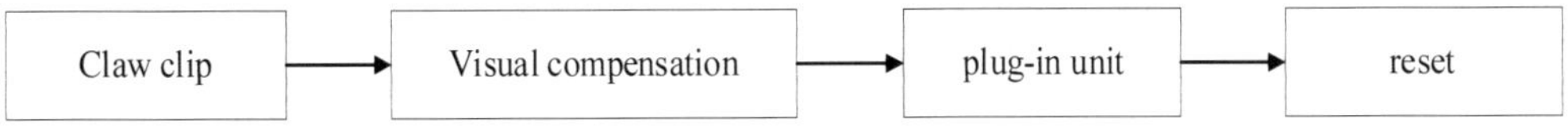

Figure 3. Flow chart of scheme 1 (Z-axis module action)

(2) The module action flow chart of scheme 2 is shown in Figure 4. The z-axis module has two small claws, which move at the same time to insert the two PCB boards. The small claws first completes the clamping and inserting of safety regulation capacitor, and then the clamping and inserting of electrolytic capacitor. This scheme requires the PCB transport line to automatically adjust the position of two PCB boards to apply the plug-in of two small claws at the same time. Two electrolytic capacitors and safety regulation capacitor feeders are required respectively to meet the needs of clamping two capacitors at the same time. On average, it takes 1.81s to complete the plug-in of one PCB.

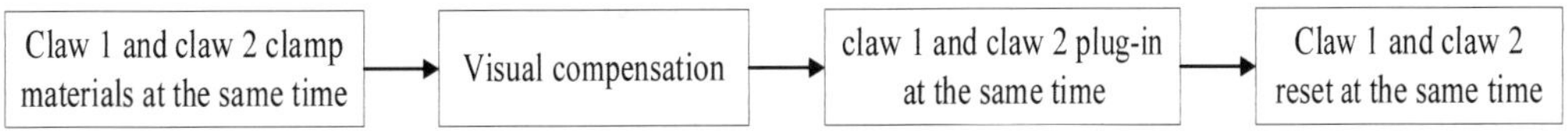

Figure 4. Flow chart of scheme 2 (Z-axis module action)

(3) The module action flow chart of scheme 3 is shown in Figure 5. The z-axis module has two interlocking claws. The two claws are installed in the same X-axis module. They move independently in the Z direction, and claw 1 and claw 2 clamp materials at the same time. Because the positions of the conductive holes of the plug-ins on the PCB of the two capacitors are inconsistent with the positions of the two small claws, the two small claws need to link the plug-ins. The small claws first completes the plug-in action of the safety regulation capacitor, and then carry out the electrolytic capacitor plug-in. On average, it takes 1.81s to complete the plug-in of one PCB.

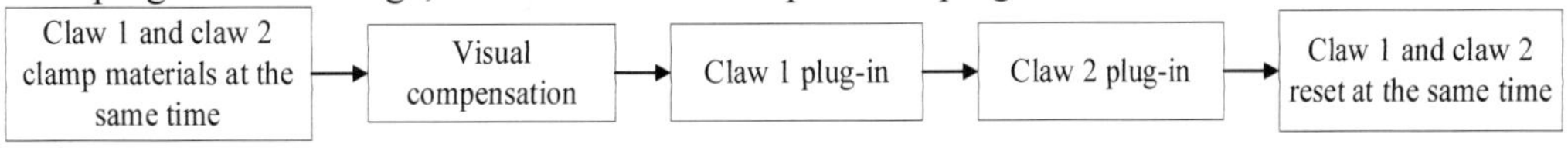

Figure 5. Flow chart of scheme 3 (Z-axis module action)

Comprehensively considering the assembly efficiency and design cost of each scheme [14-16], scheme (1) took too long to complete the assembly of a PCB board, and scheme (2) and scheme (3) met the time requirements. Scheme (2) took the least time, but two feeders are added. Because the design of PCB transportation line added automatic adjustment function, the design cost was increased. Therefore, selecting scheme (3) could not only meet the time requirements of plug-in, but also reduced the design cost. The sequence of scheme (3) is shown in Figure 6 below.

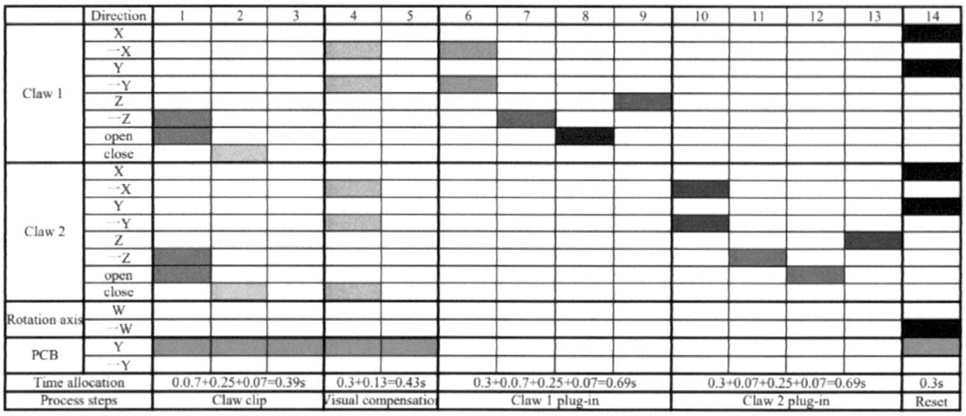

The process description:1.Claw drop; 2.Clip material; 3.Rise the two claws; 4.Two claws move to visual compensation position 5.Visual compensation;6.Move the claw 1 to plug-in position; 7.The claw 1 inserts the element; 8.Open the claw 1; 9.Rise the claw 1;10.Move the claw 2 to plug-in position; The claw 2 inserts the element; 12.Open the claw 2;13.Rise the claw 2;14.reset.

Figure 6. The Schedule of Z-axis module action

2.3. Working principle

The plug-in rectangular coordinate manipulator has four degrees of freedom, namely three degrees of freedom for linear movement in X, y and Z axes and one degree of freedom for rotation around Z axis. When the manipulator detects that the capacitor is fed in place, the z-axis ball screw spline drives the claw to move in a straight line under the drive of the motor until it drops to the grasping position. Claw 1 and claw 2 grasp the safety regulation capacitor and electrolytic capacitor at the same time, the two claws rise at the same time. During the rising process, the z-axis rotates 90 degrees clockwise. Driven by the x-axis and y-axis linear modules, the z-axis module moves to the plug-in position. The claw 1 drops, and the safety gauge capacitor is inserted into the conductive through hole of the PCB board. After the plug-in is completed, the claw 1 rises, the claw 2 falls, and the electric capacitance is inserted. After the completion of the plug-in, each module carries out reset instruction operation, thus completing a cycle process.

3. The design of each motion module

3.1. The Z-axis module of the plug-in rectangular coordinate manipulator

The SCARA manipulator is generally composed of four axes, which can realize three rotational degrees of freedom and one degree of freedom in the vertical direction. The ball screw spline shaft is the key working shaft, located at the end of the manipulator. The advantage is that the motion range is wide, the movement is more flexible, and the working range of -150~+150 degrees can be realized. The disadvantage is that the design structure is complex, and it is not easy to control, and the cost is high [17-19]. Cartesian coordinate manipulator has the advantages of simple structure, high accuracy and higher efficiency than SCARA manipulator. The disadvantage is that the movement is relatively simple and complex movement cannot be achieved [20-23]. Combining the advantages of the two different manipulators, the plug-in operation can be realized efficiently and accurately.

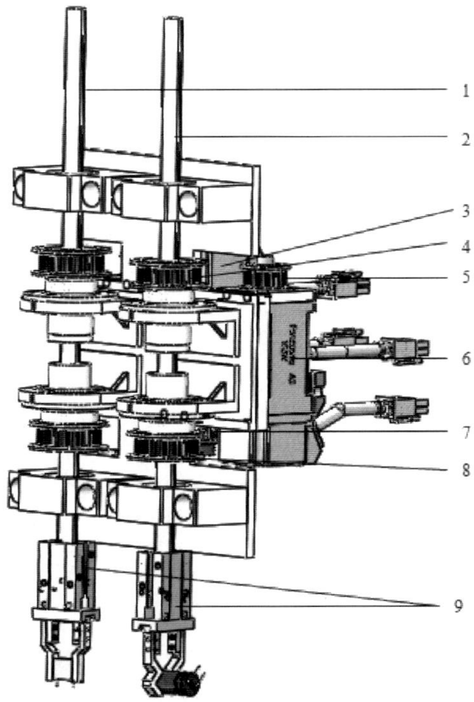

1. Z1-axis ;2. Z1-axis ;3. Spline motor ;4. Screw gear ;5. Motor gear 1;
6. Screw motor 1 ;7. Motor gear 3 ;8. Spline gear ;9. Claw

Figure 7. Z-axis module structure diagram

The structure diagram of z-axis module is shown in Figure 7. THK's BNS 1015A precision ball screw spline was used to realize linear and rotational motion. In order to improve the efficiency of the plug-in, two plug-in shafts were designed. Lead screw motor 1 is driven by synchronous belt and lead screw gear to realize the upper and lower displacement of Z1 axis. Similarly, lead screw motor 2 drives the upper and lower displacement of Z2 axis; Z-axis spline gear is connected with spline motor through synchronous belt to realize z-axis's rotation. The small claw consisted of parallel cylinder and claw fingers. The solenoid valve can control the opening and closing of the cylinder to realize the feeding and plug-in action. The claw finger of the small claw can be installed on the parallel cylinder with only two screws, which can be easily disassembled and replaced. Here the electrolytic capacitor was a column, so the claw finger was designed as a diamond, four-point contact can effectively prevent the loosening of the column. The safety capacitor is rectangular, the claws with rectangular slot in the middle can effectively fix the safety capacitor.

3.2. The Y-axis module of the plug-in rectangular coordinate manipulator

The Y-axis module was mainly composed of linear module and linear guide rail. The motion stroke of the Y-axis module is 300 mm, and the speed of the Y-axis module is up to 500 mm/s. The straight line module adopted KY55 series straight line module as the power supply end. Linear guide mainly plays an auxiliary role in motion and support, the resistance of linear guide is very small. When the Y-axis motor drives the linear module, the linear module can drive the X-axis device mounted on the Y-axis to the position of the plug-in or the pick-up. Two limit switches are installed on both sides of Y axis to protect the over-stroke movement in this direction. A reference point switch is installed in the middle of Y to reset the Y axis.

3.3. The X-axis module of the plug-in rectangular coordinate manipulator

X axis module was composed of two axes, mainly including X linear module and optical axis. The two axes are across the Y axis, one side is connected with the linear module slide block of the Y axis, the other side is connected with the slide block on the linear guide rail. The motion stroke is 300 mm, and the speed is up to 500 mm/s. The linear module of KY46 series was selected, and the motor was Siemens 1FL6 series. The motor drives the z-axis part placed on the X-axis to realize the movement in the X-direction. Two limit switches are installed on the left and right sides of the X-axis to protect the over-stroke movement in this direction. A reference point switch is installed in the middle of X to reset X axis. The biaxial design enhances the rigidity of the manipulator structure and makes the movement more stable [24-25].

4. Conclusion

This cartesian coordinate manipulator combined ball screw spline with cartesian coordinate manipulator, which simplified the structure and operation of SCARA manipulator. By the double Z-axis design, the linkage of the two manipulators improved the working efficiency of the cartesian coordinate manipulator. The addition of X-ray axis improved the synchronicity of the motion at both sides of X and enhanced the rigidity of the structure. The plug-in rectangular coordinate manipulator control system composed of PLC and servo motor had the advantages of high control precision, simple operation, smooth operation and low noise. The design of this machine could realize the plug-in speed of 2.5s/board, which was 1.5 times that of manual plug-in. It had important practical significance for enterprises.

References

[1] Lim, J. C., Ng, K. W., & Ang, M. C. (2021) "Search Operators Based on TRIZ for Optimising PCB Assembly Time",Advances in Visual Informatics 7th International Visual Informatics Conference, pp. 350-361, DOI: 10.1007/978-3-030-90235-3_31.

[2] Pandey, V., Malhotra, A., Kant, R., & Sahana, S. K. (2019) "Solving scheduling problems in PCB assembly and its optimization using ACO",Advances in Swarm Intelligence - 10th International Conference, ICSI 2019, pp.243-253. DOI: 10.1007/978-3-030-26369-0_23g.

[3] Zhang, J. L., Shao, Y. F., Wang, J. R., Bu, S. S., Niu, S., Su, M. S., Ren, J. S. (2019) Research on structural design of PCB splitter [J] Journal of Qingdao University (Engineering Technology Edition), 34 (02): 73-77. DOI:10.13306/j.1006-9798.2019.02.013.

[4] Liang, J., Ye, G., Li, J., Kuang Y and Zhang, X. (2018) "Pose Alignment for Electronic Component Insertion using Rotational Stereo Vision", 2018 IEEE International Conference on Robotics and Biomimetics (ROBIO), pp. 2421-2427, doi:10.1109/ROBIO.2018.8665293.

[5] Zheng Y. H., Wang, J. R., Wang, W. X., Liu, C. (2019) Research on improved design of manipulator of PCB automatic detection device [J] Journal of Qingdao University (Engineering Technology Edition), 34 (02): 63-67+77. DOI: 10.13306/j.1006-9798.2019.02.011.

[6] Liu, Y. S., Li, M. J., Ding, P. C. (2017) Research on control system of double SCARA special-shaped plug-in machine [J] Mechanical engineering and automation, (05): 10-12 + 15

[7] Wang, S. X. (2019) Research on automatic positioning of plug-in machine based on image [D] Harbin University of technology.

[8] Wang, D. (2018) Research on Key Technologies of robot automatic plug-in [D] Jilin University.

[9] Jiang, S., Liu, Y., Hu, J. H., Wang, X. D. (2021) Design of rectangular coordinate manipulator system based on PLC [J] Industrial instruments and automation devices, (02): 17-20. DOI:10.19950/j.cnki. cn61-1121/th. 2021.02.004.

[10] Xu, G. (2020) The dynamic characteristic analysis on Cartesian coordinate robot[J]. IOP Conference Series: Earth and Environmental Science, 474(3).

[11] Matthew, T. (2018) Wayland,Matthias Landgraf. A Cartesian Coordinate Robot for Dispensing Fruit Fly Food[J]. Journal of Open Hardware, 2(1).

[12] Ren, J. S., Wang, J. R., Su, M. S., Niu, S. (2018) Design of grasping manipulator for printed circuit board [J] Journal of Qingdao University (Engineering Technology Edition), 33 (01): 112-115. DOI:10.13306/j.1006-9798.2018.01.022.
[13] Shao, Y. F., Bu, S. S., Zhang, J. L., Wang, J. R. (2019) Research on structure design of PCB transmission identification and detection system [J] Journal of Qingdao University (Engineering Technology Edition), 34 (01): 91-94. DOI:10.13306/j.1006-9798.2019.01.016.
[14] Li, C. L. (2020) Research on balance improvement of J product plug-in production line of H company [D] Zhongbei University, DOI:10.27470/d.cnki. ghbgc. 2020.000427.
[15] Shi, L. W., Wang, J. R., Zheng, Y. H., Dai, X. W., Yang, J. Q., Li, J. (2021) Improved layout design of single-phase carrier module assembly line based on QFD and TRIZ [J] Journal of Qingdao University (Engineering Technology Edition), 36 (02): 47-53. DOI:10.13306/j.1006-9798.2021.02.008.
[16] Tian, Z. C., Wang, J. R., Wang, S. H., Zheng, Y. H., Li, J. (2020) Scheme evaluation of PCB Automatic packaging production line based on BWM and Improved Entropy TOPSIS method [J] Journal of Qingdao University (NATURAL SCIENCE EDITION), 33 (02): 52-57.
[17] Wang, Z. T., Xu, Y. T., He, Q., Fang, Z. H., Xu, G. H., Fu, J. Z. (2020) Grasping pose estimation for SCARA robot based on deep learning of point cloud[J]. The International Journal of Advanced Manufacturing Technology, 108(prepublish).
[18] Wang, L. (2020) Research on tea picking robot based on SCARA manipulator [D] University of science and technology of China, DOI:10.27517/d.cnki. gzkju. 2020.001950.
[19] Zhu, Y. N. (2021) Design and analysis of three degree of freedom palletizing and sorting manipulator based on electrical transmission [J] Papermaking equipment and materials, 50 (07): 30-31+61.
[20] Liang, F. (2020) Design of handling manipulator for automatic production line with high-precision pick-up of product target [J] Mechanical design and manufacturing engineering, 49 (02): 26-30 .
[21] Wang, S. H., Wang, J. R., Tian, Z. C., Zheng, Y. H., Li, J. (2020) Design and research of pallet handling manipulator [J] Journal of Qingdao University (NATURAL SCIENCE EDITION), 33 (02): 27-31.
[22] Ren ,B., Zhang, Z. R., Zhang, W. J., Feng, B., Gao, X., Wang, H. X., Gao, S., Fang, H. (2018) Cartesian coordinate manipulator for dispensing based on visual recognition [J] Mechanical engineering and automation, (05): 168-170.
[23] Wei, Z. L. (2017) Several methods of PLC controlling stepping servo motor to realize positioning [J] Modern manufacturing technology and equipment, (04): 160 + 164 DOI:10.16107/j.cnki. mmte. 2017.0275.
[24] Ao, M. Y. (2019) Research on fruit and vegetable robot grasping control system based on PLC [J] Guangxi Agricultural Journal, 34 (02): 34-37
[25] He, X. D., Chen, X. S. (2021) Application of PLC-Based Stroke Control in Industrial Production[J]. IOP Conference Series: Earth and Environmental Science, 632(4).

Journal of Physics: Conference Series **2387** (2022) 012016 doi:10.1088/1742-6596/2387/1/012016

Improved performance of nafion membranes by blending ultra-high molecular weight polyvinylidene fluoride

Chao Feng[1,a] and Weidong He[1,*]

[1]School of Physics, University of Electronic Science and Technology of China, Chengdu 610054, P.R. China.

*weidong.he@uestc.edu.cn; [a]chaofeng@std.uestc.edu.cn,

Abstract. Proton exchange membrane (PEM) is developing towards thin thickness and high mechanical strength for extraordinary performance proton exchange membrane fuel cells (PEMFCs). However, the commercial membrane such as Nafion can hardly satisfy the practical application of PEMFCs because of high gas crossover and low mechanical strength when the thickness is less than 20 μm. Here, a reinforced composite membrane (denoted as P110-PFSA) was synthesized by blending poly(vinylidene fluoride)(PVDF) featured with high molecular weight of M_w=1100000 g mol^{-1} into perfluorosulfonic acid resin (PFSA). The P110-PFSA with the thickness of 15 μm exhibits tensile strength of over 33 MPa because the PVDF with high molecular weight forms a higher density of hydrogen bonds with PFSA, resulting in a reinforcement of the bonding strength between PVDF and PFSA. H_2/O_2 fuel cell performance with the P110-PFSA shows more than 1170 mW cm^{-2} fed with H_2 at 70 °C and 100% RH much better than that with Nafion 211. Direct methanol fuel cell power densities of the blent PEM are 92, 61, 50, 28 and 15 mW cm^{-2} fed with 2, 6, 10,16 and 20 M methanol solution respectively at the anode.

1. Introduction

As the climate warming gradually become a global consensus, the nations of the world are vigorously developing clean energy technologies to reduce carbon emissions produced by the fossil energy. Proton exchange membrane fuel cells (PEMFCs) have become one of the promoting power generation systems in the 21st century due to its high density of energy, simple design, low operation temperature and zero-carbon emissions [1-6]. As a critical component in PEMFCs system, proton exchange membrane (PEM) works as H+ conductor and separator between oxygen and fuel [7,8]. Nafion is a commercial PEM which is widely used in PEMFCs due to superior proton conductivity under high humidity condition, satisfactory mechanical features and excellent chemical stability in severe environment. However, with the urgent demand for high power density of fuel cell, the ultra-thin PEM (~15μm) based Nafion severely restrict the further development of PEMFCs due to undesirable gas permeation and poor mechanical strength [9,10].

Incorporation of organic or inorganic nano materials is one of the promising methods to optimize Nafion membrane for operation. Zhou *et al.*, reported a blent PEM consisted of Nafion and crosslinked poly(vinylidene fluoride) (PVDF) [11]. They first fabricated Nafion/PVDF membrane by solution casting and then impregnated the membrane into the mixture crosslinker and solvent. The obtained semi-interpenetrating polymer networks (semi-IPN) modified PEM exhibits extraordinary mechanical features and enhanced structural stability. However, the composite PVDF/Nafion membranes have big thicknesses and low conductivity. Si *et al.* reported multilayer membrane with the methanol-

nonconduction layer of PVDF and two conductive layers of Nafion to prevent the methanol diffusing pass [12]. The 55 wt % PVDF, 45 wt % Nafion, 10 μm nonconduction layer sandwiched with 20 μm Nafion layers decreased the methyl alcohol permeability by 64% in comparison to the Nafion membrane of 50 μm. Pintauro and co-workers prepared single-fiber Nafion membranes blended with PVDF by electrospinning for hydrogen/bromide fuel cells [13]. The nanofiber composite membrane with 20 wt.% PVDF exhibits low Br_2/Br_3^- permeability and up to 1310 mW·cm^{-2} power, much superior than Nafion 212. Unfortunately, the nanofiber PFSA membrane still has unacceptable mechanical property even though the content of PVDF is about 40%.
In the present work, two kinds of different molecular weights PVDF/PFSA blent membrane were prepared by solution casting. And the effect of molecular weights of PVDF on properties of PEM was studied comprehensively. This study indicates the PVDF polymer with high molecular weight contributes to dramatically reinforce the mechanical strength of PFSA membrane due to high molecular weight PVDF forms a higher density of hydrogen bonds with PFSA and these hydrogen bonds net can strengthen the bonding energy between PVDF molecule and Nafion molecule. The mechanical strength of P110-PFSA (molecular weight=1100000 g mol^{-1}) membrane is over 33 MPa with a thin thickness of 15 μm. In comparison, the pure PFSA membrane (~15 μm) and commercial Nafion 211 (~25 μm) show only 11.2 MPa and 21 MPa respectively. Moreover, the composite P110-PFSA membranes display proton conductivity of up to 0.168 S cm^{-1} at 80 °C, higher than commercial Nafion 211. The power density of hydrogen fuel cells based P110-PFSA membrane is 1170 mW cm^{-2}, superior to that of Nafion 211 (1130 mW cm^{-2}) and slightly lower than that of original PFSA membrane. In the end, the blent PEM with 15 μm was also applied in direct methanol fuel cell (DMFC) and showed 92, 61, 50, 28 and 15 mW cm^{-2} fed with 2, 6, 10,16 and 20 M methanol solution respectively at the anode.

2. Experimental section

Preparation of composite membrane: PFSA powder was purchase from Kedi Chemical Technology Co., LTD (China). N-Methylpyrrolidone (NMP) (>99.9%) was purchased from BASF Co. Ltd., Mainland, China. PVDF powders (Solef®5130 (Mw=1100000 g/mol), Solef®1010 (Mw=350000 g/mol) were purchased from Solvay Co. Ltd. PFSA powder was dissolved in N-Methylpyrrolidone at 70 °C with continuously string for 3 h to obtain 15 wt.% solution. Afterwards, the PVDF was added into the PFSA solution and continued to be stirred at 80 °C for 1h. The mixed solution was cast onto a glass plate using a blade apparatus with a thickness of 180 μm. Then the wet membrane was dried in oven at 80 °C to evaporate the solvent. The obtained membrane was taken off from glass by immersing it into water. The obtained membrane was boiled in 5% H_2O_2, 1 M H_2SO_4 and deionized water for 1 h respectively before using.
Membrane characterization: Rectangular membranes (approximately 20 mm×30 mm in size) were dried at 90 °C for 24 h and then weighed using high precision balance. Then the membrane was immersed in water 25 °C for 1 day. The hydrous membranes were quickly wiped to remove surface water by tissue. The water uptake of samples was calculated by the specific value of the weight of the wet membrane and the dried one. The surface morphologies of the PEM were obtained by a scanning electron microscope. The mechanical properties of the PEMs were measured on Power Test machine (CMT6104, China) at room temperature with 50%~60% RH. The in-plane proton conductivity of the composite PEM was measured by two-probe apparatus through electrochemical impedance spectroscopic measurements (CHI760E, Chenhua Instrument Company, Shanghai, China) from 50 °C to 100 °C at 100% RH with a driving voltage of 0.1 V and a frequency range of $1 \times 10^6 \sim 1$ Hz. The proton conductivity was calculated using Eq. (1):

$$\sigma = \frac{D}{RWT} \quad (1)$$

where σ (S cm^{-1}) is proton conductivity,D (cm) represents the distance of two testing electrodes， R (Ω) represent the tested impedance of PEM, W (cm) represents the width of the hydrous PEM, and T is the thickness of hydrous PEM.

Journal of Physics: Conference Series **2387** (2022) 012016 doi:10.1088/1742-6596/2387/1/012016

Electrochemical characterization: The membrane assemble electrode (MEA) was prepared by coating catalyst on to membrane using a spray gun. The loading of Johnson Matthey Pt/C (40%) at the anode was controlled up to 0.35 mg cm^{-2} and the cathode is 0.55 mg cm^{-2}. The active areas of MEA are 4 cm^{-2}. 100% RH H_2 (100 sccm) and 100% RH O_2 (150 sccm) were supplied to the anode and cathode with a back-pressure of 14.5 PSI. The fuel cell was heated to 70 °C during the testing. The cell performance was recorded by Arbin Fuel Cell Instrument. For direct methanol fuel cells, the loadings of E-TEC Pt-Ru (Pt/Ru 40%/20%) black at the anode and Johnson-Matthey Pt black at the cathode were both 2.0 mg cm^{-2}. The active areas are 4 cm^{-2}. The anode was fed with different concentration of 2, 6, 10, 16, 20 M aqueous methanol solution (6 ml min^{-1}). The cathode of fuel cell was supplied with O_2 (120 sccm) with a back-pressure of 10 PSI. The temperature of fuel cell was maintained at 80 °C. The methanol crossovers of all samples were measured by linear sweep voltammetry (LSV) method in the voltage range from 0 and 1.0 V at the scan rate of 2 mV s^{-1}. During measurements, the anode side of the MEA was supplied with 2 M methanol solution (6 mL min^{-1}) while the humidified nitrogen was supplied to the cathode at 60 °C. The cathode of cell worked as working electrode.

3. Results and discussions

The PVDF and PFSA are immiscible based on the thermodynamic principle owing to their different dissimilar structures, but a large number of literatures have confirmed that PFSA has good miscibility with PVDF [14,15]. As shown in Fig.1, all samples show homogeneous surface morphology. No agglomeration of PVDF powder is observed from the cross-section SEM image, indicating good dispersion of PVDF in PFSA matrix. However, P35-PFSA blent membrane exhibits clear PVDF spherulite, implying severe macro-phase separation when PVDF and PFSA are blent together from Fig. 1d. For comparison, the P110-PFSA and pure PFSA membranes show uniform and interconnected morphology, implying that no macroscopic separation. This result indicates that PVDF®5130 (M_w=1100000 g mol^{-1}) owns better compatibility with PFSA.

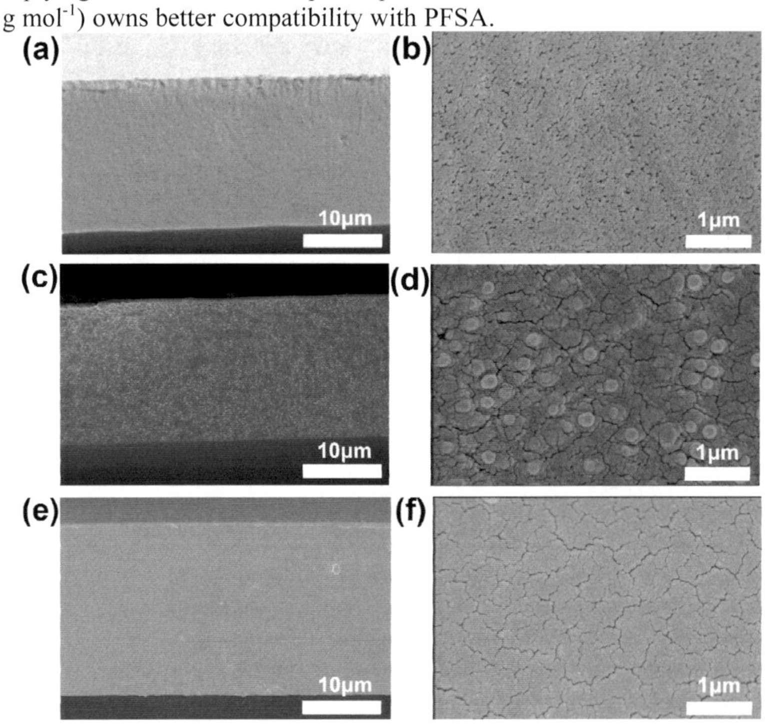

Figure 1. Cross-section SEM image: (a, b) P110-PFSA, (c, d) P35-PFSA, (e, f) Pure PFSA

Good water uptake is important for a PEM to construct proton conductivity channels due to the proton-hopping mechanism [16,17]. The table 1 shows the water uptake of the Nafion 211 and PVDF/PFSA membrane at 25 °C. The P110-PFSA membrane has lower water uptake (5.99%) compared to Nafion 211 (8.42%) and pure PFSA (30.5%) due to incorporation of hydrophobic PVDF. However, the P110-PFSA membrane shows better swelling ratio in water than those of Nafion 211 and pure PFSA owing to the rigid structure of PVDF molecule restrain the swelling of Nafion' ion clusters [18]. Furthermore, the blent membrane with high molecular weight PVDF (M_w=1100000) exhibits bigger water uptake and smaller swelling ratio than that with low molecular weight PVDF (M_w=350000) potentially because of better miscibility between PVDF 5130 and PFSA. The mechanical property of membrane is closely related its dimensional stability and durability of MEA. As shown in Fig. 2, the P110-PFSA membrane (~ 15 μm) exhibits much higher breaking strength over 33 MPa than those of pure PFSA (15 μm, 11.2 MPa), P35-PFSA (15 μm, 24.2 MPa) and Nafion 211 (25 μm, 21.5 MPa) because PVDF with high molecular weight forms a higher density of hydrogen bonds with PFSA and these hydrogen bonds net can strengthen the bonding strength between PVDF molecule and PFSA molecule.

Table 1. The water uptake and swelling ratio of PFSA/PVDF blent membranes, pure PFSA and Nafion 211.

Membranes	Liquid water uptake (25 °C)	Swelling ratio (25 °C, in water)
Nafion 211	8.42%	20%
Pure PFSA	30.5%	33.3%
P110-PFSA	5.99%	5.59%
P35-PFSA	4.46%	13.3%

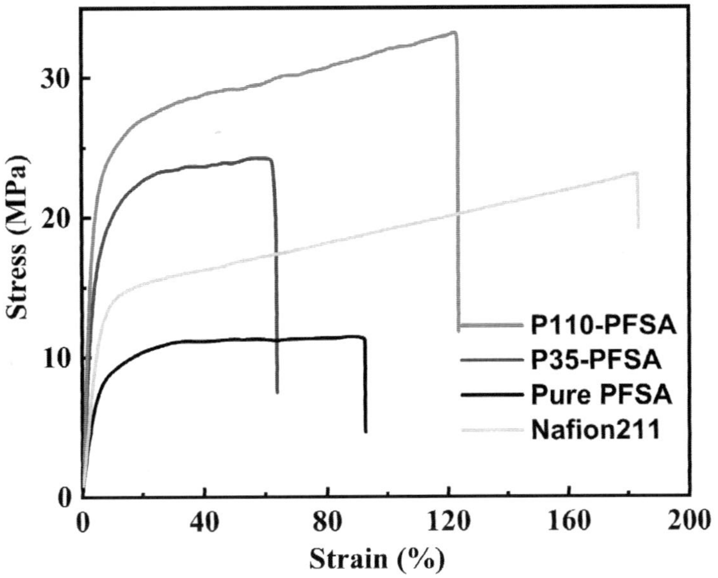

Figure 2. Stress-strain curves for Nafion 211and PVDF/PFSA blent membranes at room temperature and 50~60% RH.

The proton conductivities of Nafion 211 and PVDF/PFSA blent membranes with the varied temperature are shown in Figure. 3a. The P110-PFSA and P35-PFSA membranes exhibit lower proton conductivities than pure PFSA because of the addition of non-proton conductive PVDF. However, the proton conductivity of P110-PFSA is higher than those of Nafion 211 and P35-PFSA. This result indicates the

PVDF with high molecular weight does not break the continuous ion transport channel in PFSA because of better miscibility between PVDF molecule and PFSA molecule. This result is also consistent with the performance of the fuel cell. The MEA power density based P110-PFSA blent membranes is much higher than those of Nafion 211 and P35-PFSA due to better proton conductivity as shown in Figs. 3b and 3c. The PEMFC based on P110-PFSA blent membrane delivers up to 1170 mW cm^{-2}, much higher than that of P35-PFSA (890 mW cm^{-2}) and comparable to Nafion 211(1130 mW cm^{-2}), suggesting the membrane has great potential application in hydrogen fuel cell.

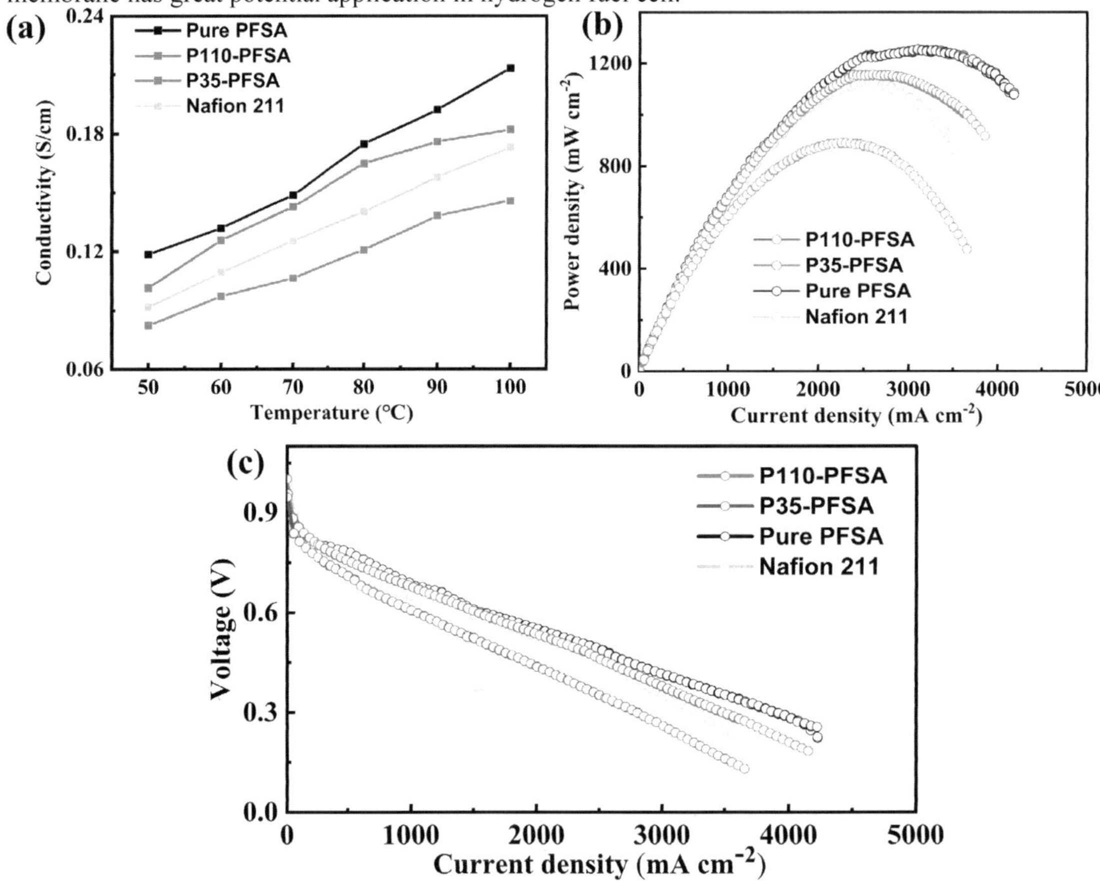

Figure 3. (a) The proton conductivities, (b) power density and (c) polarization curves of MEAs with Nafion 211 and PVDF/PFSA blent membranes.

The commercial application of the blent membrane in this work was further assessed by a DMFC at 80 °C under 10 PSI pressure. The polarization curves and power densities of P110-PFSA blent membrane, pristine PFSA and commercial PEM are exhibited in Figure 4a-4c. The thickness of P110-PFSA blent membrane and pure PFSA membrane is about 15 μm while the commercial PEM is approximately 25 μm. The P110-PFSA blent membrane displayed the lower polarization loss and higher power density under different concentration of methanol solution compared to pristine PFSA and commercial PEM. The DMFC based P110-PFSA blent membrane outputted 92 mW cm^{-2} at 0.3 V when the anode was fed with 2 M methanol solution, more than 156% and 95% of original PFSA membrane and Nafion 211 respectively. The blent PEM with high molecular weight PVDF still showed extraordinary methanol resistance, representing high power density of 15 mW cm^{-2} even though the concentration of methanol solution was approximate to net methanol. In contrast, the DMFC based the pristine PFSA and commercial Nafion 211 barely generated electricity under identical operating conditions due to high methanol permeation, resulting in severe mixed potential at cathode. The blent membrane with high molecular weight PVDF showed good performance in single cell of DMFC due to superior methanol barrier property of PVDF. Moreover, high molecular weight PVDF can effectively restrain swelling and absorption of the PFSA matrix in methanol solution because more long-chain molecules in PVDF enhance the adhesive strength of PVDF and PFSA. This result is further confirmed by the experiment of methanol crossover. The methanol crossover of P110-PFSA blent PEM, pristine PFSA membrane and commercial PEM is evaluated by the LSV. The current measured from the fuel cell is called methanol crossover, which corresponds to the oxidation of diffused methanol. The value of plateau current in LSV curve represents the methanol permeation level. The limiting current densities (25 mA cm^{-2}) of P110-PFSA blent membrane much lower than those of pritine PFSA (108 mA cm^{-2}) and commercial PEM (58 mA cm^{-2}) shown in figure 4 (d), mainly owing to narrower hydrophilic regions in P110-PFSA blent PEM.

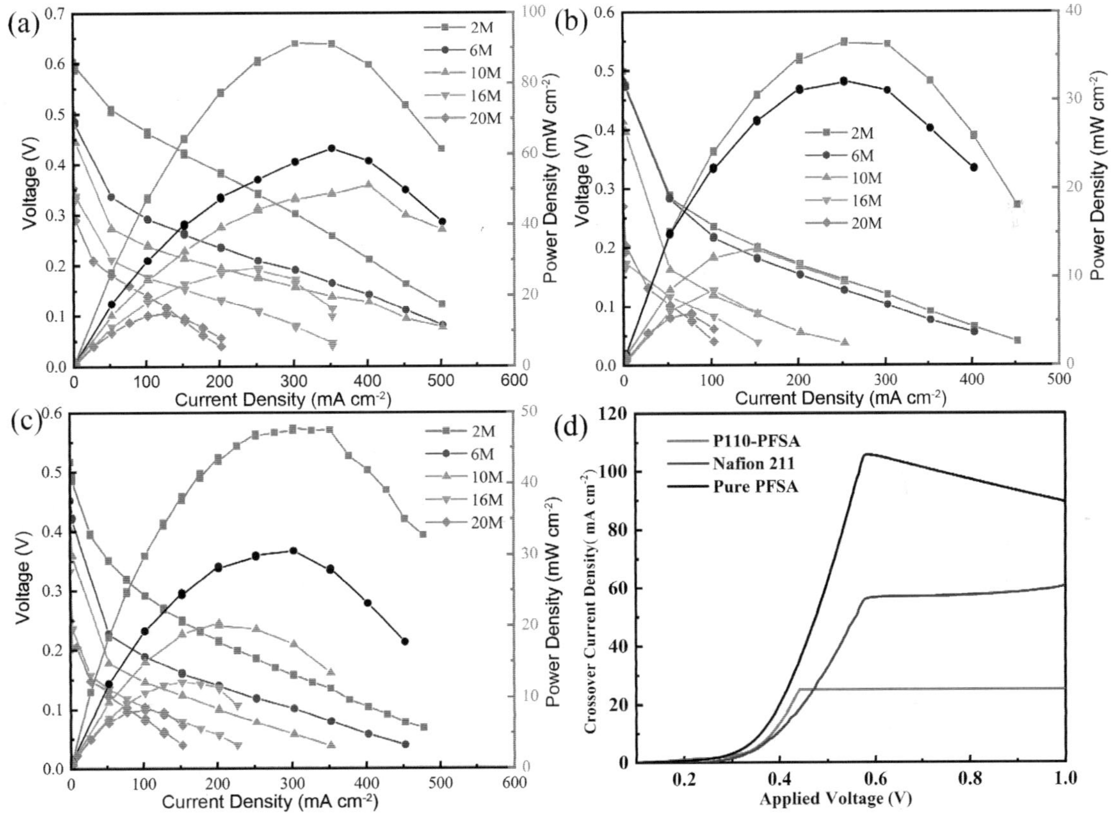

Figure 4. DMFC performances of MEA based (a) P110-PFSA membrane, (b) pritne PFSA membrane and (c) Nafion 211 fed with 2, 6, 10, 16, 20 M methanol solution at 80 °C; (d) Comparison of methanol crossover current densities of P110-PFSA, pure PFSA and Nafion 211 (methanol concentration of 2 M).

4. Conclusion

Homogeneous and high-powered PEMs are fabricated with high molecular weight PVDF (Solef®5130, M_w=1100000 g mol^{-1}) by solution-casting method. The PFSA membrane with 25% content PVDF caused decrease in water absorption and ionic conductivity because of the incorporation of PVDF. In contrast, the mechanical properties and swelling behaviours of blent membrane are improved obviously. Meanwhile, the H_2/O_2 power density of 1170 mW cm^{-2} based the blent PFSA shows better than that based Nafion 211 (1130 mW cm^{-2}). Moreover, the composite membrane with 15 μm was also applied in direct methanol fuel cell (DMFC) and showed 92, 61, 50, 28 and 15 mW cm^{-2} fed with 2, 6, 10,16 and 20 M methanol solution respectively at the anode. The blent membranes can be one of the best candidates for mass-produced PEMFCs in the future.

Acknowledgments
The work was supported by the Science & Technology Support Funds of Sichuan Province (Grant No. 2016GZ0151) and the Fundamental Research Funds for the Chinese Central Universities (Grant No. ZYGX2015Z003).

References

[1] Zhao, Z., Liu, Z., Zhang, A., Yan, X., Xue, W., Peng, B., et al. (2022). Graphene-nanopocket-encaged PtCo nanocatalysts for highly durable fuel cell operation under demanding ultralow-Pt-loading conditions. Nat. Nanotechnol., pp.1-8.
[2] Zaman, S., Wang, M., Liu, H., Sun, F., Yu, Y., Shui, J., et al. (2022). Carbon-based catalyst supports for oxygen reduction in proton-exchange membrane fuel cells. Trends Chem..
[3] Ahn, C. Y., Kim, S., Choi, H. J., Lee, J., Kang, S. Y., Kim, O. H.,et al. (2022). Effect of iridium oxide as an additive on catalysts with different Pt contents in cell reversal conditions of polymer electrolyte membrane fuel cells. Int. J. Hydrogen Energy, 47(3), pp.1863-1873.
[4] Liu L., Xing Y., Li Y., et al. (2022) Enhanced mechanical durability of perfluorosulfonic acid proton-exchange membrane based on a double-layer ePTFE reinforcement strategy. Int. J. Hydrogen Energy, 67(47), pp. 29014-29026.
[5] Liu L., Xing Y., Li Y., et al. (2022) Double-Layer expanded polytetrafluoroethylene reinforced membranes with cerium oxide radical scavengers for highly stable proton exchange membrane fuelcells. ACS Appl. Energy Mater., 5(7), pp. 8743-8755.
[6] Yang J., Tong L., Alsubaie A. S., et al. (2022) Hybrid proton exchange membrane used in fuel cell with amino-functionalized metal–organic framework in sulfonated polyimide to construct efficient ion transport channel. Adv. Compos. Hybrid Mater., pp. 1-9.
[7] Li, Y., Hui, J., Kawchuk, J., O'Brien, A., Jiang, Z and Hoorfar, M. (2019) Composite membranes of PVDF nanofibers impregnated with nafion for increased fuel concentrations in direct methanol fuel cells. Fuel Cells, 19(1), pp. 43-50.
[8] Mahimai, B. M., Sivasubramanian, G., Sekar, K., Kannaiyan, D., Deivanayagam, P. (2022). Sulfonated poly (ether ether ketone): efficient ion-exchange polymer electrolytes for fuel cell applications–a versatile review. Materials Adv..
[9] Ryu, S., Lee, B., Kim, J. H., Pak, C and Moon, S. H. (2021) High-temperature operation of PEMFC using pore-filling PTFE/Nafion composite membrane treated with electric field. Int. J. Energy Res., 45 (13): 19136-19146.
[10] Vinothkannan, M., Kim, A. R., Gnana kumar, G and Yoo, D. J. (2018) Sulfonated graphene oxide/Nafion composite membranes for high temperature and low humidity proton exchange membrane fuel cells. RSC Adv., 8 (14) 7494-7508.
[11] Zhou, B., Pu, H., Pan, H and Wan, D. (2011) Proton exchange membranes based on semi-interpenetrating polymer networks of Nafion® and poly(vinylidene fluoride) via radiation crosslinking. Int. J. Hydrogen Energy, 36 (11):6809-6816.
[12] Si, Y., Lin, J. C., Kunz, H. R and Fenton, J. M. (2004) Trilayer membranes with a methanol-barrier layer for DMFCs. J. Electrochem. Soc., 151 (3): A463.
[13] Park, J. W., Wycisk, R., Lin, G., Chong, P. Y., Powers, D., Van Nguyen, T., et al. (2017) Electrospun Nafion/PVDF single-fiber blended membranes for regenerative H_2/Br_2 fuel cells J. Membr. Sci., 541, 85-92.
[14] Lin, J. C., Ouyang, M., Fenton, J. M., Kunz, H. R., Koberstein, J. T., et al. (1998) Study of blend membranes consisting of Nafion and vinylidene fluoride-hexafluoropropylene copolymer. J. Appl. Polym. Sci., 70 (1): 121-127

[15] Kim, H. J., Kim, H. J., Shul, Y. G and Han, H. S. (2004) Nafion-Nafion/polyvinylidene fluoride-Nafion laminated polymer membrane for direct methanol fuel cells. J. Power Sources, 135 (1-2) 66-71.
[16] Lu, Y., Liu, Y., Li, N., Hu, Z and Chen, S. (2020) Sulfonated graphitic carbon nitride nanosheets as proton conductor for constructing long-range ionic channels proton exchange membrane. J. Membr. Sci., 601, 117908.
[17] Wang, H., Sun, N., Xu, X., Wang, S., Kang, W., Zhuang, X. et al. (2021) Adenosine triphosphate@graphene oxide proton channels for proton exchange membranes constructed via electrostatic layer-by-layer deposition. J. Membr. Sci., 620, 118880.
[18] Javed, R. M. N., Al-Othman, A., Tawalbeh, M., Olabi, A. G. (2022). Recent developments in graphene and graphene oxide materials for polymer electrolyte membrane fuel cells applications. Renewable Sustainable Energy Rev., 168, pp.112836.

Primary frequency modulation of microgrid based on consistent droop control method

Linlin Hu[1,*], **Long Fu**[2,a]

[1]College of Electrical and Electronic Engineering, Guangdong Technology College , Zhaoqing, China
[2]College of intelligent manufacturing, Guangdong Technology College, Zhaoqing, China

*E-mail: hll@gdlg.edu.cn; [a] fulong0819@163.com

Abstract. The rapid development of new energy sources gradually challenges the frequency stability of power systems. This article focuses on how to realize the frequency modulation control of the power grid under the fluctuation of the power grid. In order to improve the frequency stability of the microgrid system, based on the traditional droop control and consensus algorithm research, this paper proposes a microgrid primary frequency modulation based on the consistency droop control method. This control strategy improves the consistency of the progressive correction term to realize the control of the output of the energy storage unit, and improves the traditional droop controller to quickly restore the stability of the primary frequency. The root locus method verifies the effectiveness and feasibility of the control strategy. The simulation experiment verifies the frequency modulation ability of the proposed control strategy, which not only has a faster adjustment speed and response ability, but also has a strong anti-interference ability.

1. Introduction

Large scale new energy access to the grid has gradually become the development trend of modern power grid, gradually causing power grid frequency fluctuations.

Aiming at the problem of low accuracy of reactive power compensation allocation in traditional reactive power droop control, reference [1] proposes an improved droop control method with parallel inverters of different power levels to improve the adaptability of the system to inductive droop control and effectively realize the regulation of reactive power output. The traditional droop control and double loop control methods have poor power distribution accuracy and circulating current suppression effect. Reference [2] proposes a control strategy for parallel operation of low-voltage micro grid resistive inverters, which uses negative inductive reactance components to offset the impedance characteristics in the inverter output impedance, improving the power distribution accuracy and dynamic response speed.

The penetration rate of new energy into the grid has a great impact on the dynamic frequency deviation of the grid[3]. Reference [4] proposed the frequency modulation optimization of microgrid based on DFIG virtual inertia and droop expansion to further improve the frequency stability. Aiming at the problem of voltage optimal control in microgrid[5-6], reference [7] solves the optimal value of objective function, and transforms the deviation adjustment problem in proportional controller into the voltage tracking problem of MPC, so as to realize the rolling optimization of DC bus voltage dynamic response. The voltage stability and dynamic performance of the microgrid need to be improved, reference [8] improves the intercept of the droop control curve to improve the speed at which the controller releases or absorbs power, so as to provide inertial support for the microgrid system. The line resistance between the inverter and the DC bus will affect the power sharing accuracy[9]. Reference [10] uses the distributed consistency algorithm in the secondary event triggered control to achieve voltage stability and frequency recovery, and reduces the communication requirements. The inconsistency of line impedance often leads to the problems of uneven power distribution and circulating current. Reference [11] uses the consistency algorithm to dynamically correct the deviation, so as to realize the accurate distribution of reactive power. The high penetration rate of new energy grid connection[12-13] often leads to poor droop control effect. Reference [14] makes use of the influence of droop adaptive change on system circulation to reduce DC bus voltage deviation.

Based on the mathematical model of droop control and the consistency control method, this paper proposes a micro grid primary frequency modulation control strategy based on the consistency droop control method. Through the improvement of the consistency correction term, the dynamic response speed and anti-interference ability of the power grid primary frequency modulation are improved, and the power balance of the micro grid is quickly restored.

2. Consistent droop control method

2.1. Energy storage output control structure

The dynamic response speed of the energy storage battery unit is determined by the control mode of the energy storage battery converter. The virtual droop control method is often introduced into the converter control of the energy storage battery unit, which can realize the virtual droop control. The coefficient is the key parameter of droop control, which is defined by the ratio of the output of the energy storage battery unit to the frequency increment.

$$P_{EMS-1} = -K_{droop}\Delta B \tag{1}$$

$$\Delta B(s) = \frac{\Delta P_{Load}(s)}{Ns + Y + K_{gen}G(s) + K_{EMS}M(s)} \tag{2}$$

Where, K_{droop} is the coefficient, P_{EMS-1} is the output power of the energy storage battery unit, ΔB is the system frequency deviation, the inertia time coefficient is M, K_{gen} is the unit power regulation coefficient of the power generation unit, M(s) is the transfer function of the battery energy storage unit participating in frequency regulation, G(s) is the transfer function of the generator unit participating in frequency regulation, and Y is the load damping coefficient. If the virtual droop control coefficient is larger, the time-domain relative sensitivity is larger, which can greatly improve the efficiency of power grid frequency recovery.

2.2. Consistent regulation strategy

When the load and power are unbalanced, the frequency imbalance occurs to achieve the purpose of maintaining frequency stability. The core of consistency algorithm is consistency and correction term, which are very critical. Among them, the imbalance between load and power often needs to be offset by the correction term part, and maintain the state balance to achieve the consistency of consistency state variables. In general distribution network, when all state variables in the whole system tend to be consistent, the optimal solution of the system is obtained.

The state iteration equation of the consistency correction term is shown in equation (3).

$$\beta_c(f) = \beta_c(f-1) + Y_c^*(f) \tag{3}$$

$$Y_c^*(f) = \Delta\beta_c(f) = \frac{d\beta}{dP_c}\Delta P_c(f) = 2G_f\Delta P_c(f) \tag{4}$$

Where, $\beta_c(f)$ is consistency maintenance term, $Y^*_c(f)$ is consistency progressive correction term. The key of one-time frequency modulation control is to correct the output of new energy generating units by using the consistent progressive correction term according to the frequency deviation of the power grid, so as to achieve the power balance at both ends of the source load in the power system. As shown in equation (4), the consistency progressive correction term uses equation (3) to change the difference of consistency state variables. This part introduces the consistency algorithm into the traditional primary frequency modulation control, and its discrete form is shown in equations (5) - (7).

$$\beta_c(f+1)=\sum_{i=1}^{M} d_{ij}\beta_c(f)+Y_c^*(f) \tag{5}$$

$$Y_c^*(f)=A^*\Delta B(f) \tag{6}$$

$$Y_c^*(f)=A^*\left[B^*-B(f)\right] \tag{7}$$

Where, A^* is the overall progressive correction coefficient in the consistent frequency modulation algorithm, B^* becomes the rated frequency of the system, and $B(f)$ is the frequency of the f-th iteration on the power grid side. When the whole power grid meets the load power balance, the frequency value at this time is B^*.

2.3. Frequency modulation controller

Traditional primary frequency modulation is often realized by droop control, and the consistency algorithm can be applied to droop control structure. In the droop controller, B^*-B (f) can be measured by the phase-locked loop control part. The droop coefficient of the droop controller is a very critical part. Reasonable setting of the droop coefficient will help the frequency recover to a stable value as soon as possible. Its overall structure is shown in Figure 1.

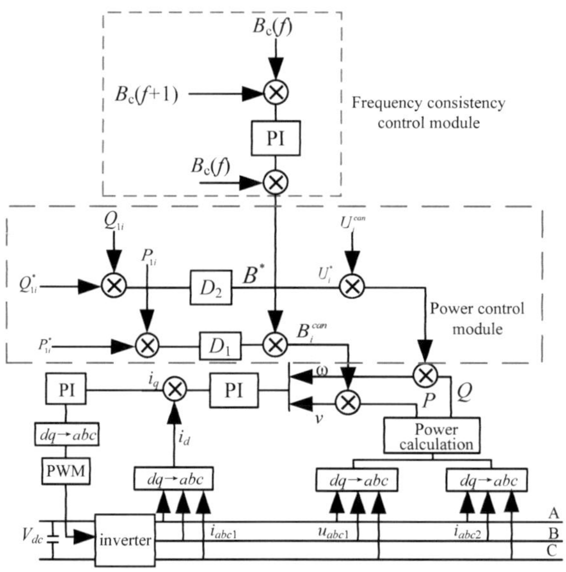

Figure 1. Energy storage participates in the overall structure of primary frequency regulation

The discrete algorithm of traditional droop control is:

$$B_i=B_i^*-D_1\left(P_{1i}-P_{1i}^*\right) \tag{8}$$

$$U_i=U_i^*-D_2\left(Q_{1i}-Q_{1i}^*\right) \tag{9}$$

Where, D_2 and D_1 are the droop control coefficients of voltage and reactive power, frequency and active power in the power grid respectively, and U^*_i and B^*_i are the given value of grid voltage and frequency in droop control respectively. When the load power imbalance occurs in the system, droop control can use droop control curve to change the system voltage and frequency. Considering that the control has the function of error free regulation, discrete PI is used to realize frequency recovery in this paper. The frequency difference can be changed into equation (10).

$$B^{*}-B(f)= K_{BL}\left(\beta_{c}(f+1)-\beta_{c}(f)\right)+K_{J}\int\beta_{c}(f+1)-\beta_{c}(f)\mathrm{d}\tau \tag{10}$$

Where, K_J and K_{BL} are the integral coefficient and proportional coefficient of the controller respectively.

In the process of coordinated operation of power grid, the power often has strong coupling, so the control of reactive power and voltage should be considered in the research of frequency modulation control. To achieve voltage equalization, the voltage connected to the distribution network at the generation side will be the reactive voltage of the consistency variable, and its discrete algorithm is shown in formula (11).

$$Q_{1i}(f+1)=Q_{1i}(f)+k\sum_{x}^{n}\left(-L_{ix}Y(f)\right) \tag{11}$$

Where, Y is the droop balance factor, the stability convergence factor is k, $Q_{1i}(f)$ is the actual reactive power signal of voltage control at the f-th iteration, $Q_{1i}(f+1)$ is the reactive power signal of voltage control calculated at the power side at the f-th iteration, and L_{ix} is the Laplace matrix, which are the controllers respectively.

2.4. Consistent optimization of frequency regulation control strategy

The specific implementation steps of the consistency algorithm proposed in this paper are as follows:

Step 1: parameter initialization. The PLL in the droop control at the power side starts to measure the frequency deviation of the system. At the same time, equations (8) - 9 start to calculate the power difference between the active power and reactive power output at the power side, and use equations (5) and (7) to calculate the consistency correction term, that is, to calculate the difference between the existing consistent state variable f and the optimal consistent state variable f^*, The state space equation and Laplace matrix are calculated by state space method and Laplace matrix respectively.

Step 2: calculate the frequency deviation. According to the calculation results of the consistency correction term, and update the system frequency deviation.

Step 3: judge the power response range of the system. If it exceeds the specified range, the power generation side needs to operate according to the minimum constraint range and update the one-time state increment. If the power response of the system still exceeds the specified range after adjusting the output at the power side, formula (10) is used for frequency error optimization frequency modulation.

Step 4: continue to optimize the power balance of the system. In the process of frequency modulation, the droop balance factor is constantly updated iteratively to achieve the convergence speed of the algorithm, and always maintain the system frequency fluctuation within 0.2Hz, so as to achieve the rapid convergence of the consistent optimized frequency modulation algorithm.

3. Stability analysis

In order to study the stability and reliability of the microgrid primary frequency modulation based on the consistent droop method, this paper establishes a small signal model of the control system, combined with the change of the characteristic root curve, to verify the feasibility and superiority of the control algorithm proposed in this paper. In this paper, a small signal model of primary frequency modulation of microgrid based on consistent droop method is established, and its discrete form is shown in formula (12).

$$\Delta M^{*}=Z\Delta M \tag{12}$$

Where, ΔM is the system state variable and Z is the system characteristic root.

Figure 2 shows the root locus curve of the control system under the change of droop balance factor, the red arrow indicates the characteristic root curve of the control system when the droop balance factor increases from 100 to 150. In the process of increasing the droop balance factor, z_1, z_2 and z_3

will move rapidly to the left real part, and the convergence speed of their respective dominant state variables will also gradually accelerate. Figure 2 shows that the larger the droop balance factor is, the faster the convergence speed of the control system is and the stronger the stability is.

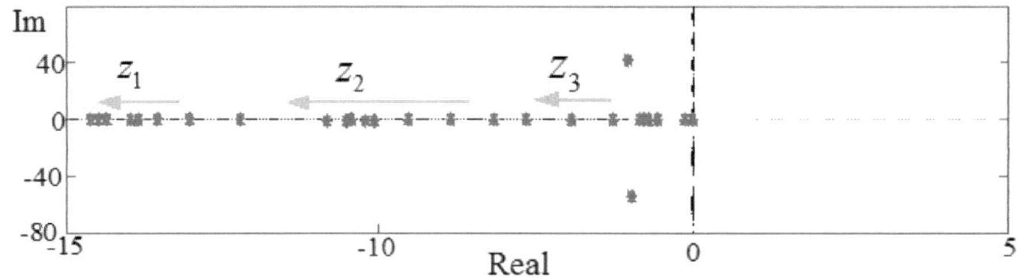

Figure 2. The root locus curve of the control system under the change of the droop balance factor

Figure 3 shows the root locus curve of the control system under the change of the stability convergence factor, the red arrow indicates the characteristic root curve of the control system when the stability convergence factor increases from 200 to 400. In the process of increasing the stability convergence factor, Z4, Z5 and Z6 will move rapidly to the right real part direction, and the convergence speed of their respective dominant state variables will also gradually slow down. Figure 3 shows that the larger the stability convergence factor, the smaller the convergence speed of the control system and the weaker the stability. Therefore, when selecting the parameters of the stability convergence factor, it is necessary to select a smaller stability convergence factor to achieve the purpose of strong stability of the control system.

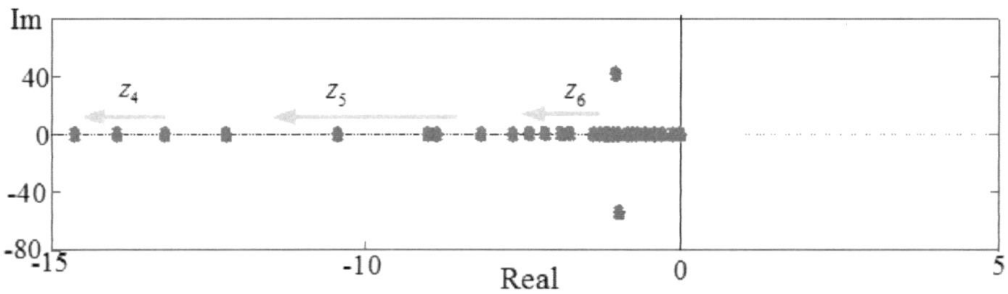

Figure 3. Root Locus Curve of Control System with Stable Convergence Factor Changes

4. Simulation analysis

In order to verify the effectiveness and feasibility of the microgrid primary frequency modulation based on the consistent droop method, this paper builds an IEEE3 machine 9-node system on the MATLAB/Simulink simulation experiment platform. Figure 4 shows the structure of the energy storage system connected to the 3-machine 9-node system, in which G1, G2 and G3 are all set to 30MW, and load1, load2 and load3 are 28MW respectively. The energy storage capacity of the battery is 15MW, and the upper and lower limits of SOC are set to 0.9 and 0.3 respectively. This paper will carry out the simulation experiment under the load step fluctuation, and compare the traditional droop control and the traditional consistency algorithm in this process to verify the feasibility and effectiveness of the control strategy proposed in this chapter.

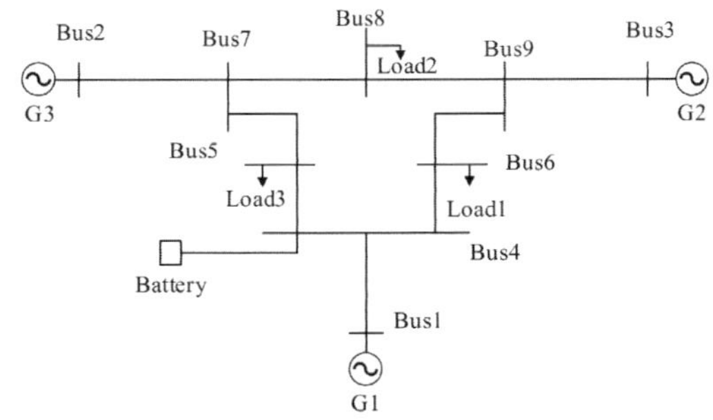

Figure 4. The energy storage system is connected to a 3-machine 9-node system

In this paper, 15MW load step disturbance is applied to the system, and the simulation time length is set to 100s. The system frequency change curve under load step is shown in Figure 5. In the simulation process of three control strategies, the algorithm proposed in this paper is superior to the other two control algorithms. The root mean square value and the maximum value of frequency deviation of the traditional droop control algorithm in the process of system frequency change are the maximum values, and the unit climbing rate in the process of frequency recovery is low. At the same time, the frequency recovery speed of the traditional consistency algorithm is slow, so the algorithm proposed in this chapter has faster dynamic response speed and anti-interference ability.

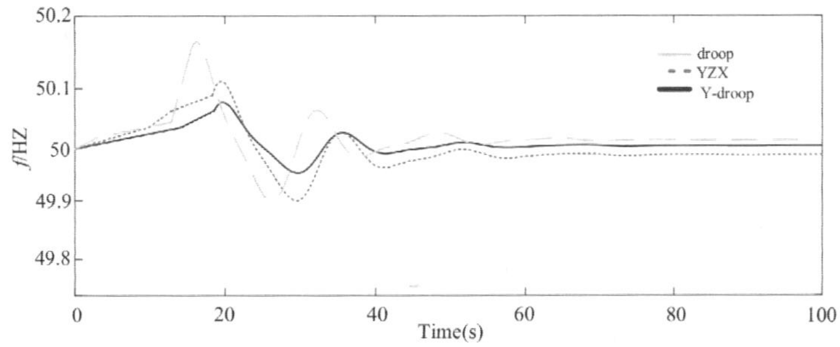

Figure 5. System frequency change curve under load step change

The SOC curve of energy storage with load step change is shown in Figure 6. Compared with the traditional consistency algorithm and the traditional droop control algorithm, the SOC of the frequency modulation strategy proposed in this paper decreases slowly, and the SOC target value remains relatively stable. Under the influence of droop balance factor, the charging and discharging of battery energy storage system will be affected by the change of active power of generator set.

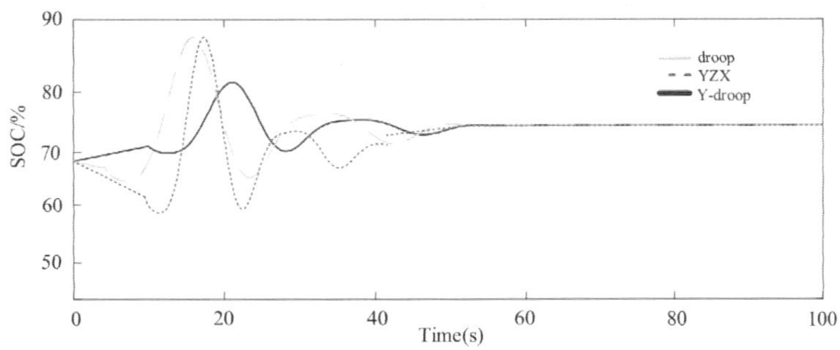

Figure 6. SOC curve of energy storage under load step chang

Figure 7 shows the G2 active power change curve of load step change. The control strategy proposed in this chapter can improve the frequency response of the system to the greatest extent and improve the power utilization at the generation side, so as to realize the rapid recovery and stability of the system frequency.

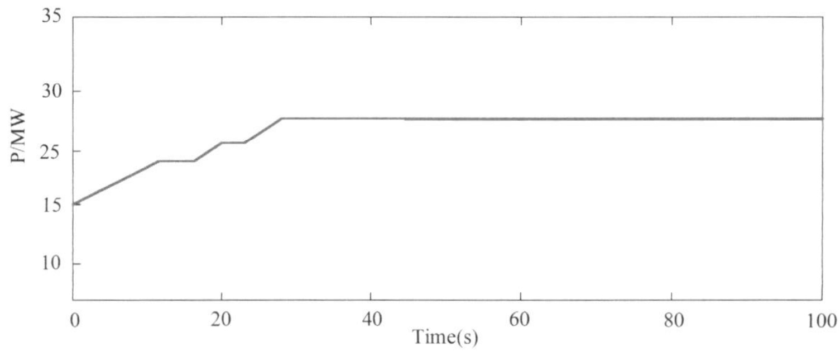

Figure 7. G2 active power change curve under load step change

5. Conclusions

Aiming at the problems of poor stability and long regulation time of primary frequency regulation in power grid, this paper proposes a micro grid primary frequency regulation based on consistent droop control method. The control strategy first proposes a consistent progressive correction term to control the output of the energy storage unit, and improves the traditional droop controller, which can quickly restore the stability of the primary frequency. The root locus method verifies the effectiveness and feasibility of the control strategy. Simulation experiments verify the frequency modulation ability of the proposed control strategy, which not only has faster regulation speed and response ability, but also has strong anti-interference ability.

Acknowledgments

Supported by the 2020 Guangdong Provincial Department of education young innovative talents project in Colleges and Universities (2020KQNCX113); "Key cultivation project" of Guangdong Technology College in 2021(2021GKJZD001).

References

[1] Ma, W. T., Wang, J. M., Mao, H. T., Yuan, X. W., Wang, Z. Y. (2021) Research on Improved Droop Control Strategy for Paralleled Inverters with Different Power Levels[J]. Acta Energiae Solaris Sinica, 42(08): 16-22.

[2] Zhang, J. H., Zhao, R., Liu, Y. F., Yang, P. H., Wu, Z. K., Zhang, X. (2022) Resistive Inverters Output Impedance Parallel Operation Strategy in Low-voltage Microgrid[J]. High Voltage Engineering, 48(01): 136-146.

[3] Jing, L., Huang, X., Wu, X. Z. (2014) Research on Improved Microsource Droop Control Method[J]. Transactions of China Electrotechnical Society, 29(02): 145-152+184.

[4] Bian, X. Y., Yin, L. Y., Ding, Y., Zhou, Q. B., Li, D. D., Zhao, J. (2021) Micro-grid Frequency Modulation Optimization Based on DFIG Virtual Inertia Droop Control[J]. Acta Energiae Solaris Sinica, 42(09): 452-460.

[5] Zheng, Y. W., Chen, M. Y., Li, C., Xu, R. L,, Xu, X. (2013) A Microgrid Control Strategy Based on Adaptive Drooping Coefficient Adjustment[J]. Automation of Electric Power Systems, 37(07): 6-11.

[6] Song, Q., Zhang, H., Sun, K., Wei, Y. L. (2017) Improved Adaptive Control of Inertia for Virtual Synchronous Generators in Islanding Micro-grid With Multiple Distributed Generation

Units[J]. Proceedings of the CSEE, 37(02): 412-424.

[7] Zhu, X. R., Ma, Y. Q., Zhao, C. H. (2021) Distributed voltage secondary control strategy of multi wind-storage DC microgrid based on model predictive control algorithm[J]. Electric Power Automation Equipment, 41(10): 184-191.

[8] Wang, H., Zhao, S. Q., Meng, J. H., Wang, C., Tian, Y. J. (2021) Adaptive Virtual Inertia Control for DC Microgrid Based on Droop Curve Intercept Adjustment[J]. Automation of Electric Power Systems, 45(24): 97-105.

[9] Xie, L. L., Shi, B., Hua, G. Y., Wen, F., Yang, L. L., Dong, Q. L. (2013) Parallel Operation Technology of Distributed Generations Based on Improved Droop Control[J]. Power System Technology, 37(04): 992-998.

[10] Zheng, J. L., Wang, J., Sun, Z., Wu, F. (2022) Improved Droop Control of PV Energy Storage DC Microgrid Based on Consensus Algorithm[J]. Proceedings of the CSU-EPSA, 34(04): 116-125.

[11] Zhang, Y., Meng, R. Q., Wang, Z. A., Han, X. Q., Wang, J. H., Liu, Y. Z. (2021) An improved droop control strategy based on a consensus algorithm[J]. Power System Protection and Control, 49(14): 104-111.

[12] Liu, Q., Zhang, T. Y., Tian, Y. J. (2021) Adaptive Segmented Droop Control for Flexible Output of Photovoltaic Converter in DC Distribution Network[J]. Electric Power Construction, 42(10): 60-70.

[13] Kan, Z. Z., Zhang, C. J., Xue, H. F., Dong, J., Wu, W. Y. (2011) A Novel Droop Control of Three-phase Inverters in Wireless Parallel Operation in Microgrid[J]. Proceedings of the CSEE, 31(33): 68-74.

[14] Guo, L., Feng, Y. B., Li, X. L., Wang, C. S., Li, Y. W. (2016) Stability Analysis and Research of Active Damping Method for DC Microgrids[J]. Proceedings of the CSEE, 36(04): 927-936.

Design of broadband electromagnetic metamaterials absorption in double-layer resistance film

Guanfang Zhu*, Yifan Kang, Yanru Jiang, Aixia Wang

Fundamentals Department, Air Force Engineering University, Xi'an710051, China

*zhugf2006@163.com

Abstract. This paper designs a lossy double-layer resistance film absorption structure, which consists of two layers of resistance film and PFA medium alternately superimposed, the bottom of the Cu conductive film is used as the reflection bottom plate, and the metamaterials absorber thickness is 4.267mm. Simulation results show that an absorption rate greater than 90% is achieved in the 6.4-21.1GHz frequency interval, with a maximum absorption peak of 98.5%, a bandwidth reaching 14.6GHz, and a relative bandwidth of up to 106.2%. The wave absorption structure presents polarized insensitive absorption and has good wide-angle incident absorption characteristics. The single-layer resistance film realized periodic and multi-resonant peak narrow-band absorption, with a maximum absorption rate of 99.9%. Based on the software simulation results, the current density field distribution map and impedance characteristics, and the absorption mechanism were analysed.

1. Introduction

Since 2008, when Landy et al of Boston University designed a "perfect" metamaterial with an absorption rate close to 100% in a microwave frequency band and metamaterials have attracted great attention in wave absorption technology [1]. The absorbing structures designed by resistive materials have the advantages of wide absorption bandwidth [2][3], polarization insensitivity [4][5], and wide incidence angle [6][7]. The resistive film is considered an attractive microwave absorber. For FSS, researchers have made great achievements in the design and theoretical analysis of broadband absorbers. A perfectly conductive FSS layer loaded with lumped resistors can be employed to synthesize thin and wideband absorbers without the limitation of one quarter-wavelength thickness [8][9]. However, the use of massive lumped resistors leads to complexity of the manufacturing and high cost of high-frequency resistors.

Structures with multiple resonant frequencies can further expand the absorber bandwidth. It can be introduced by using different resistance values of or multi-loop topology [10][11] in a single layer. The absorbers of double [12], three, and even four-layer [13] are also proposed to discuss more resonant modes. Based on the Kramers-Kronig relationship, the Ran Lixin team of Zhejiang University pointed out that the introduction of loss in the metamaterial structure can regulate the dispersion, to achieve the impedance matching [14] in the wider frequency band. Since the absorber principle is based on electromagnetic resonance, the key solution to increase the absorption bandwidth is to increase the number of resonant frequencies. One approach is to introduce multiple resistive layers, Jooman [15] uses multilayers to introduce additional resonances and increase bandwidth. However, However, the use of multilayers would result in thicker thicknesses and more complex processing. In addition, the use of multi resonant frequency selection faces (FSSs) can also greatly increase the absorption bandwidth. In addition, the use of multi resonant frequency selection surfaces (FSSs) can also greatly increase the absorption bandwidth [16-20] due to the advantages of thin thickness and strong design ability, it is also conducive to the realization of electromagnetic metamaterials based on broadband absorption.

This paper designed a double layer resistance film metamaterials, by two different resistance films and medium substrate superposition, radar wave broadband high absorption can be realized by adjusting the resonator size and ITO square resistance, the structure in the 6.4-21.1GHz frequency range absorption more than 90%, 14.6GHz bandwidth, relative bandwidth as high as 106.2%, has strong practicability.

2. Design of the double-layers resistance film absorber

Ran Lixin's team from Zhejiang University pointed out based on the Kramers-Kronig relationship that introducing loss into the metamaterial structure can regulate the dispersion relationship between the complex dielectric constant and the magnetic conductivity in a certain band, to realize the impedance matching [21] in the wider frequency band. The response function of any linear passive system, actually the relationship between the part and the virtual part is called the Kramers-Kronig relationship (K-K relationship). Metamaterials consist of sub-wavelength structural units with strong Lorentzian dispersion characteristics [22][23]. whose constitutive parameters (such as equivalent dielectric constant and equivalent magnetic conductivity) satisfy the K-K relationship. The real part of the complex magnetic conductivity and the complex dielectric constant represents the dispersion, and the imaginary part represents the loss. By introducing loss materials in metamaterial structures with multiple Lorentz resonance characteristics, the real part can be adjusted by controlling the imaginary part to achieve a relatively flat dispersion relationship in a certain frequency band. The structural design is shown in Figure 1. a, consisting of two layers of resistance film resonant unit, medium and copper bottom plate. Resistance film material is ITO / PET (Indium tin oxide / Polyethylene). ITO membranes due to the thickness of the nano grade, ITO film can be set to 0, the relative dielectric constant of PET substrate ε_{PET}=3.0 (1-j 0.06), and the thickness is 0.125mm. The dielectric substrate is made of PFA (Perflfluoroalkoxy) material with its relative permittivity ε_{PFA}=2.03 (1-j 0.0002). The structural shape and parameters of the resonant unit of the two layers are shown in Fig. 1 (b) and (c), and the resistance film has a central symmetric shape (Fig .1).

3. Simulation analysis

3.1. Simulation results

Modeling calculations were performed using the commercial electromagnetic software, CST Microwave Studio. unit cell boundary conditions were used in the x and y directions, and two excitation ports are set in the ± z-direction, one for the transmitting port and the other for the receiving port. Through CST software, the cell cycle size p=14.5mm, the upper PFA medium substrate thickness h_1=3mm, h_2=1mm, the lower medium substrate thickness h_2 = 1mm, the upper resistance film seam width t_1=1.1mm, the lower resistance film seam width t_2=0.3mm, R=6.15mm, the upper and lower resistance film surface resistance R_1=250 / sq and R_2=35/ sq, respectively. The overall thickness of the metamaterial absorber is 4.267mm.

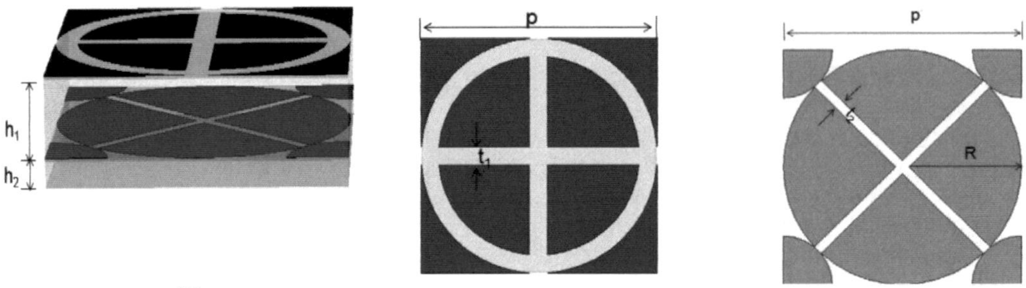

(1) (2) (3)

Figure1. Double-layer electromagnetic metamaterials wave absorber (1): overall structure; (2) upper layer resistance film structure; (3) lower laye resistance film structure.

Fig. 2 gives the simulation results of the absorption rate of electromagnetic wave vertical incidence, the black line for the TE absorption curve, and red for the TM wave absorption curve. Because the design shape of the resistive membrane resonant unit is centrally symmetrical, the absorption rate of TE waves and TM waves is unchanged, which maintains good broadband absorption characteristics, indicating that the metamaterials absorbing device has polarization-insensitive absorption characteristics. The structure achieves a broadband wave absorption rate above 90% in the 6.4-21.1G GHz frequency band, with a relative bandwidth as high as 106.2%, and a resonant absorption peak at 7.60 GHz and 15.40 GHz, with an absorption rate of 96.9% and 98.5%, respectively.

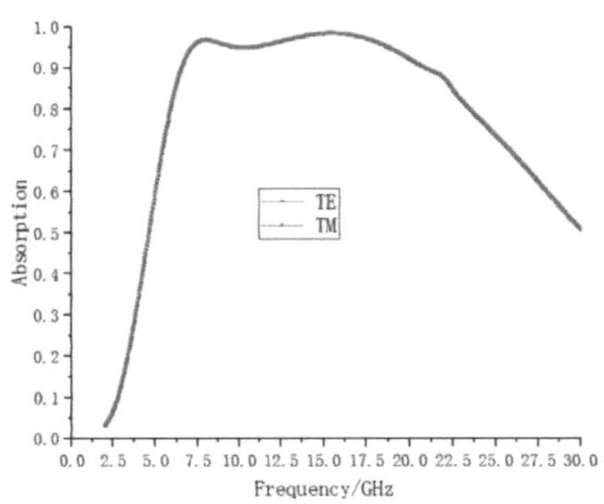

Figure 2. Simulated absorption spectra of the absorbing structure.

To study the effect of wide-angle incidence on metamaterials absorber, the variation of metamaterials absorber absorption at different incidence angles is simulated (Fig.3.).As can be seen from Fig. 3 (a), when the TE mode is an oblique incident, the absorption rate gradually decreases as the incidence angle increases, and the absorption bandwidth moves to the high frequency. When the incidence angle is increased to 50°, the structure can always maintain a broadband electromagnetic absorption rate above 80% in the absorption frequency band of 6.8 - 25.4 GHz.When the TM mode wave incident, when the incidence angle increases to 60°, the absorption rate in the 9.03-29.4 GHz band always remains above 90%; Therefore, the metamaterial absorber has good wide-angle incident absorption characteristics, and the TM mode is better than the TE mode.

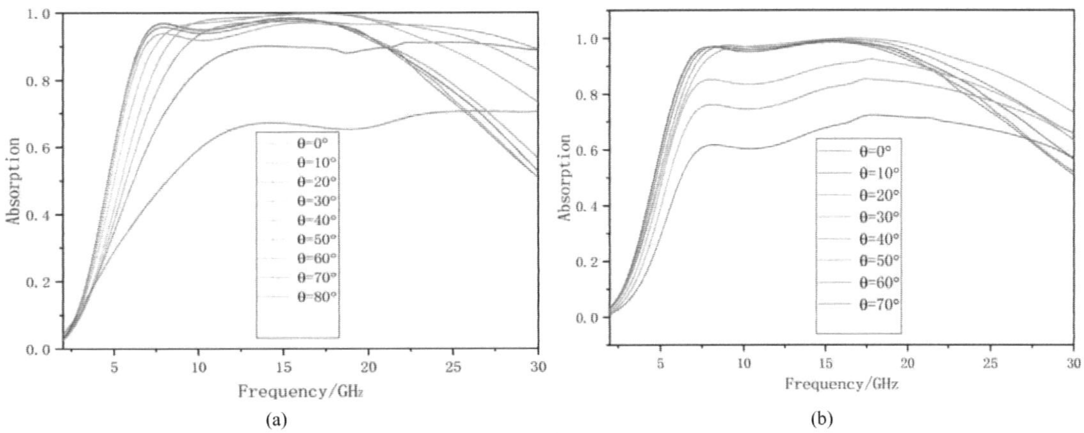

Figure 3. (a) Simulated absorption at different incident angles of TE;(b) Simulated absorption at different incident angles of TM.

3.2. Influence of resistance film structure on absorption performance of absorbing structure

To analyze the broadband wave absorption properties of the resistance film structure of each layer, the wave absorption properties of the single-layer resistance membrane unit structure were studied by keeping the total thickness of the PFA medium substrate unchanged at 4mm. Therefore, the upper resistance film and the lower resistance film are used as periodic resonant units of the single-layer absorption structure, the 4mm thick, the Cu / PET conductive film at the bottom, and other parameters remain unchanged for numerical simulation. The structure is shown in Fig. 4. The following three Models are discussed separately. Model 1 in Fig. 4 is the absorbing structure discussed above. The Model 2 structural unit in Figure.4 is composed of an upper resistance film as an independent resonant unit, with a 0.15mm thick PET medium and a 4mm PFA and a copper back plate. Model 3 mainly consists of the lower resistance film as an independent resonant unit, consisting of a 0.15 mm thick PET medium, a 4mm PFA and a copper back plate.

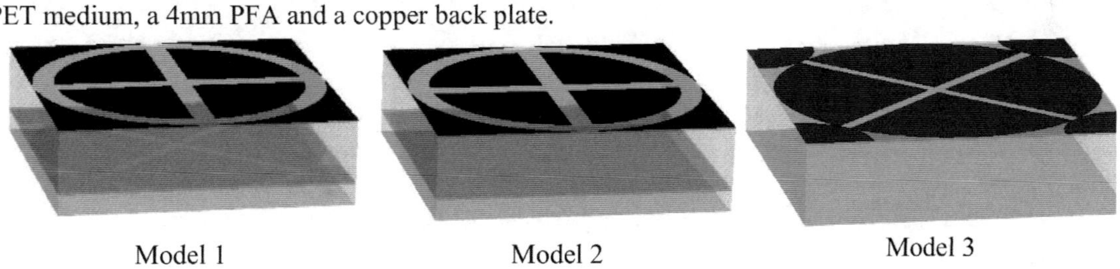

Figure 4. Absorbing structure of monolayer and double-layers.

The relevant structure parameters and material parameters of the two models are exactly consistent. The results of the absorption rate simulation are shown in Fig. 5. (a), Where the black curve is the absorption curve of Model1, the red curve shows the absorption curve of Model2, and the blue curve shows the absorption curve of Model3. Comparing the results of the simulation, the absorbing wave in the low-frequency band is caused by the lower resistive film structure unit, the reflective backplane, and the medium, The high-frequency band is mainly caused by the second resistance film, the reflection backplane, and the medium; When the first and second resistive membranes act together, the strong loss occurs simultaneously in the high and low-frequency bands, which makes the absorption performance stable and efficient in the 6.4-21.1GHz frequency interval, and the multi resonant peak narrow band absorption is converted into broadband absorption.

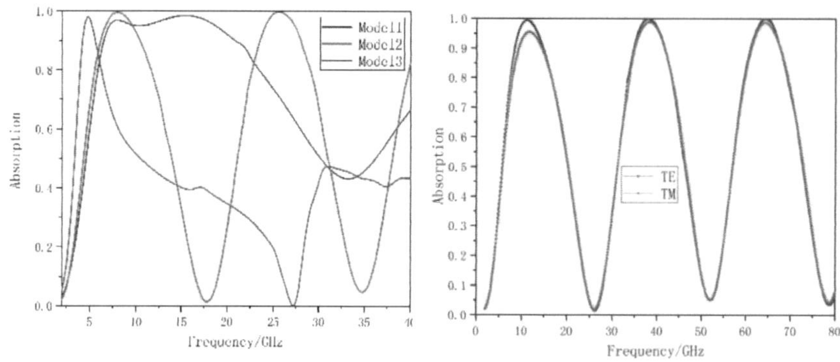

(a) (b)

Figure 5. (a). Absorption curves of the different models;(b) Simulated absorption results of TE AND TM WAVE.

3.3. *Wave absorption analysis of resistance film loss characteristics*

To better analyze the structural properties of the resistance membrane units in each layer, the surface current and impedance properties of Model 2 and Model 3 were analyzed below, respectively. As can be seen in Fig.5. (b), the absorption rate of the Model 2 structure is above 99% at both the 11.169GHz,37.958GHz, 64.438GHz, and 64.438GHz frequency points. Based on the above results, the surface current distribution map of model 2 at the above frequency point is presented by a field analysis, as shown in Fig. 6.

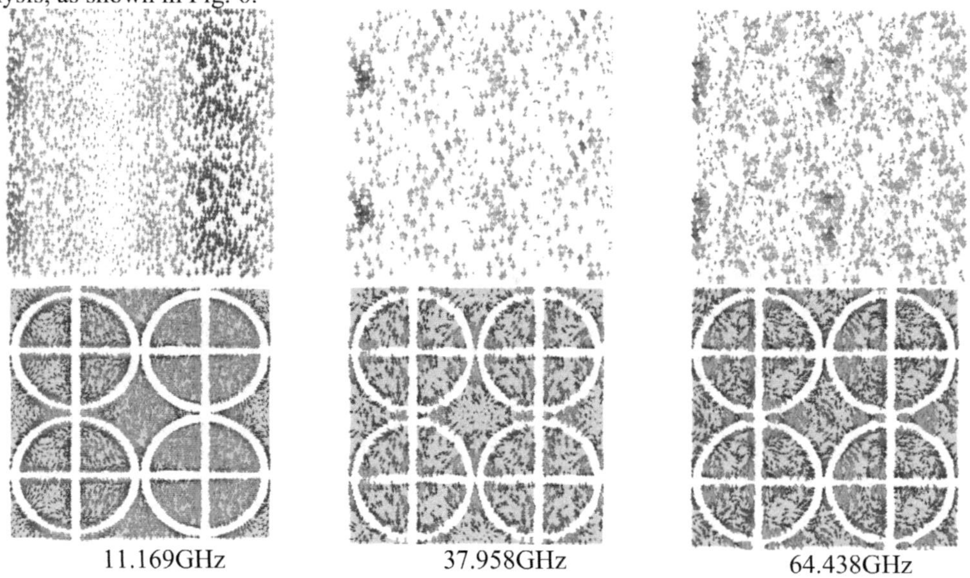

11.169GHz 37.958GHz 64.438GHz

Figure 6. Surface current density map of the resonance point at11.169GHz,37.958GHz, 64.438GHz.

According to the figure, the surface current of the upper layer resistance film structure unit at the four frequency points has the opposite direction of the surface current at the corresponding position of the backplane, which forms a ring current and forms a magnetic resonance. Based on the medium equivalence theory, the Smith circle of the equivalent relative impedance Z_{eff} was shown in the Fig.7. As can be seen from Fig.7, the impedance of model 2 at the frequency corresponding to the three absorption peaks is closest to the relative characteristic impedance value of the air（blue match point）. At this time, the solid Re (Z_{eff}) is close to 1, and the virtual Im (Z_{eff}) is close to 0, ensuring the matching degree of impedance with the air, so the absorption rate can reach more than 99%. However, in other frequency bands and the air impedance matching is poor, therefore, the broadband absorption characteristics are not reflected. The relative air impedance value Z_0=1 ensures the impedance matching degree with the air so that all electromagnetic waves can enter the material absorber for loss, so the absorption rate can reach more than 99%. However, in other frequency bands and air impedance matching is poor, therefore, model 2 does not exhibit the broadband absorption characteristics.

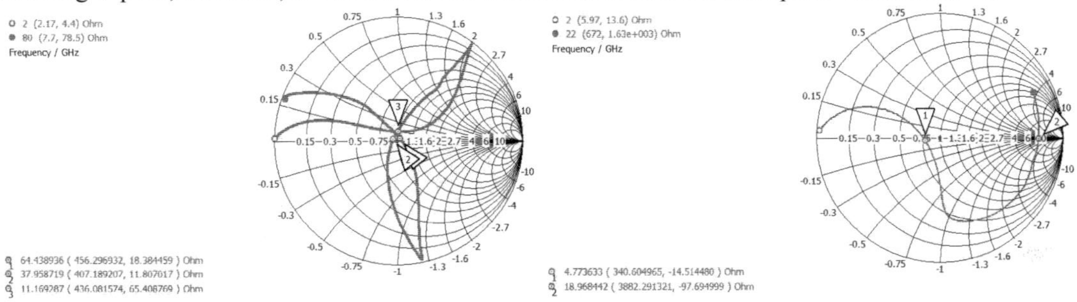

(a) (b)

Figure 7. (a). The impedance of Model2 within 2-80GHz; (b) Impedance of Model2 within 2–22GHz.

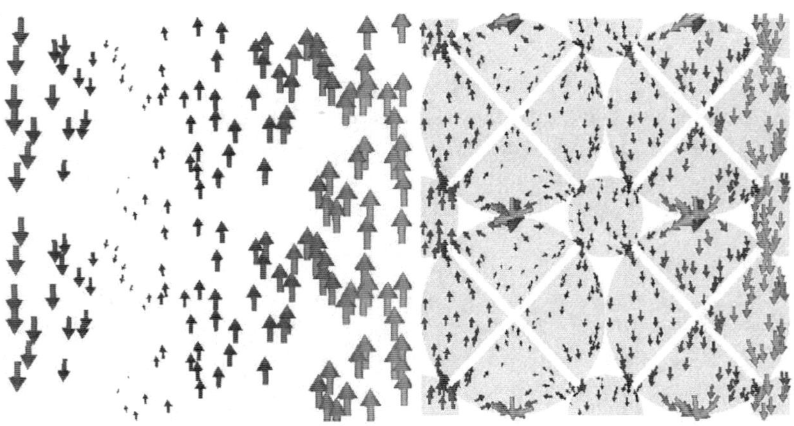

Figure 8. Surface current density map of the Model3 at 4.77GHz.

Fig. 8 shows that in Model 3, the resistance film and the bottom plate generate a current in the opposite direction at the resonance frequency of 4.77GHz inspired by an incident electromagnetic wave forming magnetic resonance. Fig. 7(b) is an impedance-matched Smith circle diagram of Model 3 in the 2-22GHz frequency band. As can be seen from the figure, In the 2-4.77GHz frequency interval, the structure is inductive, and the 4.77-18.9GHz frequency interval structure is capacitive. When the frequency is 4.77GHz, that is, the 1 point in the figure and the air impedance match are relatively good. At this time, the real part of the impedance is 340.6, and the imaginary part ImZ=-14.5, which matches the corresponding frequency of the peak absorption rate of Model3, indicating that the impedance matching enables the electromagnetic waves to enter the metamaterial structure for loss.

To contrast the three model impedance properties, Smith circle plots of Model 1, Model 2, and Model 3 in the frequency range of 6-22GHz are shown in Fig. 9 (a), (b), and (c), respectively. As can be seen from the figure, Model2 is inductive in the absorption band range and Model3 is capacitance in most frequency bands. The impedance matching of Model1 structure in the range of 6-22GHz range is significantly better than Model 2 and Model 3, making Model 1 absorption performance stable and efficient in the frequency range of 6 to 22 GHz, and the multitenancy peak narrow band absorption of Model 2 is converted into the broadband absorption of Model1.

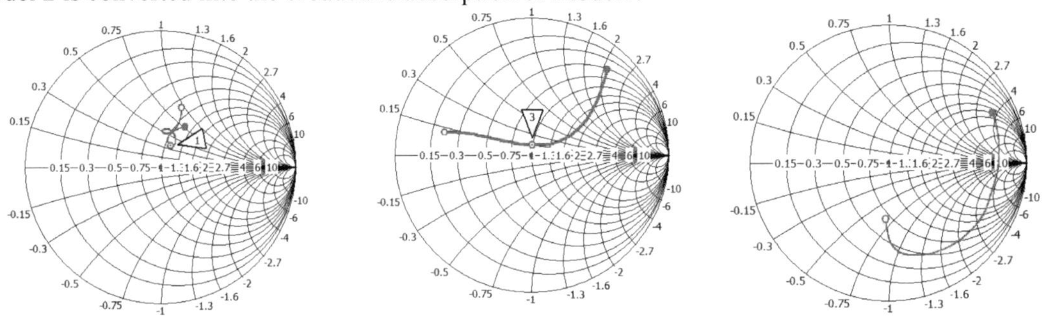

(a) (b) (c)

Figure 9. The impedance matching situation of the three models.

By comparison, under the premise of not increasing the medium thickness, in the metamaterial absorption structure composed of the double resistance film, the two resistance film resonance units affect each other, so that the equivalent relative impedance of the metamaterial absorption structure is smoother and more stable in the wide range. And the air relative characteristic impedance value Z_0=1 improves the impedance matching with air, the corresponding absorption peak also high-frequency offset, absorption performance, realize the efficient absorption of a wider band.

By comparison, without increasing the thickness of the medium, the resonance units of the two layers affect each other. This makes the equivalent impedance of metamaterial absorption structure better match the air impedance in the broadband range, and enhances the absorption performance. At the same time, it is found that the combination of double-layer resistance film moves the absorption bandwidth to high frequency, realizing the efficient absorption of a wider frequency band.

4. Conclusion

In this paper, an absorbing structure composed of a double-layer resistance film unit is designed to realize the broadband absorption of wide-angle radar waves. The absorption rate is greater than 90% in the 6.4-21.1GHz frequency interval and the polarization is insensitive. When the TM mode wave increases the incidence Angle to 60°, the absorption rate of the metamaterial structure always remains above 90% in the broad band of 9.03 to 29.4 GHz. When the TE mode wave incident and the incidence angle increases to 50°, the structure can maintain absorption above 80% in the frequency range of 6.8 to 25.4GHz.Simulation results of monolayer and double-layers resistance film structures showed that the equivalent impedance of double-layer resistance film structures matched better with the air impedance in the wide range. The results also show that a broadband radar absorber can be designed using a structure with a multi resonant frequency. This design has the advantages of a wide absorption band and moderate thickness, which has a certain guiding significance for the preparation of a new broadband electromagnetic metamaterials absorber, so it has potential application value.

Acknowledgment

The financial support was received from the Shaanxi Province Natural Science Basic Research Program (Grant No.2022JQ-075) and the National Natural Science Foundation of China (Grant No. 62001504).

References

[1] Smith, D. R., Smith, D. R., Vier, D. C,, et al. (2000) Direct calculation of permeability and permittivity for a left-handed metamaterial[J]. Applied Physics Letters, 77(14):2246-2248.
[2] Costa, F., Monorchio, A. (2012) A Frequency Selective Radome With Wideband Absorbing Properties[J]. IEEE Transactions on Antennas & Propagation, 60(6):2740-2747.
[3] Costa, F., Monorchio, A., Manara, G. (2007) Ultra-thin absorbers by using high impedance surfaces with resistive frequency selective surfaces[C]// Antennas & Propagation Society International Symposium. IEEE.
[4] Tayde, Y., Saikia, M., Srivastava, K. V., Ramakrishna, S. A. (2018) Polarization-Insensitive Broadband Multilayered Absorber Using Screen Printed Patterns of Resistive Ink[J]. IEEE Antennas and Wireless Propagation Letters, 17(12):2489-2493.
[5] Ling, X., Xiao, Z., Zheng, X . (2018) A Three-Dimensional Broadband Infrared Metamaterial Absorber Based on the Plasmonic and Dipole Resonance Responses[J]. Plasmonics.
[6] Long, L., Lv, Z . (2017) Ultra-wideband polarization-insensitive and wide-angle thin absorber based on resistive metasurfaces with three resonant modes[J]. Journal of Applied Physics, 122(5):055104.
[7] Sui, S., Ma, H., Wang, J., et al. (2015) Topology optimization design of a lightweight ultra-broadband wide-angle resistance frequency selective surface absorber[J]. Journal of Physics D: Applied Physics, 2015, IEEEMTT-SInternationalMicrowaveSymposium (21):215101.
[8] Liu, T and Kim, S. (2016) Design of wide-bandwidth electromagnetic wave absorbers using the inductance and capacitance of a square loop-frequency selective surface calculated from an equivalent circuit model [J]. Optics Communications, 359,372-377.
[9] Liu, H. T., Cheng, H. F., Chu, Z. Y., Zhang, D. Y. (2007) Absorbing properties of frequency selective surface absorbers with cross-shaped resistive patches[J]. Materials & Design, 28(7):2166-2171.
[10] Guo, R. Z., Pei, H. Z., Hui, B. Z., et al. (2013) Analysis and design of triple-band high-impedance surface absorber with periodic diversified impedance[J]. Journal of Applied Physics, 114(16):17-395.
[11] Mei, L., Xiao, S. Q., Bai, Y. Y., et al. (2012) An Ultrathin and Broadband Radar Absorber Using Resistive FSS[J]. IEEE Antennas & Wireless Propagation Letters, 11:748-751.
[12] Sheokand, H., Ghosh, S., Singh, G., et al. (2017) Transparent broadband metamaterial absorber based on resistive films[J]. Journal of Applied Physics, 122(10):105105.
[13] Sudhendra, C., Madhu, A. R., Pillai, A., et al. (2013) A novel ultra-wideband Radar Absorber based on hexagonal resistive patch FSS[C]// 2013 IEEE Applied Electromagnetics Conference (AEMC). IEEE.
[14] Ye, D., Wang, Z., Xu, K., et al. (2013) Ultrawideband Dispersion Control of a Metamaterial Surface for Perfectly-Matched-Layer-Like Absorption[J]. Physical Review Letters, 111(18):187402.
[15] Knott, G., Tuley, M., Shaeffer, J. (2004) Radar Cross Section[M]. IET Digital Library, 2004.
[16] Kundu, D., Mohan, A., Chakrabarty, A. (2016) Single-Layer Wideband Microwave Absorber Using Array of Crossed Dipoles[J]. IEEE Antennas & Wireless Propagation Letters, 1589-1592.
[17] Shang, Y., Shen, Z., Xiao, S. (2013) On the Design of Single-Layer Circuit Analog Absorber Using Double-Square-Loop Array[J]. IEEE Transactions on Antennas & Propagation,

61(12):6022-6029.

[18] Yao, Z., Xiao, S ., Jiang, Z., et al. (2020) On the Design of Ultrawideband Circuit Analog Absorber Based on Quasi-Single-Layer FSS[J]. IEEE Antennas and Wireless Propagation Letters, PP(99):1-1.

[19] Li, S., Gao, J., Cao, X., et al. (2014) Wideband, thin, and polarization-insensitive perfect absorber based on the double octagonal rings metamaterials and lumped resistances[J]. Journal of Applied Physics, 116(4):207402.

[20] Yuan, W., Cheng, Y. (2014) Low-frequency and broadband metamaterial absorber based on lumped elements: design, characterization, and experiment[J]. Applied physics, A. Materials science & processing, 1915-1921.

[21] Ye, D., Wang, Z., Xu, K., et al. (2013) Ultrawideband Dispersion Control of a Metamaterial Surface for Perfectly-Matched-Layer-Like Absorption[J]. Physical Review Letters, 111(18):187402.

[22] Pendry, J. B., Holden, A. J., Stewart, W. J, et al. (1996) Extremely Low-Frequency Plasmons in Metallic Mesostructures[J]. Physical Review Letters, 76(25):4773-4776.

[23] Pendry, J. B., Holden, A. J., Robbins, D. J., et al. (1999) Magnetism from conductors and enhanced nonlinear phenomena[J]. IEEE Transactions on Microwave Theory Techniques, 47.

Partial discharge location method in GIS coaxial bus structure based on multi-normal photoelectric array

Chen Li[1,*], Ming Ren[1], Yue Wang[1], Haobin Guan[1], Changjie Xia[1], Jiahe Yu[1], Tao Zhang[2] and Jin Miao[2]

[1] State Key Laboratory of Electrical Insulation and Power Equipment, Xi'an Jiaotong University, Xi'an, Shaanxi, 710000, China.
[2]Wuxi Electric Power Supply Company, State Grid Jiangsu Electric Power Co, LTD, Wuxi, Jiangsu, 214000, China.

*Corresponding author's e-mail: lc19991212@stu.xjtu.edu.cn

Abstract. Partial discharge is an important reason for the failure of GIS equipment, and its detection can determine the type and position of the defect in GIS. In the discharge inspection, accurate positioning of defects can provide an important reference for maintenance. In response to this, this paper proposes a discharge optical location method based on multi-normal photoelectric array. Based on the SiPM sensor, a multi-normal array is designed and built into the coaxial structure of GIS equipment. Measuring the responses of each unit in the array to get the response ratio of each element and calculate the direction of the discharge light in space. This paper firstly introduces the array structure, and proposes a solution algorithm in space, which uses the proportional relationship of multi-normal elements to solve the discharge direction. Secondly, the feasibility and error source of this method are verified in the simulation software. Finally, the application effect of this method in actual power equipment is verified by actual measurement, and the results show that this method can accurately locate the partial discharge in GIS.

1. Introduction

Gas insulated switchgear (GIS) has been widely used because of its advantages of small floor area and convenient installation. However, the failure rate of GIS was much higher than recommended by IEC[1]. Most of the faults were insulation faults. The insulation faults caused by abnormal surface of basin insulator, metal particles, loose shielding cover and burr of high voltage guide rod are particularly prominent [2].

In order to monitor discharge faults in GIS, various detection methods based on the characters of current, electromagnetic and acoustic have been proposed, which have become an important means of fault prevention at present [3]. However, the practical problems such as complex electromagnetic or acoustic interference, propagation path and multi-source discharge have seriously affected their application effect. Although scholars have conducted a lot of research and exploration in interference suppression, photoelectric joint diagnosis, acoustic and electric fault location and other aspects [4], the complex field application environment still brings great challenges to PD monitoring confidence and fast location.

Considering the limitations of existing schemes, a new partial discharge optical location method of GIS equipment for photoelectric sensor array is proposed r. The faults are located by detecting intensity relationship between each unit of detection array. In this paper, a new type of solid-state silicon photoelectric semiconductor sensor is used to realize the optical detection of internal discharge in the equipment, and the problem of built-in sensor in GIS equipment is solved. The sensor is a new type of low-light detector, with a small size and a square shape of 3.16mm side length, which can be flexibly installed inside GIS equipment. Secondly, the multi-normal vector localization algorithm based on the multi-normal SiPM array is introduced. Based on the linear correlation between the response intensity of the sensor to the light coming from different directions, and the cosine value of the light coming angle, the light direction tracing based on the response ratio relationship of each element of the array is realized. Finally, the location effect of sensor array in actual equipment is simulated. The results show that the 3d array positioning technology can accurately locate the discharge points in GIS, which provides a reference for fault diagnosis and maintenance of GIS equipment.

2. Location principle

The SiPM(Silicon photomultiplier) sensor is small in size (millimeter level), which can be ignored compared to GIS.[5] When installed on the inner wall of the housing, the impact on the electric field is extremely small, and the light generated by the partial discharge defect inside the GIS hits the sensor. It can be seen as parallel light coming from a distance. Since light from different directions will produce different angles with respect to the normal vector of the photosensitive surface of the sensor, the luminous flux received by the sensor surface will be different when the same light is irradiated from different directions [6]. The response intensity of the solid-state silicon photoelectric semiconductor sensor is proportional to the luminous flux received by its surface, and the direction of the partial discharge source can be judged by the geometric angle relationship of the normal vector and the light intensity. In order to orient the discharge source, three solid-state silicon photoelectric semiconductor sensor arrays can be used, and the three sensors are fixed on different planes as shown in Fig.1(a). The position of the partial discharge source can be determined through the geometric angle relationship of the normal vector and the ratio of the received light intensity. The ultraviolet transposing device is composed of fluorescent glass and bracket, which will be placed in the front of photoelectric sensor to change the wavelength, to enhance the efficiency of partial discharge photometry.

Under the assumption that the incident light is parallel light, the luminous flux that the solid-state silicon photoelectric semiconductor sensor can receive is related to its equivalent area in parallel light, that is, the projected area of the sensor on the plane perpendicular to the incident light, as shown in the figure 1 shown. In order to realize the partial discharge localization based on the detection intensity of the three-dimensional array, the normal vector translation and angle calculation are firstly performed, as shown in Figure 1(b), that is, the normal vectors $\mathbf{n_1}$ and $\mathbf{n_2}$ in the sensor array are translated so that the starting point coincides with the **O** point. In the plane γ formed by $\mathbf{n_1}$ and $\mathbf{n_2}$, a ray i is incident to point O, and any point A is selected on the ray **I**. Next, as shown in Fig 1(c), through point **A**, draw a line perpendicular to the plane γ to point **B**. On the plane γ, draw a line perpendicular to the normal vector n1 through the point **B** and intersect the normal vector $\mathbf{n_1}$ at the point C. Similarly, draw a line perpendicular to the normal vector $\mathbf{n_2}$ through point **B**, and the line intersects the normal vector $\mathbf{n_2}$ at point **D**. Point **A**, point **C**, and point **D** are located on ray **I**, normal vector $\mathbf{n_1}$, and normal vector $\mathbf{n_2}$, respectively. The included angles between the vector $\boldsymbol{OA}$ and the normal vector n1 and the normal vector $\mathbf{n_2}$ are α_1 and α_2 respectively, and the included angles between the vector $\boldsymbol{OB}$ and the normal vector n1 and the normal vector $\mathbf{n_2}$ are β_1 and β_2 respectively. According to the knowledge of solid geometry, it can be concluded that the angles α_1, α_2, β_1 and β_2 have the following relationship:

$$\begin{cases} \dfrac{\cos(\alpha_2)}{\cos(\alpha_1)} = \dfrac{\boldsymbol{OC}/\boldsymbol{OA}}{\boldsymbol{OD}/\boldsymbol{OA}} = \dfrac{\boldsymbol{OC}}{\boldsymbol{OD}} \\ \dfrac{\cos(\beta_2)}{\cos(\beta_1)} = \dfrac{\boldsymbol{OC}/\boldsymbol{OB}}{\boldsymbol{OD}/\boldsymbol{OB}} = \dfrac{\boldsymbol{OC}}{\boldsymbol{OD}} \end{cases} \rightarrow \frac{\cos(\beta_2)}{\cos(\beta_1)} = \frac{I_2}{I_1} \tag{1}$$

Combining the relationship between the angles β_1, β_2 and β ($\beta_1+\beta_2=\beta$), the size of the angle β_1 and the angle β_2 can be calculated, so as to further calculate the value of $\boldsymbol{OB}$, and then obtain the vector $\boldsymbol{BA}$ perpendicular to the γ plane. value. Therefore, we can determine a plane γ_1 on which the incoming light rays lie. Similarly, another plane γ_2 where the incoming light rays are located can also be determined. The direction of the light is the intersection of the two planes γ_1 and γ_2, as shown in Figure 1(d). The final ray direction can be solved by equation (2).

$$\begin{cases} \dfrac{\cos(\beta_2)}{\cos(\beta_1)} = \dfrac{I_2}{I_1} \\ \beta_{11} + \beta_{12} = \beta_{01} \\ \dfrac{\cos(\beta_{22})}{\cos(\beta_{21})} = \dfrac{I_3}{I_1} \\ \beta_{21} + \beta_{22} = \beta_{02} \end{cases} \tag{2}$$

In the formula, β_{11}, β_{12}—the included angle determined by sensor A and sensor B; β_{21}, β_{22}—the included angle determined by sensor A and sensor C. In this experiment, because of its simple fabrication and good light transmittance, the fluorescent glass was prepared by the phosphor coating method, and the UV transposition device was built

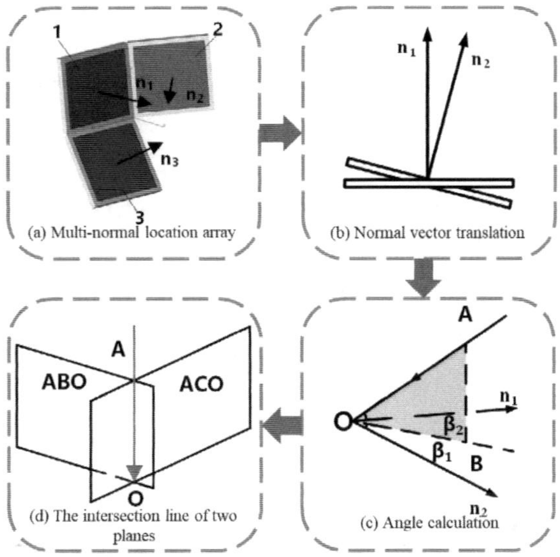

Figure 1. PD location method based on 3D array detection intensity

3. Simulation process

In this paper, TracePro software is used to simulate the research content. This paper takes the 550kV GIS coaxial bus section as an example to carry out further simulation analysis on the three-dimensional photoelectric plane positioning method. In this paper, the actual size of 550kV GIS is used to construct the simulation geometric model[7]. Due to the extremely small size of the PD light source and the short luminescence quenching time, similar to the pulsed point light source, this paper adopts a uniform spherical light source with a small diameter to simulate the discharge light signal. In order to use discrete single beam sets to simulate the continuous light field in space, and comprehensively consider the calculation amount of the simulation, the total number of rays is set to 780,000.

The simulation structure is shown in Figure 2, in which Figure 2(a) is the GIS simulation model. The location of the PD light source are shown in Figures 2(b),. The occlusion effect of the rod on the light path divides the internal space of the GIS into bright field and dark field areas. Make a tangent to the inner guide rod through the center of the sensor on the inner wall of the busbar shell to obtain the bright field area as shown in the figure, which is the effective field of view that can satisfy the discharge light source directing the sensor. The simulations in this paper are all aimed at the detection of discharge faults in the bright field range. For the faults in the dark area, the detection can be carried out by increasing the number of sensors and expanding the detection range. The sensor arrays (A,B,C) are fixed on the GIS enclosure and the coordinates are set to (0,496). The discharge light source simulates the common guide rod burr defect model. Three groups of defects are set in the axial direction of the inner guide rod, and the distances from the center of the guide rod are 0mm, 200mm and 400mm respectively; each group of light sources is set with the center of the guide rod and the center of the sensor. Four burr defect light sources with angles of 0°, 15°, 30° and 45°.

The simulation results of light propagation inside the GIS are shown in Figure 3(a)[8]. In order to better show the light propagation process, the number of rays in Figure 3(a) is extracted by 0.5% for display, as shown in Figure 3(b). Red means the light energy is attenuated to 100%~66% of the original, green means the light is attenuated to 66%~33% of the original, and blue means the light is attenuated to 30% of the original. As can be seen from the figure, when the light source is located in front of the guide rod, the light entering the sensor is divided into direct light[9] and reflected scattered light; when the light source is located on the guide rod and behind the guide rod, the light entering the sensor is both reflected light and scattered light. The reflected light and scattered light entering the sensor will undoubtedly bring some errors to the sensor. Table 1 shows the energy distribution of the 7512 rays of light received on the sensor surface C when the discharge defect is at position 1. It can be seen that after refraction and reflection, most of the light energy is distributed below 10%, and the direct light responds to the energy of the sensor. The proportion of 89.1%, most of the energy is still caused by direct light. Therefore, the sensor response, which includes reflected and scattered light energy, still reflects the approximate location of the light source to a certain extent[10].

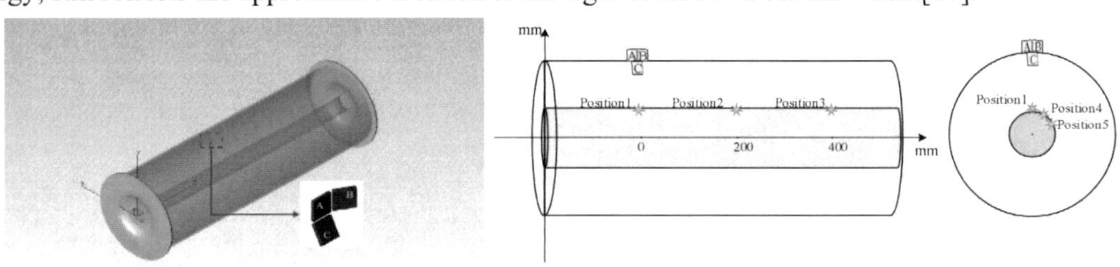

(a) GIS coaxial bus structure (b) The location of the PD light source

Figure 2. Light source positions in guide rod

(a) Light path of light source in GIS (b)Light path inside GIS when the light quantity is 1%

Figure 3. Simulation results of light propagation in GIS

Figure 4 shows the irradiance distribution map and candela plot on the three surfaces of A, B, and C in the sensor array when the discharge defect is at position 1. The left side of the picture represents the irradiation energy corresponding to each color. It can be seen from the figure that the irradiance distribution on the A sensor with direct light is relatively uniform, while the irradiance distribution on the B sensor and the C sensor varies significantly with the coordinate change. Figure 5 shows the radiation distribution diagram on sensor A when the discharge defect and sensor array are on the same cross section of GIS and the included Angle between the discharge defect position and sensor array is 0°, 30°, 45° (positions 1, 5 and 6). It can be seen that the farther the light source is from the sensor or the larger the Angle between the light source and the sensor, the smaller the luminous flux on the sensor

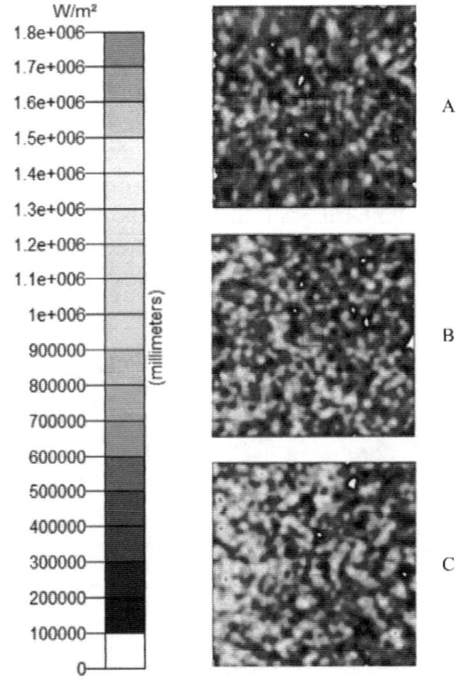

Figure 4. Radiation distribution diagram of discharge defects on A, B and C array at position 1

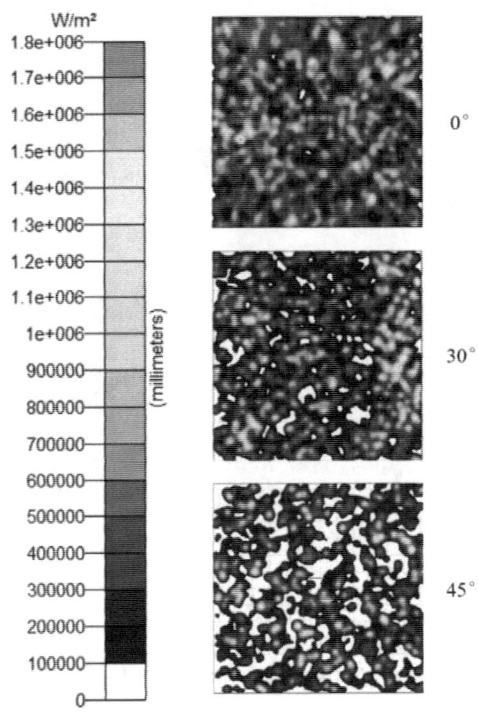

Figure 5. Irradiance analysis of sensor a when the angle between discharge defect location and sensor array changes

The discharge direction and the actual incoming light direction calculated according to the simulation results are shown in Table 3. Inside the GIS, when the discharge point is closer to the sensor array, the orientation error is smaller. In the structure set in this paper, when the discharge point is 0mm and the angle with the sensor is 0°, the orientation error is the smallest, which is 7.2977°; when the discharge point is 200mm and the angle with the sensor is 45°, the orientation error is the largest, which is 15.6527°. This is because when the distance is short and the discharge point is facing the sensor array, the incident light is mainly direct light from the discharge point. When the distance becomes farther and the sensor has a certain angle, the proportion of reflected light increases, resulting in a larger orientation error. . Overall, the method proposed in this paper can localize PD points.

Table 1. Simulation results and actual values of light source positioning (distance unit: mm)

Distance	0mm			
Angle	0°	15°	30°	45°
Actual direction	(1,0,0)	（1,-0.079,0）	(1,-0.128,0)	(1,-0.178,0)
Simulation direction	(1,-0.1,-0.08)	（1 -0.12 0.16）	(1,-0.01,0.1)	(1,0.03,0.1)
Error	7.2977°	9.3190°	8.8110°	10.1234°

Distance	200mm			
Angle	0°	15°	30°	45°
Actual direction	(1,0,0.49)	(1,-0.079,0.482)	(1,-0.128,0.461)	(1,-0.178,0.43)
Simulation direction	(1,0.12,0.65)	(1,-0.28,0.32)	(1,-0.11,0.24)	(1,0.12,0.41)
Error	8.9845°	13.4111°	11.1966°	15.6527°

Distance	400mm			
Angle	0°	15°	30°	45°
Actual direction	(1,0,0.98)	(1,-0.079,0.964)	(1,-0.128,0.922)	(1,-0.178,0.86)
Simulation direction	(1,0.22,0.82)	(1,-0.25,1.51)	(1,0.11,1.14)	(1,-0.12,1.31)
Error	10.8935°	13.2955°	11.2890°	12.3927°

The discharge direction and the actual incoming light direction calculated according to the simulation results are shown in Table 3. Inside the GIS, when the discharge point is closer to the sensor array, the orientation error is smaller. In the structure set in this paper, when the discharge point is 0mm and the angle with the sensor is 0°, the orientation error is the smallest, which is 7.2977°; when the discharge point is 200mm and the angle with the sensor is 45°, the orientation error is the largest, which is 15.6527°. This is because when the distance is short and the discharge point is facing the sensor array, the incident light is mainly direct light from the discharge point. When the distance becomes farther and the sensor has a certain angle, the proportion of reflected light increases, resulting in a larger orientation error. . Overall, the method proposed in this paper can localize PD points.

4. Conclusion

In this paper, a method for localization of partial discharges in GIS based on the principle of multi-normal photoelectric location principle and SiPM sensor is proposed. At the internal spatial scale of GIS equipment, the array can satisfy the cosine response relationship, which verifies the rationality of applying the principle of multi-normal vector location to the internal fault location of GIS equipment. Using the multi-normal vector positioning principle proposed in this paper, 12 groups of GIS internal discharge simulation faults are simulated and analyzed for direction traceability, and the direction vectors of the discharge source relative to the sensor at different positions in the bright field area detected by the sensor are obtained.

The analysis results and the actual position error Within 16°, the results show that the analytical solution can reflect the position of the discharge point inside the GIS to a certain extent, and the accurate source traceability of the discharge point can be realized by reversely tracing the source to find the intersection of the equipment space, and the deviation between the fault position and the actual position can be located. The maximum distance is 1.3cm, and the average deviation is 0.62cm, which can meet the internal positioning requirements of GIS equipment. The orientation accuracy of the discharge point distribution in the GIS equipment is related to the fault location. The closer the discharge point is to the sensor array and the smaller the angle between the discharge point and the sensor, the smaller the orientation error.

The principle of the multi-normal photoelectric array proposed in this paper can realize the accurate orientation of the discharge source inside the GIS equipment. In order to further reduce the error, the structure of the sensing array needs to be optimized in the follow-up research, and the optimal parameters such as the number of cells and the angle between cells are obtained. And cooperate with experiments to optimize the response curve of the sensor to light in different directions, to achieve accurate positioning and even imaging in practical applications.

Acknowledgments

The work was supported by the Science and Technology Program of the State Grid Corporation of China (5500-202218131A-1-1-ZN).

References

[1] Lin, L. C., Chen, D. (2021) The Research of GIS Discharge Influence Based on Particle Properties in Installation Environment[J]. Journal of Physics: Conference Series, 2121(1).

[2] Tang, Q., Li, G., Wang, J., et al. (2021) GIS partial discharge defect diagnosis system and method based on Extreme Learning Machine[J]. IOP Conference Series: Earth and Environmental Science, 632(4):042005 (6pp).

[3] Zhi, Y., Min, L., Li, X., et al. (2021) Study on UHF electromagnetic wave propagation characteristics of partial discharge in GIS[C]// Society of Photo-Optical Instrumentation Engineers (SPIE) Conference Series. SPIE.

[4] Kitamura, Y., Hirabayashi, S. (2016) Partial discharge deterioration of epoxy resin for electronic parts[C]// Conference on Electrical Insulation & Dielectric Phenomena - Annual Report 1985. IEEE.

[5] Ren, M., Zhou, J., Miao, J. (2020) Adopting spectral analysis in partial discharge fault diagnosis of GIS with a micro built-in optical sensor[J]． IEEE Transactions on Power Delivery, PP(99):1-1.

[6] Sassaroli, A., Fantini, S., et al. (2004) Comment on the modified Beer–Lambert law for scattering media[J]. Physics in Medicine and Biology, 49(14):N255.

[7] Zang, Y., Qian, Y., Wang, H., et al. (2021) Method of GIL partial discharge localization based on natural neighbour interpolation and ECOC cm LP log VM using optical simulation technology[J]. High Voltage.

[8] Ren. M., Li, C., Wang, B., et al. (2022) Corona Discharge Optical Location Method Based on Lambert Scattering Principle， [J/OL]． Proceedings of the CSEE， 1-10 [2022-06-13]. http: //kns.cnki.net/kcms/detail /11.2107.TM.20220511.1217.002.html

[9] Xia, C., Ren, M., Chen, R., et al. (2022) Multispectral optical partial discharge detection, recognition, and assessment[J]. IEEE Trans. Instrum. Meas..

[10] Li, X., Xu, Y., Ding, D., et al. (2022) Influence of Temperature on Partial Discharge and Flashover Characteristics of Metal Particle on GIS Insulator Surface[J]． Proceedings of the CSEE, 42(01):406-415.DOI:10.13334/j.0258-8013.pcsee.210451.

Research on PID control of electric traction DC motor speed based on MATLAB

Huarong Wang[1,*], Qiyan Yan[1]

[1]Guangdong University of Science and Technology, Dongguan Guangdong, 523083 China

*wanghuarong@gdust.edu.cn

Abstract. With the development of science and technology, digital control systems are more widely used, which is more conducive to the development of intelligent power locomotives. In the speed control process of electric traction DC motor, considering the stability, rapidity, accuracy and other requirements of speed control, digital position type PID control strategy is adopted.This study first models the power traction DC motor and analyzes the performance of the system. In order to improve the performance index of the system, a digital PID control strategy is adopted.MATLAB simulation results show that the controller has good speed and stability for the speed control of electric traction DC motor.

1. Introduction

In urban transportation, electric traction motor has been widely used because of its advantages such as large traction power, high transmission efficiency and no pollution.At present, traction motors are used on railway trunk motor locomotives, subway electric vehicles, urban trams, industrial and mining electric locomotives and other vehicles.There are many kinds of traction motors, which mainly study DC traction motors. The electric traction motor transmission system takes the traction motor as the control object to adjust the traction force and speed of the motor to meet the needs of vehicle traction and braking characteristics [1-2].With the development of electric locomotives, the improvement of the control performance of the traction transmission system has always been complementary and mutually reinforcing. Therefore, the electric traction transmission and control system plays an important role in the field of transportation.

At present, in terms of speed control of traction motors, many scholars have put forward many control methods, which have achieved good results in practical engineering applications.Chen Zaigang made a systematic overview of the dynamics of the rail transit traction power transmission system, especially the key components or devices of the core of "stuck neck", key scientific problems, key technologies and so on.Based on the modeling of rotor dynamics system, for the control of the traction motor, four types of control strategies based on the steady-state model of the motor (scalar control), vector control, direct torque control and adaptive control are summarized. Through these control methods, the speed of the motor is controlled [3].Zuo Yun adopts speed sensor-free technology to solve the problems that the mechanical speed sensor of the traction motor will increase costs.In the speed control strategy based on the motor model, the phase-locked loop (PLL) speed identification method is adopted to extract the speed and phase from the rotor magnetic flux, and an improved q-type PLL speed control method is proposed to make up for the shortcomings of traditional q-PLL. It has good performance in a variety of industrial states, and the structure is simple and easy to implement [4].Wang Huimin proposed a high-performance speed identification control based on the backstepping control theory and MRAS for the speed sensorless control of the linear traction motor of medium and low-speed maglev trains [5]. Chen Ming design and application on the speed sensor of traction motor of rail vehicles, using dual Hall type integrated circuit components, design speed sensor and PCB board.It realizes non-contact installation, zero-speed measurement, high reliability, and solves the problem of high price of such imported products [6]. You Bingtao considered the power loss of traction motors in the energy-saving speed of high-speed trains, and proposed model-based predictive control and dynamic programming methods to ensure the speed tracking accuracy, reasonably allocate the number of traction motors, and reduce the energy consumption of high-speed trains [7]. To sum up, this paper aims to digitize the electric traction motor. The traction speed of DC motor is controlled by position-type PID, and the speed control is realized in the digital control system, which has better performance of the analog system.

This study uses professional knowledge such as automatic control principles, as well as more mathematical, physical, electronics, motor, mechanical and other discipline knowledge, to establish a perfect and feasible control system model, mathematically model it and use MATLAB simulation for dynamic and steady-state performance analysis, which is conducive to combining theoretical knowledge with practical engineering applications to deal with and solve engineering problems.

2. Electric traction DC motor

The working principle of DC traction motor is the same as that of ordinary DC motors, but it has special working conditions and many design and structural requirements.The block diagram of this electric traction motor control system is shown in Figure 1.This system uses sensor devices such as speed generators. Using the principle of generators, voltages that change with the measured speed are collected, the input is fed back to form a deviation signal with the input signal, and then the deviation signal is sent to the power amplifier and the traction DC motor [8][9].This motor adopts an armature-controlled motor, and its transfer function is

$$\frac{\omega(s)}{u(s)}=\frac{5}{s^2+1.25s+0.75} \tag{1}$$

The power amplifier is nonlinear and can be roughly represented by an exponential function $v_2=2e^{3v_1}$. Its normal working point is 1.5V, and the transfer function of the power amplifier is $\frac{v_2(s)}{v_1(s)}=540$ [9]. The transfer function of the speed generator is 0.1. From literature [9], we can know that the closed-loop transfer function of electric traction motor is:

$$G(s)=\frac{\omega(s)}{\omega_{\mathrm{d}}(\mathrm{s})}=\frac{2700}{s^2+(1.25s+2700.75)} \tag{2}$$

then its open-loop transfer function can be approximated is

$$G_{\mathrm{k}}(S)=\frac{2700}{s(s+1.25)} \tag{3}$$

$\omega_d(s)$,$\omega(s)$ are the given input (that is the expected angular velocity) and angular velocity output of the electric traction DC motor.

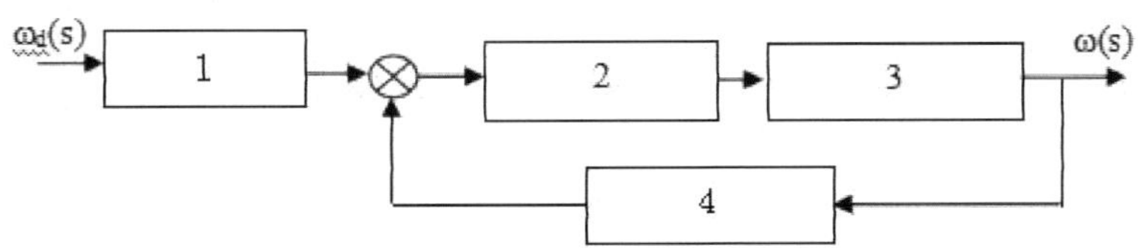

Figure 1. Block diagram of electric traction motor control system

Given device; 2-Power amplifier; 3-Electric traction machine;4-Speed measuring generator

3. PID digital control algorithm

3.1. Characteristics analysis of electric traction DC motor

The dynamic characteristics of the electric traction DC motor can be derived from the closed-loop transfer function of the electric traction motor, $G(s)=\frac{\omega(s)}{\omega_{\mathrm{d}}(\mathrm{s})}=\frac{2700}{s^2+1.25s+2700.75}$. MATLAB software is used to simulate the system [10]. Through the unit step response, observe its dynamic characteristics and performance indicators.

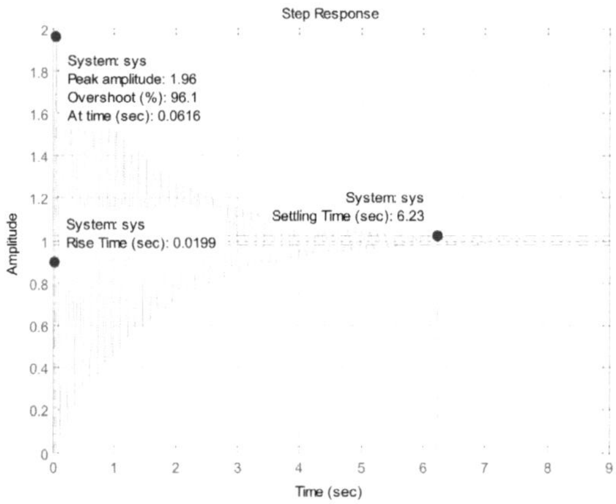

Figure 2. Unit step response curve of the original system

According to the unit step response dynamic curve, although the system will be stable eventually, its dynamic performance index is not ideal. The overshoot reached 96.1%, and the stationarity was very poor. The adjustment time is 6.23s, which is long and fast.The adjustment time is 6.23s, which is long, and speed is not good.

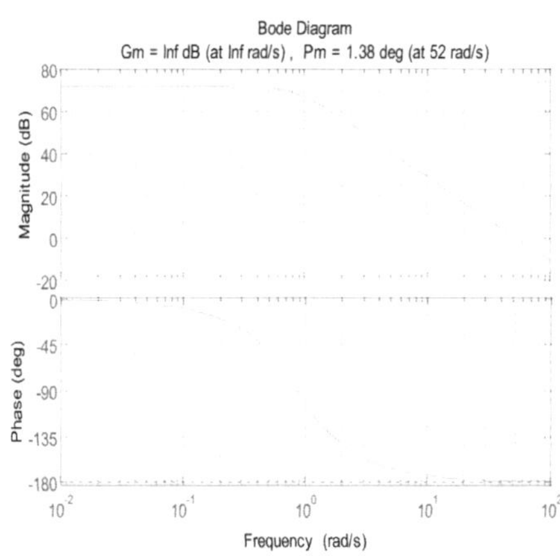

Figure 3. Bode diagram of the original system

From the BODE figure, we can see that the phase angle margin of the system is γ=1.38°, which is very small, far less than 30°~60°, so the performance index is not good.

3.2. Principle of digital PID control for speed control system

In order to submit the performance index of the speed control system of the electric traction DC motor, a regulator, namely a PID controller, must be added to the system. If a continuous controller is used, the new control system block diagram is shown in Fig. 4.

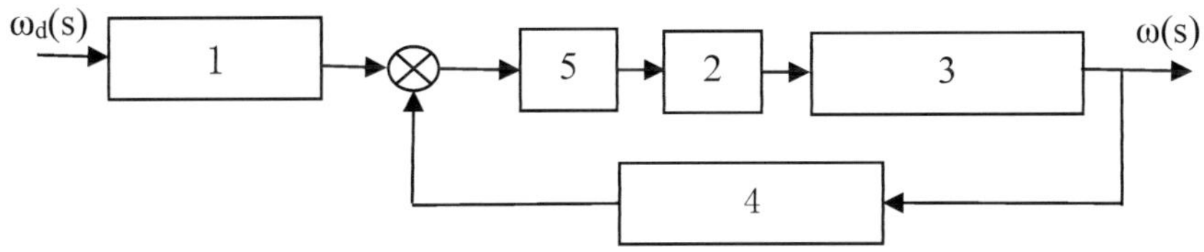

1-Given device; 2-Power amplifier; 3-Electric traction machine;4-Speed measuring generator;5-PID control

Fig. 4. Block diagram of the new control system

The system uses a sensor device to collect the voltage that changes with the measured speed, and then sends it to the digital PID controller through analog-to-digital conversion. The speed control adopts an improved PID digital control algorithm.In terms of system implementation, microprocessors such as single-chip microcomputer, DSP or STM32 are used to realize the digital PID control algorithm. The detection angular speed is compared with the expected angular speed, and then the system control signal is calculated through the digital PID control algorithm, and then amplified by power to act on the traction motor to achieve the purpose of controlling the speed. In MATLAB software, variables ωd(k) and ω(k) are used as the given input (expected angular velocity) and angular velocity output at time k, as shown in Fig. 5.

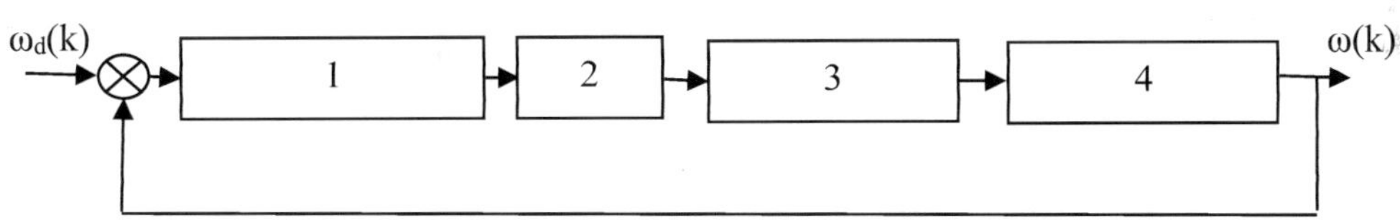

1-Digital PID control; 2-D/A converter; 3-Power amplifier; 4-Electric traction machine

Figure 5. Digital PID control principle

3.3. Regulation law of speed control system

In MATLAB software, variables rin(k), yout(k) are used as the given input at time k (that is the expected angular velocity) and angular velocity output. It should be emphasized that the input signal of the regulator refers to the deviation signal compared between the measured signal (the regulated parameter) y(t) and the given signal r (t), denoted by e(t).

$$e(t) = r(t) - y(t) \tag{4}$$

The output signal of the regulator refers to the change of the corresponding output signal after the regulator receives the deviation signal e(t), which is represented by u(t).

Although the addition of integral adjustment can eliminate the static difference, the cost is to reduce the response speed.In order to speed up the control process, it is necessary not only react to the deviation immediately at the moment of occurrence or change, but also to the change of deviation, or control it according to the trend of deviation change, so that the deviation can be eliminated in the bud.In order to achieve this goal, differential adjustment can be added to the PI adjustment to obtain the following PID control rules.

$$u(t) = k_P\left(e(t) + \frac{1}{T_I}\int_0^t e(t)dt + T_D\frac{de(t)}{dt}\right) + u_0 \tag{5}$$

In the formula, TI is called integral time, and TD is called differential time.

3.4. Digital PID controller

In this control system, the PID control adopts the digital PID position control algorithm regularly. When the sampling period is quite short, summation is used instead of integration, and backward difference is used instead of differential, so that the analog PID discretization becomes a difference equation.

It is believed that the initial control amount is zero, and the formula 4 is transformed into a difference equation. For this reason, the following approximation can be made.

$$\int_o^t e(t)dt = \sum_{j=o}^{k} e(jT) = T\sum_{j=o}^{k} e(j) \tag{6}$$

$$\frac{de(t)}{dt} = \frac{e(kT) - e((k-1)T)}{T} = e(k) - e(k-1) \tag{7}$$

In the formula, T is the sampling period, k is the sampling serial number, and t=kT(k=0,1,2,...).
The available digital PID position control formula is

$$u(k) = k_p(e(t) + \frac{T}{T_1}\sum_{j=o}^{k} e(j) + T_D(e(k) - e(k-1))) \tag{8}$$

$$u(k) = k_p(e(t) + k_i\sum_{j=o}^{k} e(j)T + k_d\ (e(k) - e(k-1))) \tag{9}$$

In the formula, Kp=1/δ is called the proportional gain, $k_i = \frac{k_p}{T_1}$ is called the integral coefficient, $k_d = k_p T_D$ is called the differential coefficient, $e(k)$ and $e(k-1)$ is the deviation signal obtained at the k time and k-1 time respectively.

3.5. PID parameter adjustment

PID parameters are manually adjusted by MATLAB simulation. First, let ki=0 and kd=0, and then slowly increase the value of proportional gain Kp until the output of the closed-loop system oscillates, that is, the system reaches a critical stable state.After determining this value of the proportional gain Kp, the value of Kp is reduced so that the system output reaches the so-called 25% amplitude attenuation state.That is to say, the amplitude output of the closed-loop system can be reduced to about 25% of the maximum value within an oscillating period.According to experience, the proportional gain kp can be reduced to half of the critical stability value first. The next step is to increase the values of ki and kd to enable the closed-loop system to produce the expected step response [9]. In addition, according to the influence of proportional gain Kp, integral gain Ki and differential gain Kd on the step response performance of the system (If the proportional gain Kp increases, the overshoot increases, the steady-state error decreases, and the adjustment time has little effect. If the integral gain Ki increases, the overshoot increases, the adjustment time increases, and the steady-state error is zero. If the differential gain Kd is increased, the overshoot and adjustment time are decreased, and the steady-state error has no effect)[9], After several tests, adjusting each parameter to observe the effect of step response, Finalize kp=1, ki=10, kd=0.03.

3.6. Programming of digital PID controller

In MATLAB software, variables rin(k), yout(k) are used as the given input at time k (that is the expected angular velocity) and angular velocity output. The corresponding flow chart of control is shown in Fig. 6.

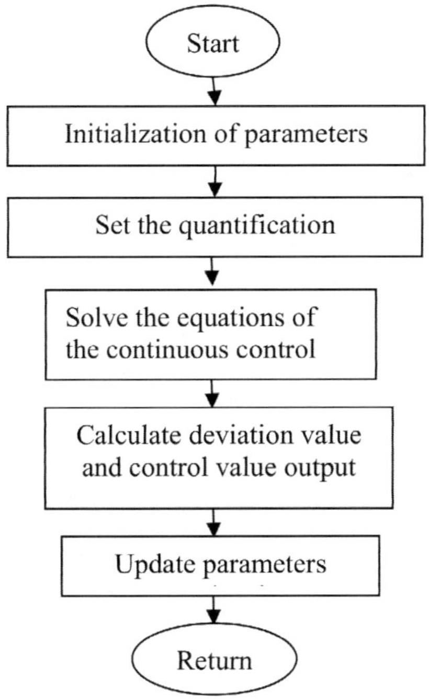

Figure 6. Flowchart of digital position-type PID control algorithm

The unit step response curve of PID simulation is shown in figure.7(a). It can be seen that the overshoot is 22.3% and the adjustment time is 0.0464s. Figure.7(b) is the unit step response curve of digital position-type PID control, and its dynamic overshoot performance is better than that of analog PID simulation.

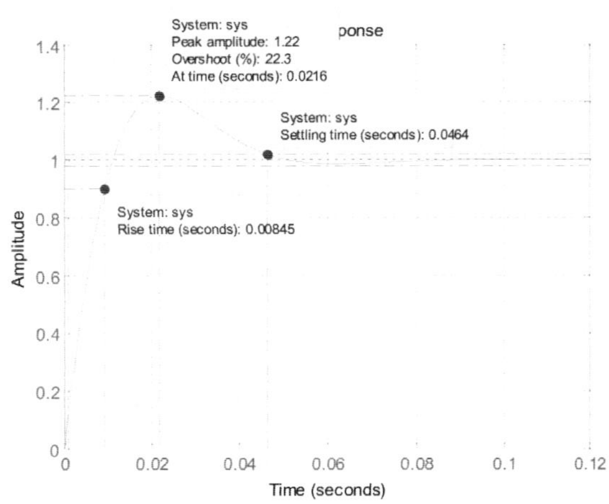

(a) Simulation PID control

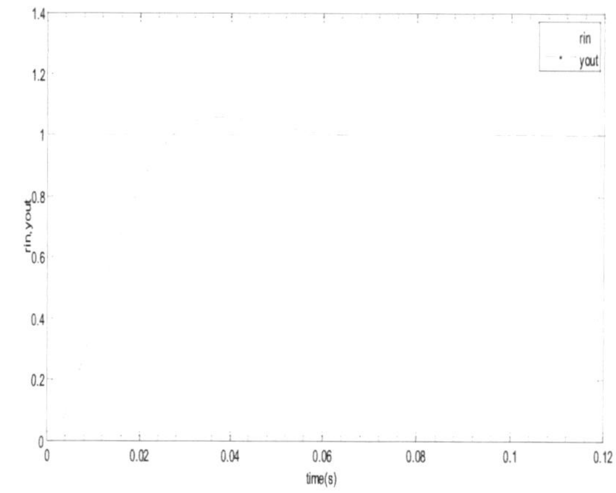

(b) Digital position type PID control

Figure 7. Unit step response curve of PID control

The BODE diagram of the new system, shown in Fig.8, shows a good phase margin.

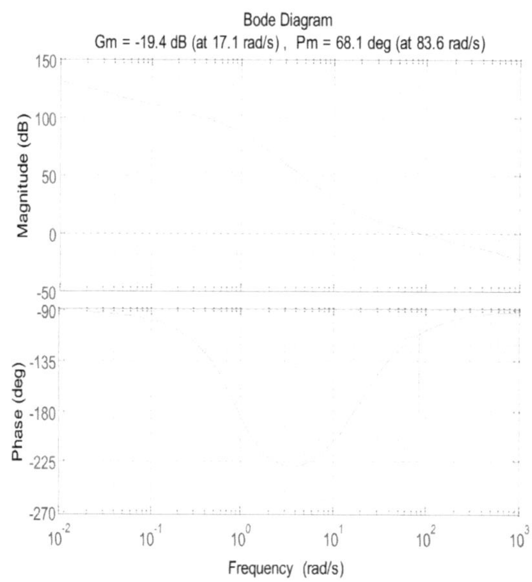

Figure 8. Bode diagram of the new system

4. Conclusion

This article in view of the electric traction precise control of dc motor of high power and rotational speed, and realize the demand of the application system of digital, intelligent, analogue controller is converted into digital controller, in order to satisfy the functional requirement of the electric traction motors and speed performance requirements, and we compare the performance of the control system. PID regulator is a classical control system. In the speed control system of electric traction DC motor, the PID regulator of digital position type can get satisfactory control effect. It is widely used in industrial and transportation systems.

Acknowledgment
2020 Guangdong University of Science and Technology scientific research project "Speed Control System Design of electric traction DC Motor based on Digital PID control(GKY-2020KYYBK-31).

References

[1] Bao, G. Q., He, T., Liu, X. B. (2018) Permanent magnet synchronous motor finite set model for direct torque control prediction [J]. Electric machines and control applications, 45(02):19-24.
[2] Zhou, M. L., You, X. J., Wang, C. C. (2011) Vector control of traction drive system for electric locomotive [J]. Transactions of China electrotechnical society, 6(09): 110-115+129.
[3] Chen, Z. G., Liu, Y. Q., Zhou, Z. W., Ning, J. Y. (2021) Review on Dynamics of traction powertrain in rail transit[J].Journal of Traffic and Transportation Engineering, 21(6):31-49.
[4] Zuo, Y., Ge, X. L., Li, S. T., Wang, H. M., Yan, P. L. (2021) Speed sensorless control of traction motor based on improved Q-PLL[J].Proceedings of the Chinese Society for Electrical Engineering, 41(1):383-391.
[5] Wang, H. M., Xie, D., Yao, Bo, Ge, X. L. (2018) Linear traction motor based on Backstepping theory[J].Proceedings of the Chinese Society for Electrical Engineering, 38(22):6711-6722.
[6] Chen, M., Wei, T. (2022) Design and application of speed sensor for traction motor of rail vehicle[J]. Innovation in science and technology, 20:42-45.
[7] You, B. T., Zhao, Z. N., Feng, J. H., Fu, C. C. (2021) Research on energy saving speed tracking control of high speed train considering traction motor power loss[J].Modern computer, 7:26-32.
[8] Hu, S. S. (2020) Principle of Automatic Control (7th Edition)[M]. Beijing: Science Press, 9
[9] Duo, E. F., Bi, X. P. (2015) Modern Control System (12th Edition)[M]. Beijing: Publishing House of Electronics Industry, 2.
[10] Zhang, L., Ren, X. Y. (2018) MATLAB and Control System Simulation [M]. Beijing: Publishing House of Electronics Industry, 8.

Stage division strategy for gas insulation surface discharge based on multi-physical information

Ke Zhao[1], Yujie Li[1,*], JingTan Ma[1], Hanyan Xiao[1], Tianxin Zhuang[1] and Yongfei Liu[1]

[1] State Grid Jiangsu Electric Power Company Research Institute Nanjing 211103, China

First author's e-mail: 15105168884@163.com

*Corresponding author's e-mail: liyj7@js.sgcc.com.cn

Abstract. The internal insulation defects in gas insulated equipment leading to partial discharges will affect the insulation performance of the equipment. Among them, surface discharges are a typical type of partial discharge, which often occurs at basin-type insulators. To realize effective discharge detection and discharge degree determination, a refined discharge stage division strategy for gas-insulated creeping discharge based on multi-physical information is investigated in this study. First, an experimental platform for gas-insulated multi-physical signal detection is established, of which high frequency current transformer (HFCT), silicon photomultiplier (SiPM), ultra-high frequency (UHF) and acoustic emission (AE) sensors are applied to measurement the multi-physical energy releases in the process of creeping discharge. Then an artificial surface discharge defect is developed following the actual surface discharge of basin insulator. Subsequently, the relative optical radiation power, electromagnetic radiation power and ultrasonic radiation power produced in the process of discharge are deduced from the multi-physical information. Meanwhile, the relationships among the three energy releases are carried out for variation of different energy with the applied voltage increases. Finally, development path of multi-physical energy ratio for surface discharge is proposed to evaluate the severity of discharge.

1. Introduction

Gas insulated switchgear (GIS) usually uses SF6 and other gases as the insulating medium, which has the advantages of high electrical strength, small floor area and less maintenance workload, so it has been widely used [1,2]. However, due to the influence of some inevitable factors in the production, transportation and assembly, long-term service and other processes of GIS equipment, there are sometimes insulation defects in the equipment, such as metal spikes, abnormal surface of insulating parts, suspended potential and so on, which bring varying degrees of potential safety hazards to the operating equipment. Among them, surface discharge is a common fault in GIS, which usually occurs in the position of the pot insulator of the equipment, and is generally caused by foreign metal objects or the defects on the surface of the insulator itself, and may even cause serious accidents such as flashover and breakdown, threatening the safe and stable operation of the power system.

At present, scholars at home and abroad have carried out a lot of research work in the field of surface discharge detection, and have made remarkable achievements in signal-to-noise separation, feature extraction, fault recognition and other aspects of discharge signals by using different measurement methods [3–6], which have been applied to the condition inspection and repair work at the production site. However, when condition based maintenance equipment is widely used, new problems also appear. Due to different measurement principles, various detection methods have great differences in detection sensitivity, effectiveness and scope of application for different working conditions, partial discharge types and discharge intensity [7]. For example, literature [8] found in experiments that if partial discharge occurs on the surface of basin insulator, due to the weak absorption of epoxy material for ultrasonic wave and the excitation effect of electromagnetic wave of this type of partial discharge, The sensitivity of UHF and ultrasonic detection method is significantly lower than that of pulse current method. Literature [6] compared the partial discharge detection characteristics of optical detection method and UHF method in switchgear, and found that optical method has higher sensitivity for the detection of surface discharge. On the other hand, existing research shows that although different detection methods have their own detection advantages, any single detection method has detection blind spots and cannot cover all types of partial discharge defects [9–11]. Therefore, how to effectively reuse the detection sensitivity range of different detection methods and realize the reasonable cooperation of various detection equipment in the application is the focus and difficulty in the field of partial discharge detection at present.

In view of the above problems, in order to make full use of the various physical energy information released by the surface discharge, reasonably configure the confidence intervals of various monitoring methods, and improve the sensitivity and accuracy of joint detection, a gas insulation simulation experiment platform is built, with built-in surface discharge defect simulation electrodes, and multi physical signal acquisition is realized through high frequency current transformer (HFCT) sensor, silicon photomultiplier (SiPM) sensor, ultra high frequency (UHF) antenna, narrow band acoustic emission (AE) sensor. The discharge patterns and time sequence characteristics of four PD measurements are analyzed. Then a refined discharge stage division strategy for the surface discharge of gas insulation equipment based on the multiple physical information is proposed in this study.

2. Experimental steps

2.1. Experimental platform

In order to study the change law of typical partial discharge types and phase related physical phenomena in gas insulated switchgear equipment, an experimental platform as shown in Figure 1 is built in this paper. The platform is composed of test transformer, SF6 inflation system, GIS simulation test chamber, typical partial discharge defect model, multi physical signal measurement system. It can simulate the internal gas environment and typical fault types of gas insulated electrical equipment. Besides, the high frequency current signal, optical signal, UHF signal and AE signal will be synchronously acquired by the multiple channels digital acquisition card. In the test circuit, a 1:2000 resistance capacitance voltage divider is used to measure the applied voltage on the test chamber, and the coupling capacitor is a 9.96nf standard capacitor with a withstand voltage of 100kV.

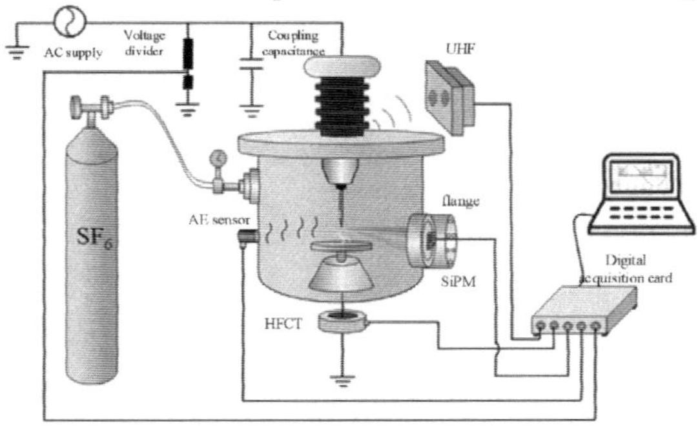

Figure 1. Gas insulation experimental platform with multiple sensors.

To our knowledge, surface discharge often occurs at the gas-solid interface of insulating medium such as the surface of pot insulator. Hence, we developed an artificial surface discharge defect, where the defect electrode model of surface discharge is shown in Figure 2. The defect model is composed of two parts: rod and plate, brass electrode and epoxy resin plate. The diameter of rod electrode is 6mm. The parameters of the plate electrode are the same as those of the corona discharge model; The epoxy resin plate is 2mm thick and 80mm in diameter, between the clamp and the rod and plate electrode.

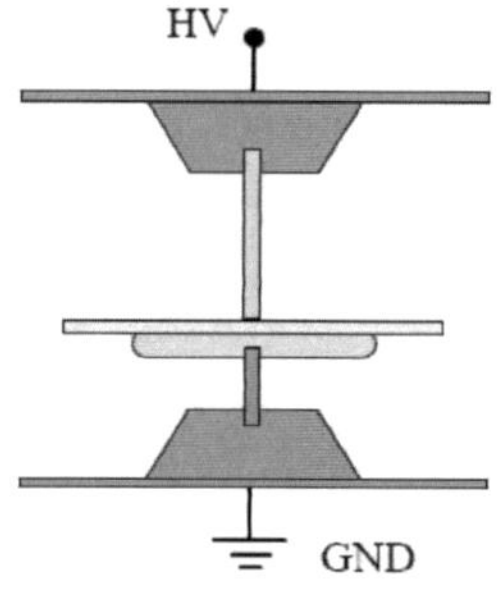

Figure 2. Surface discharge defect electrode model.

2.2. Sensor settings

The SiPM sensor is developed based on microFJ-30035-TSV SiPM units. Through the independent development of SiPM array composed of peripheral circuits, the sensor can effectively respond to optical signals in the 200-900mm band. UHF sensor adopts double helix structure, where the working bandwidth of receiving antenna is 300MHz-1.5GHz, and the standing wave ratio in the bandwidth is <1.5:1. Due to the high frequency of the original UHF signal, the detection and amplification module is connected to the rear end of the antenna, so that the original discharge pulse signal only retains the amplitude and phase information. Moreover, the narrow band AE sensor possesses detection frequency range is 20kHz-180kHz, and the peak sensitivity can reach more than -75db.

3. Research on characteristics and stage division of multi-physical information

3.1. Relationships among the multi-physical energy releases

The relationships among the relative light intensity, UHF PD magnitude and AE PD magnitude deduced by the PD signals within 0.2 s are illustrated in Figure 3. Besides, the cluster centers of the scatters under different applied energy are denoted by white markers. It can be observed that the relative intensity of each PD measurement represents a positive correlation exponential relationship with the remaining ones.

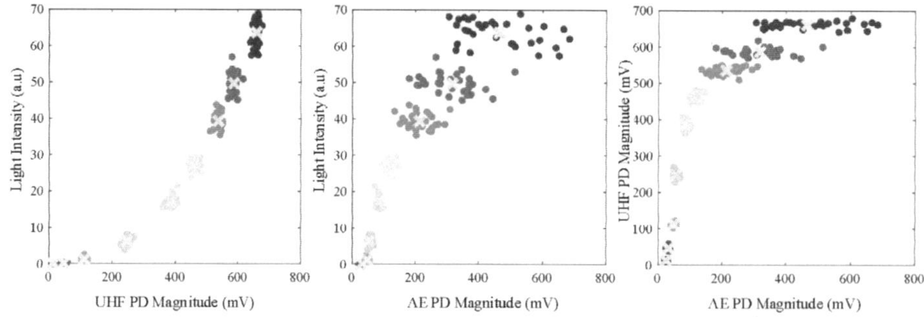

Figure 3. Relationships among the relative intensity of multi-physical detections.

To further reveal the correlation in the aspects of the multi-physical energy releases, the optical radiation power, electromagnetic radiation power and relative acoustic power are carried out with following formulas:

$$P_{LI} = \frac{\sum_i A_{\mathrm{LI}_i}}{\Delta t} \quad (1)$$

$$P_{UHF} = \frac{\sum_i A_{\mathrm{UHF}_i}^{\ 2}}{\Delta t} \quad (2)$$

$$P_{AE} = \frac{\sum_i A_{\mathrm{AE}_i}^{\ 2}}{\Delta t} \quad (3)$$

where A_{LI_i}, A_{UHF_i} and A_{AE_i} are the amplitudes of PD pulse with SiPM, UHF and AE sensors, Δt denotes the fixed sampling time for PD detection, which is 0.2 s in this study.

3.2. Discharge energy analysis

The development law and phased changes of surface discharge in the process of voltage rise, as well as the differentiation and analysis results of different severity of discharge are shown in Figure 3. It can be seen that in the initial development stage of surface discharge, the energy of each part increases exponentially with the increase of voltage, indicating that the electron avalanche intensity of partial discharge in this stage increases exponentially, and the discharge form is the same as corona discharge, which basically continues to 20kV. The apparent discharge and the relative power of acoustic radiation continue to change exponentially with the increase of voltage after 20kV, while the relative power of optical radiation and electromagnetic radiation changes linearly. This is because the discharge form has changed from the previous corona discharge form to the brush like sliding flash discharge. In this stage, the reverse electric field effect of the positive space charge generated by the electron avalanche and remaining on the gas-solid interface surface becomes more and more obvious, and leads to the gradual forward shift of the discharge phase.

At the same time, the polarity effect caused by space charge also leads to the shape feature that the positive half cycle pulse amplitude is greater than the negative half cycle in the pulse current spectrum. The phenomenon that the discharge amplitude increases with phase in the half cycle of AC voltage in the spectrum is also related to the recovery effect caused by residual positive ions in the streamer channel after discharge [12]. In addition, according to literature [13], there are obvious differences in the morphology of the positive and negative half cycle discharge channels along the surface discharge. There are multiple discharge branches in the positive half cycle discharge channel, while the negative half cycle discharge area is relatively concentrated. Therefore, the photons generated by the negative half cycle discharge in the test are easier to be detected by the sensor, resulting in the phenomenon of larger amplitude of the negative half cycle light pulse in the optical signal spectrum. In the final stage of exponential growth of discharge energy, the intensity of discharge increases much faster with voltage than before, so most of the increment of discharge intensity in the whole discharge development process occurs in the range of 27KV and above.

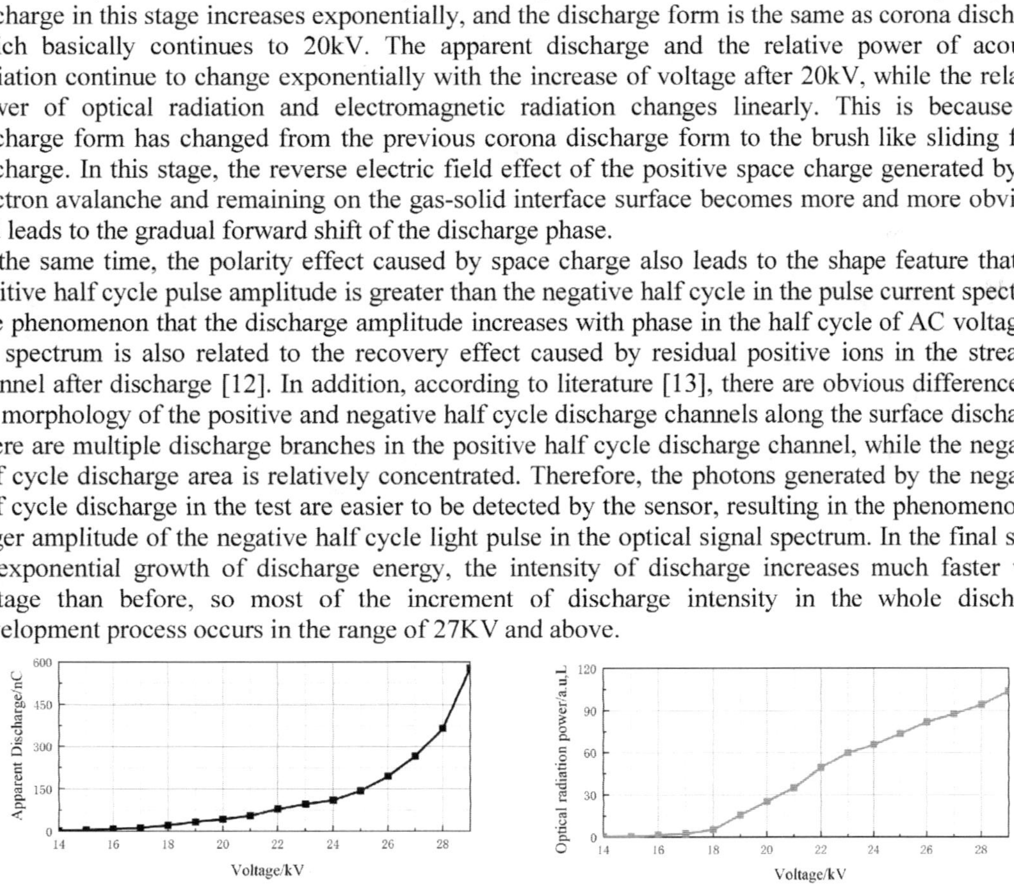

(a) Apparent discharge (b) Optical radiation power

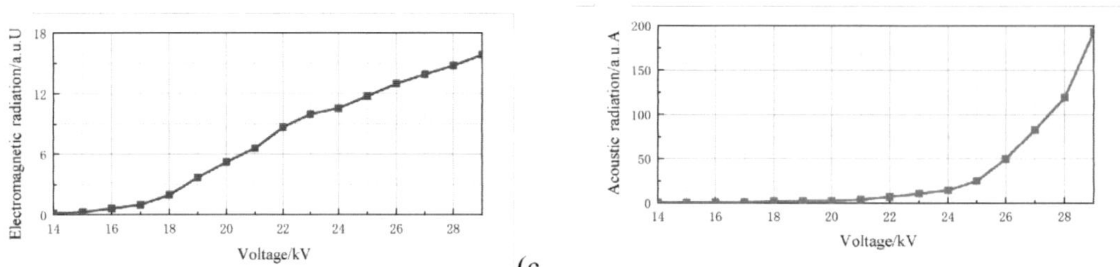

(c) Electromagnetic radiation power (d) Acoustic radiation power

Figure 4. Relative physical energy change.

3.3. Stage division

To sum up, the development process of surface discharge in the test can be divided into three stages: 14kv-20kv, 21kv-26kv and 27kv-29kv, which are called type I surface discharge, type II surface discharge and type III surface discharge respectively.

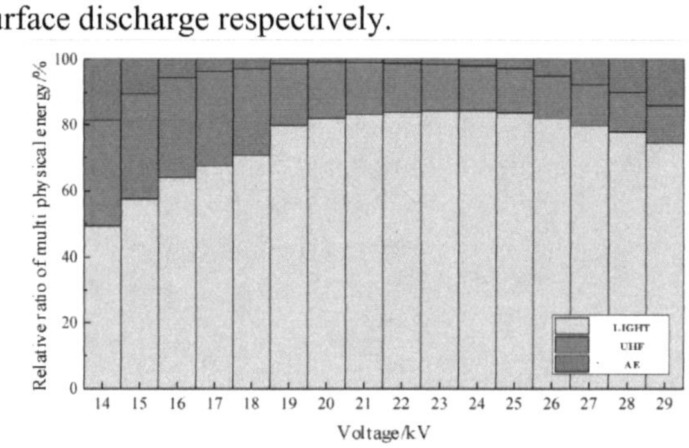

Figure 5. Relative proportion of multi physical energy.

Furthermore, ternary diagram is employed to better visualize the relationship between the three relative energy releases power in two-dimensional rectangular coordinates, as illustrated in Fig. 6. With which, the relative proportion of physical energy in the discharge development process can be clearly expressed, and the proportion of relative energy and power in the process of surface discharge has an obvious development path. Additionally, it can be seen that in type I surface discharge, the growth rate of optical radiation power is faster, while the growth rate of electromagnetic radiation and acoustic radiation is slower, resulting in a decrease in the proportion. In type II surface discharge, the growth rates of the three physical energies are relatively consistent, so the relative proportion remains basically unchanged. In type III surface discharge, the growth rate of acoustic radiation power is higher than the other two.

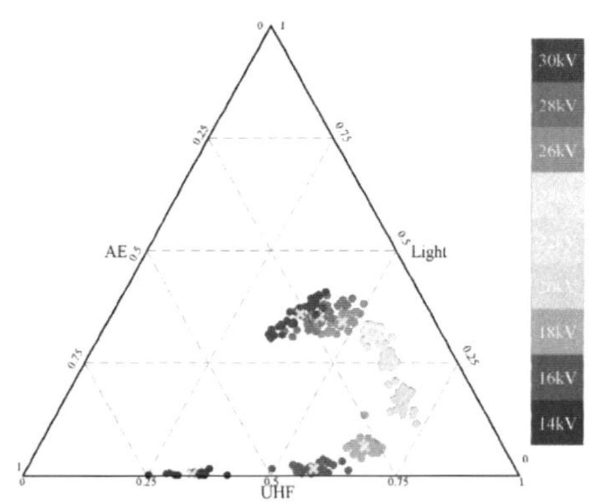

Figure 6. Ternary diagram for the proportion of multi-physical relative power.

4. Conclusion

In this paper, based on the developed experimental platform, defect simulation and sensor setting, the multi physical signal acquisition of surface discharge is realized. Then the characteristics of multi physical signals are analysed and their energy changes are calculated, with which the discharge stages are divided based on the energy analysis results, and the fine diagnosis of surface discharge is realized by simplifying the feature data set. The specific conclusions are as follows:

(1) An experimental platform for gas-insulated multi-physical signal detection is established, where the artificial surface discharge defect is established according to actual discharge conditions. Based on the platform, the multi-physical PD detections including high-frequency current, optical, ultra-high frequency, and ultrasonic emission can be synchronously acquired.

(2) With the increase of applied voltage, the phase patterns of high-frequency current, optical and UHF signals appear to increase in amplitude, pulse forward and phase, and the peak frequency of the ultrasonic signal spectrum also moves forward with the applied voltage.

(3) The proportion of optical radiation power, electromagnetic radiation power and acoustic radiation power in each stage of discharge development was analysed. In the initial stage of the discharge, the energy intensity of each relative power increased exponentially with the applied voltage. In the later stage of the discharge, the relative power of light and electromagnetic waves turned into a linear trend; according to the proportion of power distribution in the discharge process, the surface discharge was divided into three stages.

Acknowledgments

The work was supported by the Science and Technology Program of the State Grid Corporation of China (J2021067).

References

[1] Zheng, S., Wu, S. (2021) Detection Study on Propagation Characteristics of Partial Discharge Optical Signal in GIS. IEEE Trans. Instrum. Meas., 70:7005912.

[2] Ren, M., Zhang, C., Dong, M., He, Z. (2020) Fault prediction of gas-insulated system with hypersensitive optical monitoring and spectral information. Int. J. Electr. Power Energy Syst., 119: 105945.

[3] Burow, S., Straumann, U., Koehler, W., Tenbohlen, S. (2014) New Methods of Damping Very Fast Transient Overvoltages in Gas-Insulated Switchgear. IEEE Trans. Power Deliv., 29: 2332–2339.

[4] Piccin, R., Mor, A. R,, Morshuis, P., Girodet, A., Smit, J. (2015) Partial Discharge Analysis of Gas Insulated Systems at High Voltage AC and DC. IEEE Trans. Dielectr. Electr. Insul., 22:218–228.

[5] Gupta, N., Ramu, T. S. (1998) Estimation of partial discharge parameters in GIS using acoustic emission techniques: A theoretical approach. J. Comput. Acoust., 6: 403–419.

[6] Ren, M., Song, B., Zhuang, T., Yang, S. (2018) Optical partial discharge diagnostic in SF6 gas insulated system via multi-spectral detection. ISA Trans., 75: 247–257.

[7] Okabe, S., Ueta, G., Wada, H., Okubo, H. (2010) Deterioration Characteristics Due to Partial Discharges in Insulating Structure Constituting Gas-filled Power Transformers. IEEE Trans. Dielectr. Electr. Insul., 17: 1204–1213.

[8] Xing, Y., Wang, Z., Liu, L., Xu, Y., Yang, Y., Liu, S., et al. (2022) Defects and failure types of solid insulation in gas - insulated switchgear: In situ study and case analysis. High Volt., 7: 158–164.

[9] Reid, A. J,, Judd, M. D., Duncan, G. (2012) Simultaneous measurement of partial discharge using TEV, IEC60270 and UHF techniques. In: 2012 IEEE Int. Symp. Electr. Insul. San Juan. pp. 439–442.

[10] Reid, A., Judd, M. D,, Fouracre, R. A., Stewart, B., Hepburn, D. (2011) Identification of

simultaneously active partial discharge sources using combined radio frequency and IEC60270 measurement. IET Sci. Meas. Technol., 5: 102–108.

[11] Reid, A. J., Judd, M. D., Fouracre, R. A., Stewart, B. G., Hepburn, D. M. (2011) Simultaneous measurement of partial discharges using IEC60270 and radio-frequency techniques. IEEE Trans. Dielectr. Electr. Insul., 18: 444–455.

[12] Lu, S., Chai, H., Sahoo, A., Phung, B. T. (2020) Condition Monitoring Based on Partial Discharge Diagnostics Using Machine Learning Methods: A Comprehensive State-of-the-Art Review. IEEE Trans. Dielectr. Electr. Insul., 27: 1861–1888.

[13] Song, H., Zhang, Z., Tian, J., Sheng, G., Jiang, X. (2022) Multiscale Fusion Simulation of the Influence of Temperature on the Partial Discharge Signal of GIS Insulation Void Defects. IEEE Trans. Power Deliv., 37: 1304–1314.

Adopting integration of multispectral time resolved analysis and graph-based deep learning method in partial discharge type identification

Changjie Xia[1*], Ming Ren[1], Kai Wang[1], Hongyuan Zhang[1], Haobin Guan, Ming Dong[1], Tao Zhang[2] and Jin Miao[2]

[1] State Key Laboratory of Electrical Insulation and Power Equipment, Xi'an Jiaotong University, Xi'an, Shaanxi, 710049, China

[2] Wuxi Electric Power Supply Company, State Grid Jiangsu Electric Power Co, LTD, Wuxi, Jiangsu, 214062, China

*Corresponding author's e-mail: xcj19960521@stu.xjtu.edu.cn

Abstract. Partial discharge (PD) is treated as one of the major threats for gas insulated switchgear (GIS). By using the new generation multispectral detection sensor named as SiPM-based multispectral discharge sensor (SMDS), the time resolved partial discharge with multispectral information (named as MTRPD) for creeping discharge, suspension discharge and tip discharge, respectively. It indicates that the MTRPD for the specific discharge defect perform unique spectral fingerprints in discharge mode. Based on the graph characteristics of MTRPD, we introduced the convolution neural network (CNN) to implement PD type identification whose overall accuracy of Δqi-Δqi+1 and Δti-Δti+1 were 99.7% and 98.9%, respectively. This paper provides a new technique tool for fine diagnosis of PD independent of phase analysis.

1. Introduction

With the advantages of compact structure, high reliability and strong security, gas insulated switchgear (GIS) has been widely applied in power system [1,2]. However, the insulation defects will be inevitably formed and cause the partial discharge (PD) which further aggravates the insulation deterioration during the operation process of GIS [3,4]. At the same time, PD detection also takes an important role in insulation examination as well as fault diagnosis. Following the accompanying phenomena of PD, e.g. charge transfer, thermal expansion, electromagnetic radiation and light radiation, scholars have developed pulse current method, acoustic emission method, ultra-high frequency method and light detection method for PD detection [5]. Nevertheless, the electromagnetic and sound interference bring challenges for the methods based on electromagnetic wave and sound wave detection [6,7]. Light radiation occurs in the whole PD process of emission, ionization, attachment, recombination and dissipation. As an intrinsic and intuitive means of PD characterization, optical method performs the excellent outstanding advantages in anti-interference, characterization, and confidence. To realize the PD fine diagnosis via optical method, there were two problems needed to be solved: (1) How to realize hypersensitive detection of the PD light radiation? (2) How to build the PD fine diagnosis method based on the light radiation characteristics?

For the first problem, the optical pluses of PD with low radiation intensity could only last few picoseconds even nanoseconds, which bring the performance challenges for optical detection sensors [8]. Limited by the size, driving voltage, working life and quantum efficiency of the traditional optical detection sensors, optical method mainly stays in the laboratory research stage, let alone the PD spectral detection in the filed [9]. Benefiting from the rapid development of a new generation of solid-state silicon photoelectric sensors, the integration of the multi-channel silicon photomultiplier (SiPM) and optical coated filters, named SiPM-based multispectral discharge sensor (SMDS), has been adopted to implement the efficient and practical PD spectral detection in the previous works [10–12].
For the second problem, due to the rich information of PD light radiation in wavelength dimension, besides the traditional statistic characteristics, e.g. phase resolved partial discharge patterns (PRPD), PD spectrum brings a new technique approach to explore the physical process of discharges. Based on the SMDS, literature [8,13] proposed the ternary spectral bands analysis method and multispectral ratio analysis method in PD fine diagnosis, respectively.
In this paper, we are committed to developing an integration method of the statistic method and multispectral analysis method to describe PD process. On this basis, a convolutional neural network was adopted to implement the pattern recognition of PD types.

2. Test methods

2.1. Test facilities

To realize the synchronous acquisition for multispectral information of PD, literature [10,13] introduced a SMDS based on the SiPM which integrates thousands of single-photon avalanche diode (SPAD) which stay in Geiger avalanche mode. When the photon enters the SPAD, the avalanche diode will be activated and converting the incident light into avalanche current. In this paper, referring to the design of the SMDS in reference [13], a SMDS with ultraviolet (UV)-, visible (VIS)-, near infrared (NIR)- and full- band were employed as PD detection sensor. The basic structure of SMDS system is showed in Fig. 1. And Table 1 lists the filter transmittance of each spectral channel.

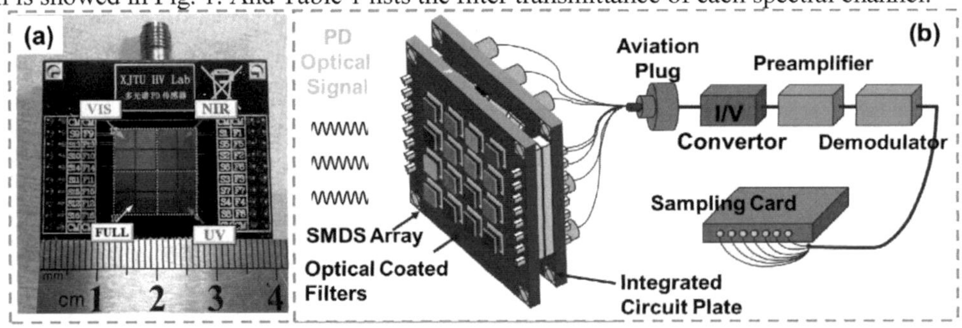

Figure 1. Basic structure of SMDS system. (a) Real picture of SMDS. (b) Block diagram of SMDS system.

Table 1. Filter Transmittances of SMDS.

Spectral Band	Central Wavelength/nm	Central Transmittance/%	Spectral Range/nm
UV	325	>73	251~398
VIS	560	>80	300~710
NIR	>700	>90	710~900
Full	550	-	200~900

2.2. Test platform

The established PD test platform consists of the high voltage loop and the signal acquisition system, which is shown in Fig. 2 (a). A test transformer with the voltage ratio of 380V/100kV was employed as the power supply. The protective resistance was used to limit the loop current and protect the equipment. A resistance capacitance divider with voltage ratio of 1:2000 was applied to measure the output voltage of the power supply. A coupling capacitance provided a low impedance circuit. Three typical PD defects in GIS were simulated in a gas insulation switch module with 0.3 MPa SF_6, that is creeping discharge, suspension discharge and tip discharge, as shown in Fig. 2 (b)-(d). The signal acquisition system consists of a high frequency current transformer (HFCT) with sensitivity of 5V/A and bandwidth of 400 kHz~250MHz, as well as a SMDS were adoptd to detect the optical signal and pulse current signal for PD, respectively.

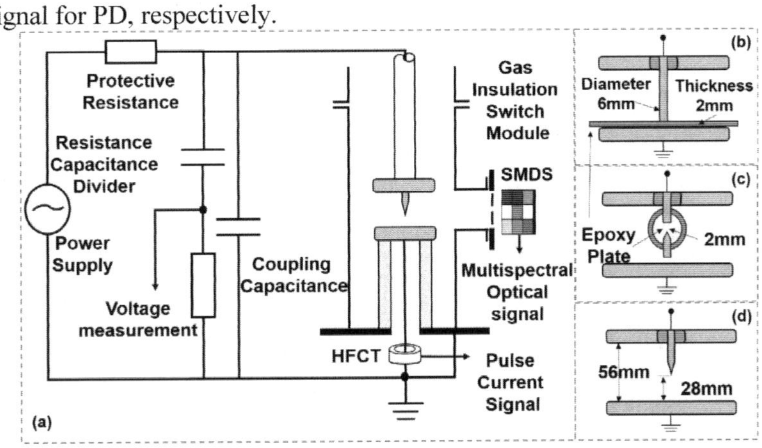

Figure 2. PD test system. (a) Optic-electric synchronous test platform. (b) Creeping discharge. (c) Suspension discharge. (d) Tip discharge.

2.3. Test procedure
To implement the rapid acquisition of a good deal of discharge data in a short time, step voltage rise method was carried out in this paper. The specific test procedures are described as following:

- Determination of noise level for each spectral channel and removal noise via thresholds.
- Measurement of PD initiation voltage (PDIV) and breakdown voltage U_b.
- Acquisition of the PD multispectral data at the applied voltage in the range from 1.1 PDIV to U_b with the increased 0.1 PDIV step.

3. Multispectral time resolved partial discharge characteristics and type identification

3.1. Multispectral time resolved partial discharge
TRPD, the abbreviation for time resolved partial discharge, which reveals the development of PD physical process with time series. As with the PRPD, TRPD can characterize discharge parameters by characteristic quantity map, while TRPD is a phase independent method. Reference [10,14] has proved that the SMDS performs an excellent liner relationship between the response values of PD light pulses and the apparent discharge capacity, therefore the apparent discharge capacity can be characterized by the light intensity. Fig. 3 displayed the time series of PD multispectral pulses in two power frequency cycles for tip discharge. It can be found that, although the intensity of light pulses varies in each band, which is determined by the discharge mechanism and development process, the pulses of each band perform excellent synchronization. Therefore, the multispectral characteristics of discharge pulses can be acquired by extracting the light intensity of each band at the same time.

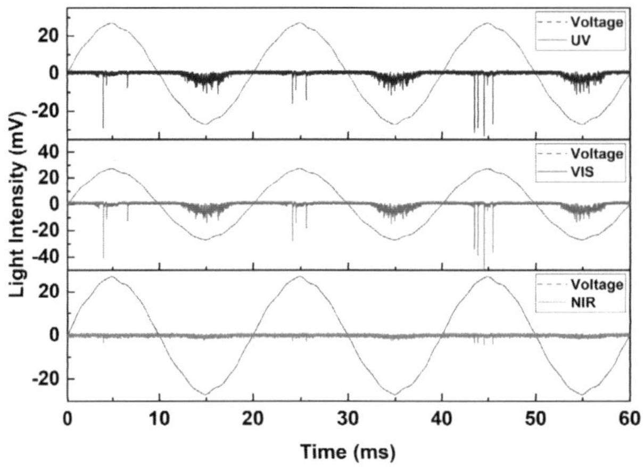

Figure 3. The time series of PD multispectral pulses in two power frequency cycles for tip discharge.

According to the classical partial discharge theory, the two basic conditions for the occurrence of PD are the high enough applied voltage that is higher than minimum breakdown voltage U_{min} of the dielectric to maintain self-sustaining discharge, and the existence of initial electrons in defects. As the external applied voltage increase, the voltage on the defects U_d will increase. The first PD occurs when U_d is higher than U_{min} and initial electrons are present. Then, after the first PD, U_d will drop below U_{min} and rise again under the applied voltage. The second PD will occur when U_d is higher than U_{min} again and initial electrons are again present. As the matter of fact, because of the changing applied voltage and polarity reversal, the actual discharge process is more complicated. Due to the difference in each charge time and the time to meet the initial electronic condition, U_d is not equal each time PD occurs which leads to the discharge amount difference of each PD pulse. It can be found that, PD characteristics are highly correlated with time series, therefore, two TRPD characteristics i.g. Δq_i-Δq_{i+1} and Δt_i-Δt_{i+1} were extracted, respectively, and fused them with multispectral information (named Multispectral TRPD, MTRPD) in this paper. Δq_i-Δq_{i+1} is defined as the discharge amplitude difference between the last discharge-current discharge and current discharge-next discharge, and Δt_i-Δt_{i+1} is defined as time interval of last discharge-current discharge and current discharge-next discharge. MTRPD of three discharge models were formed multispectral statistical analysis spectrograms in this paper.

The normalized Δq_i-Δq_{i+1} multispectral statistical analysis spectrograms for three were established in Fig. 4. Fig. 4 (a) to (c) reveal the Δq_i-Δq_{i+1} for creeping discharge (1.5 PDIV), suspension discharge (1.5 PDIV) and tip discharge (1.5 PDIV), respectively, and (-i) to (-iii) represent the Δq_i-Δq_{i+1} in UV, VIS and NIR, respectively. As Fig. 4 (a) shown, for the creeping discharge, Δq_i-Δq_{i+1} spectrograms perform a tendency to radiate from the center to the upper left, lower and right, whose shapes like "triangles". And the points are relatively more scattered in VIS. As Fig. 4 (b) shown, for the suspension discharge, Δq_i-Δq_{i+1} spectrograms were shaped like 'turtlebacks', which forms five groups. As Fig. 4 (c) shown, for the tip discharge, Δq_i-Δq_{i+1} spectrograms were shaped like 'stars'. It is worth mentioning that, although the images in each spectral band are very similar, there are still differences, especially in the distribution behaviour of points, which also provides a new feature quantity for the subsequent discharge type identification.

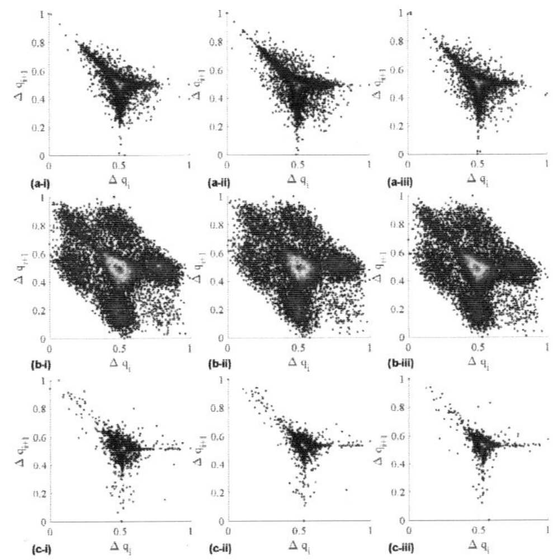

Figure 4. Normalized Δq_i-Δq_{i+1} multispectral statistical analysis spectrograms. (a) to (c) reveal the Δq_i-Δq_{i+1} for creeping discharge, suspension discharge and tip discharge, respectively, and (-i) to (-iii) represent the Δq_i-Δq_{i+1} at UV, VIS and NIR, respectively.

The normalized Δt_i-Δt_{i+1} multispectral statistical analysis spectrograms for three were established in Fig. 5. Fig. 5 (a) to (c) reveal the Δt_i-Δt_{i+1} for creeping discharge (1.5 PDIV), suspension discharge (1.5 PDIV) and tip discharge (1.5 PDIV), respectively, and (-i) to (-iii) represent the Δq_i-Δq_{i+1} in UV, VIS and NIR, respectively. As Fig. 5 (a) shown, for the creeping discharge, Δt_i-Δt_{i+1} spectrograms mainly concentrate at three groups where the coordinates are (0, 0), (0, 0.8) and (0.8, 0). And the points are relatively more scattered at UV and VIS. As Fig. 5 (b) shown, for the suspension discharge, Δt_i-Δt_{i+1} spectrograms are mainly concentrated in the triangular region of coordinates (0, 0), (0, 0.6) and (0.5, 0). Similar with the creeping discharge, as Fig. 4 (c) shown, Δt_i-Δt_{i+1} spectrograms of tip discharge also perform the clustered distribution. It is worth mentioning that, there are few points distribute at coordinate (1, 1) in UV and VIS, but in NIR band, it is obvious that four groups are mainly formed and their distribution coordinates are significantly different from those of the first two bands.

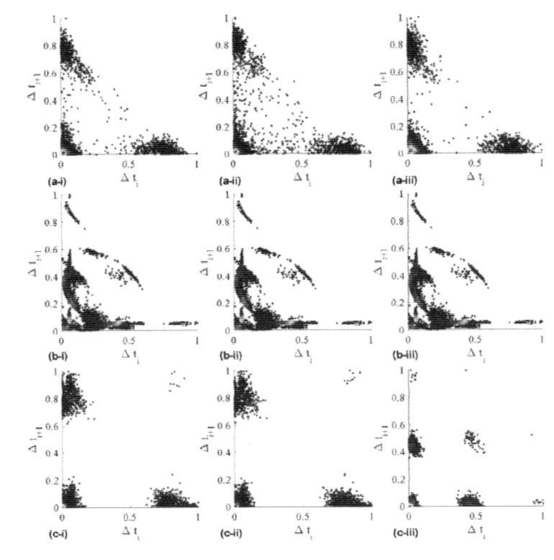

Figure 5. Normalized Δt_i-Δt_{i+1} multispectral statistical analysis spectrograms. (a) to (c) reveal the Δt_i-Δt_{i+1} for creeping discharge, suspension discharge and tip discharge, respectively, and (-i) to (-iii) represent the Δq_i-Δq_{i+1} at UV, VIS and NIR, respectively.

3.2. Pd type identification

The identification of PD type is very important for PD diagnosis and has attracted a lot of researchers' attention. Feature parameter extraction and pattern classification method are the keys to PD pattern recognition. In this paper, the aforementioned MTRPD spectrograms of each discharge show the excellent specificity and separability, therefore, the key of the study is how to establish a pattern classification method based on MTRPD. The development of deep learning can learn the deep feature information of samples from data, and has achieved good results in image recognition, speech recognition and other fields [15–17]. Therefore, considering the graph characteristics of MTRPD proposed in this paper, we introduced a convolution neural network (CNN) to realize the PD type identification.

CNN is a kind of feedforward neural network with deep structure and convolution computation, which has the ability of representation learning. CNN can shift-invariant classification of input information according to its hierarchical structure. In this paper, taking MTRPD as input, a CNN network with two convolutional layers, two pooling layers and two full connection layers was established for discharge type identification as shown in Fig. 6. The parameters of PD type, MTRPD type, training set and test set are listed in Table. 2.

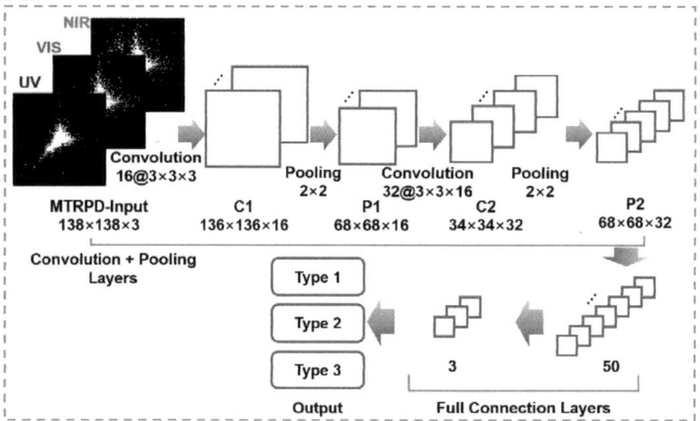

Figure 6. CNN Network Architecture.

Table 2. The parameters of PD type, MTRPD type, training set and test set.

PD Type	MTRPD Type	Training Set	Test Set
Creeping Discharge	Δq_i-Δq_{i+1}	200	300
	Δt_i-Δt_{i+1}	200	300
Suspension Discharge	Δq_i-Δq_{i+1}	200	300
	Δt_i-Δt_{i+1}	200	300
Tip Discharge	Δq_i-Δq_{i+1}	200	300
	Δt_i-Δt_{i+1}	200	300

The trained CNN was used to classify and predict the test set, and the confusion matrix of classification results of CNN was obtained, as shown in Table. 3.

Table 3. The confusion matrix of classification results of CNN

MTRPD	Type	Creeping	Suspension	Tip
Δq_i-Δq_{i+1}	Creeping	298	0	2
	Suspension	0	300	0
	Tip	1	0	299

Δt_i-Δt_{i+1}	Creeping	296	0	4
	Suspension	0	300	0
	Tip	6	0	294

Through the calculation of data in Table. 3, compared with Δt_i-Δt_{i+1}, Δq_i-Δq_{i+1} performs the higher identification accuracy, the overall accuracy of Δq_i-Δq_{i+1} and Δt_i-Δt_{i+1} were 99.7% and 98.9%, respectively. Among them, suspension discharge achieved 100% classification accuracy both in two MTRPD CNN models. And there were a few misclassifications in creeping discharge and tip discharge.

4. Conclusion

In this study, we proposed an integration method of TRPD and multispectral information to characterize the three typical PDs and applied CNN to implement PD type identification. At first, by using a hypersensitive SMDS, the MTRPD, i.g. Δt_i-Δt_{i+1} and Δq_i-Δq_{i+1} for three PDs were analyzed. The results showed that MTRPD with PD types performed specificity and separability, which provides a basis for pattern recognition. Based on the graph characteristics of MTRPD, we introduced the CNN to implement PD type identification whose overall accuracy of Δq_i-Δq_{i+1} and Δt_i-Δt_{i+1} were 99.7% and 98.9%, respectively. The focus of subsequent studies will be on expanding the data set, trying to introduce new TRPD characteristic quantities, and at the same time carrying out research on PD risk assessment.

Acknowledgments

The work was supported by the Science and Technology Program of the State Grid Corporation of China (5500-202218131A-1-1-ZN).

References

[1] Ren, M., Zhou, J and Miao, J. (2021) Adopting Spectral Analysis in Partial Discharge Fault Diagnosis of GIS With a Micro Built-in Optical Sensor IEEE Trans. Power Delivery 36 1237–40

[2] Kontani, Y., Kobayashi, T., Tamura, Y., Buhr, T and Thompson, P. M. (2012) Gas insulated transformer application for an environmentally-friendly power station upgrade PES T&D 2012 2012 IEEE/PES Transmission and Distribution Conference and Exposition (T&D) (Orlando, FL, USA: IEEE) pp 1–4

[3] Ren, M., Song, B., Zhuang, T and Yang, S. (2018) Optical partial discharge diagnostic in SF6 gas insulated system via multi-spectral detection ISA Transactions 75 247–57

[4] Yin, X., Dong, L., Wu, H., Zhang, L., Ma, W., Yin, W., Xiao, L., Jia, S and Tittel, F. K. (2019) Highly sensitive photoacoustic multicomponent gas sensor for SF6 decomposition online monitoring Opt. Express, OE 27 A224–34

[5] Yaacob, M. M., Alsaedi, M. A., Rashed, J. R., Dakhil, A. M and Atyah, S. F. (2014) Review on partial discharge detection techniques related to high voltage power equipment using different sensors Photonic Sens 4 325–37

[6] Xia, C., Liu, L., Shu, L., Hu, Q and Qiao, X. (2018) Evaluation of Operating Insulator's Hot-Water Deicing Safe Distance under 110 kV Proceedings of the International Conference on Information Technology and Electrical Engineering 2018 ICITEE '18 (New York, NY, USA: Association for Computing Machinery) pp 1–7

[7] Li, Y., Ren, M., Wang, B., Zhang, C and Dong, M. (2021) Research on Acousto-optic Joint Detection and Visual Diagnosis Method of Power Equipment 2021 International Conference on Networking, Communications and Information Technology (NetCIT) 2021 International Conference on Networking, Communications and Information Technology (NetCIT) pp 327–31

[8] Ren, M., Zhang, C., Dong, M and He, Z. (2020) Fault prediction of gas-insulated system with hypersensitive optical monitoring and spectral information International Journal of Electrical

Power & Energy Systems 119 105945

[9] Baug, A., Choudhury, N. R., Ghosh, R., Dalai, S and Chatterjee, B. (2017) Identification of single and multiple partial discharge sources by optical method using mathematical morphology aided sparse representation classifier IEEE Transactions on Dielectrics and Electrical Insulation 24 3703–12

[10] Ren, M., Zhou, J., Song, B., Zhang, C., Dong, M and Albarracín, R. (2017) Towards Optical Partial Discharge Detection with Micro Silicon Photomultipliers Sensors 17 2595

[11] Peng, Z., Yan, B., Cheng, Z., Yue, D., Ma, Y., He, S., Hong, Z., Wang, S and Xia, C. (2020) Study on the Validity of SiPM for Partial Discharge Detection in Power Equipment 2020 IEEE International Conference on Artificial Intelligence and Computer Applications (ICAICA) 2020 IEEE International Conference on Artificial Intelligence and Computer Applications (ICAICA) pp 1026–30

[12] Peng, Z., Yan, B., Cheng, Z., Yue, D., Ma, Y., He, S., Hong, Z., Li, X and Xia, C. (2020) Consistency Verification of SiPM and HFCT for detecting partial discharges 2020 Asia-Pacific Conference on Image Processing, Electronics and Computers (IPEC) 2020 Asia-Pacific Conference on Image Processing, Electronics and Computers (IPEC) pp 340–3

[13] Xia, C., Ren, M., Chen, R., Yu, J., Li, C., Chen, Y., Wang, K., Wang, S and Dong, M. (2022) Multispectral Optical Partial Discharge Detection, Recognition, and Assessment IEEE Transactions on Instrumentation and Measurement 71 1–11

[14] Ren, M., Zhou, J., Yang, S., Zhuang, T., Dong, M and Albarracín, R. (2018) Optical Partial Discharge Diagnosis in SF6 Gas-Insulated System With SiPM-Based Sensor Array IEEE Sensors Journal 18 5532–40

[15] Xia, C., Ren, M., Wang, B., Dong, M., Xu, G., Xie, J and Zhang, C. (2021) Infrared thermography - based diagnostics on power equipment: State - of - the - art High Voltage 6 387–407

[16] Wang, B., Dong, M., Ren, M., Wu, Z., Guo, C., Zhuang, T., Pischler, O and Xie, J. (2020) Automatic Fault Diagnosis of Infrared Insulator Images Based on Image Instance Segmentation and Temperature Analysis IEEE Transactions on Instrumentation and Measurement 69 5345–55

[17] Xia, C., Ren, M., Wang, B., Dong, M., Song, B., Hu, Y and Pischler, O. (2021) Acquisition and analysis of hyperspectral data for surface contamination level of insulating materials Measurement 173 108560

Study on superconductivity of strongly correlated electronic systems by high voltage method

Chang Cheng*

School of Materials Science and Engineering, Anhui University, Hefei, China, 230601

*b19201042@stu.ahu.edu.cn

Abstract. The study of condensed matter system with strongly correlated electron characteristics shows that the strongly correlated electron leads the system to have macroscopic quantum characteristics, which has multiple degrees of freedom in practical application and strong coupling relationship between them, resulting in abundant and peculiar quantum phenomena in the system. Nowadays, the research on unconventional superconductors is more and more in-depth, which not only expands the topics and ideas of practical research, but also makes excellent achievements. Therefore, on the basis of understanding the strongly correlated electron system, this paper studies the deep analysis of Fermion superconductor, copper oxide superconductor and iron based superconductor by means of high pressure experiment, so as to provide effective information for better understanding the microscopic mechanism on the basis of mastering the relevant unconventional superconductor refined research methods.

1. Introduction

Unconventional superconductors have strong electron correlation, and the complex interaction between charged particles contained in them will produce a large number of quantum states and macroscopic quantum phenomena, which can not only help researchers better understand the strongly correlated electron system, but also provide a new basis for practical research. Since the traditional B C S microscopic theory is unable to comprehensively analyze the core mechanism of unconventional superconductors, the key understanding of superconductivity in condensed matter physics theory is faced with major challenges, as shown in Figure 1 below [1-2].

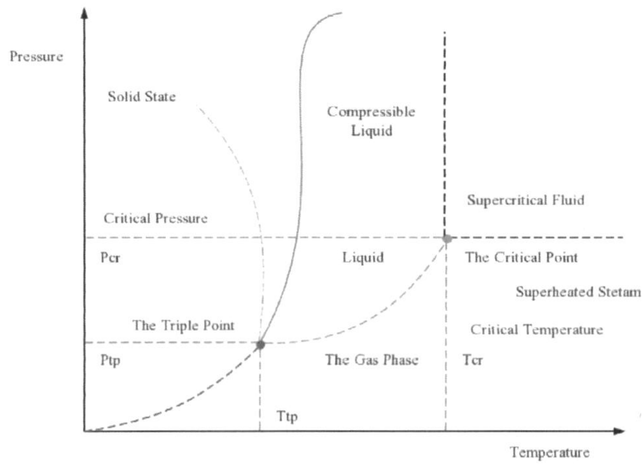

Figure 1. Theory of condensed matter physics

In the late 1970s, the discovery of CeCu2Si2, a superconductor with heavy cost, expanded the research field of unconventional superconductivity. In the mid-1980s, the copper oxide high temperature superconductor was proposed, which deepened the research intensity of high temperature superconductor and provided effective basis for the material research of high temperature superconductor. In 2008, iron based superconductor was proposed not only to expand the research direction of high temperature superconductor, [3-4] but also to create a new opportunity for the application of unconventional superconductor. In the development of economic construction in practice, people pay more attention to study the performance of the three types of unconventional superconductors, and expects to have more important information about the unconventional superconductors incentive, thus put forward a lot of research in practice analysis technology and measurement method, and based on the research needs to build a diversified superconductivity experiment, Controllable loading under comprehensive extreme conditions such as ultra-high pressure is one of the most important contents in the study of superconductivity. According to the analysis of the research results of fermion superconductor, copper oxide superconductor and iron based superconductor, it can be seen that the corresponding mechanism researches all involve strongly correlated electron system. In typical solid electronic theory, the use of landau theory of Fermi liquid can effectively simplify the solid body problem, all the atomic valence electrons can be freedom movement, and after integralization again, ignore the effects of the coulomb repulsion between them and transport properties of systems can be seen as almost independent quasiparticle weak interaction, The electron energy band theory based on nearly independent quasiparticle approximation gives a comprehensive analysis of solid systems. However, after the introduction of new materials such as pang magnetoresistance and high temperature superconductivity, it is found that their extraordinary physical properties can not be analyzed by traditional theories. In these materials, the neighboring electrons can be regarded as the nearest neighbors, and the interaction between them is strong, so they cannot be considered as separate electrons. In this process, the electronic - electronic interaction occupies the dominant position, and the differences in the traditional sense of the free electrons way, so these systems will be as strong associated electronic systems, although related materials with relatively broad application development space, but also provides scientific research the technology research of scholars with higher requirements. Therefore, starting from fermion superconductor, copper oxide superconductor and iron based superconductor, this paper discusses the superconductivity of strongly correlated electronic system deeply by using high pressure experimental method, so as to provide effective basis for practical technical research [4-5].

Journal of Physics: Conference Series **2387** (2022) 012023 doi:10.1088/1742-6596/2387/1/012023

2. Method

2.1. Fermion superconductors

This content belongs to the representative strongly correlated electron system, and the materials included are divided into two categories, one is the conductive electron with itinerant property, the other is the magnetic cation with f electron or D electron locally. Different types of heavy Fermi subsystems are the result of the interaction between travelling electrons and local electrons, which can eventually construct more abundant ground states and exhibit diversified quantum criticality behaviors. [6-8]After nearly 40 years of research and development, the current international scholars have put forward more than 40 heavy fermion superconductors, and most of all with characteristics and antiferromagnetic long program, in other words superconducting phase diagram of superconductivity is usually reflected in the boundary of the antiferromagnetic perimeter, therefore antiferromagnetic customize fluctuation can be regarded as one of the driving mechanism of Fermi system is the longest in.

On the one hand, the study of the piezoinduced superconductivity of CeRhGe3 and the superconductivity of CeTX3 system shows that this kind of compound is likely to have singlet and triplet mixing phenomenon due to the lack of antisymmetric spin-orbit coupling caused by the space inversion center, so it can be regarded as having p-wave superconducting paired component. In this study, Wang et al. conducted a deep study of CeRhGe3 compound using a variety of high-pressure measurements. According to the experimental results of high pressure in-situ resistance and AC susceptibility, CeRhGe3 compound can produce superconducting transition at 19GPa pressure, which is the last compound in CeTX3 family to achieve superconducting transition. According to the phase diagram analysis following pressure changes as shown in FIG. 2 below, in the range of 19GPa ~ 21GPa, the antiferromagnetic transition temperature does not change with pressure rise, but the superconducting transition temperature Tc will achieve the maximum value[9-10].

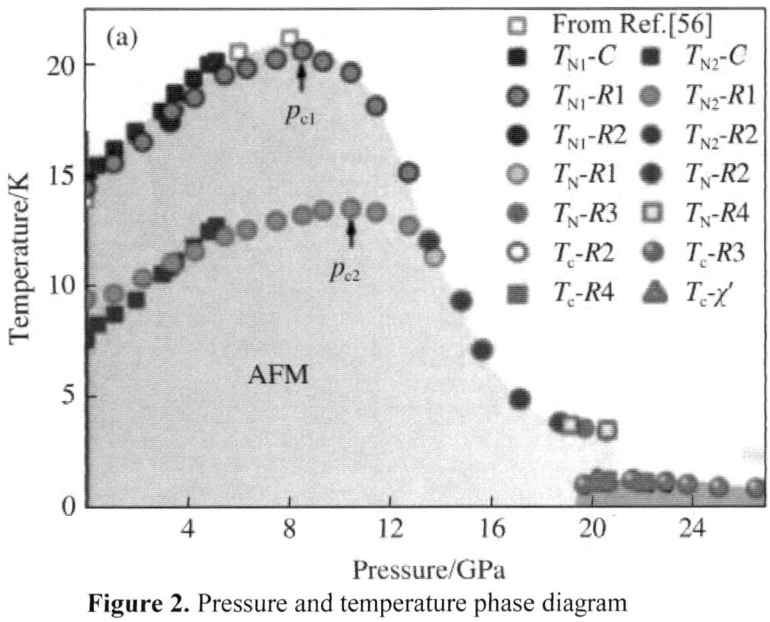

Figure 2. Pressure and temperature phase diagram

On the other hand, the analysis of the relationship between the antiferromagnetic image presented by the heavy Fermion compound CeRhGe3 and the superconducting phase shows that there is an interaction between the two. Combined with the research experience of scholars from various countries in recent years, although effective research has been carried out from this perspective, it is still a difficult problem in the field of condensed matter physics research. According to the analysis of the high-pressure phase diagram shown in FIG. 3 below, the antiferromagnetic transition temperature does not continuously decrease to zero and does not change between 19GPa and 20GPa, and suddenly disappears at about 21.5gpa, but the transition temperature can reach the maximum value. According to the analysis of the normal fitting results of the resistance curve of CeRhGe3 compound by Wang et al., it is found that the state of residual resistance and exponent N changing with pressure can be accurately understood. According to the change of resistivity, it can be clearly found that the superconductivity is directly related to the instability of electricity price in the compound, and the change of valence state under pressure is relatively low. Therefore, it is difficult to guarantee the effectiveness of practical operation based on the current high voltage absorption spectrum measurement technology.

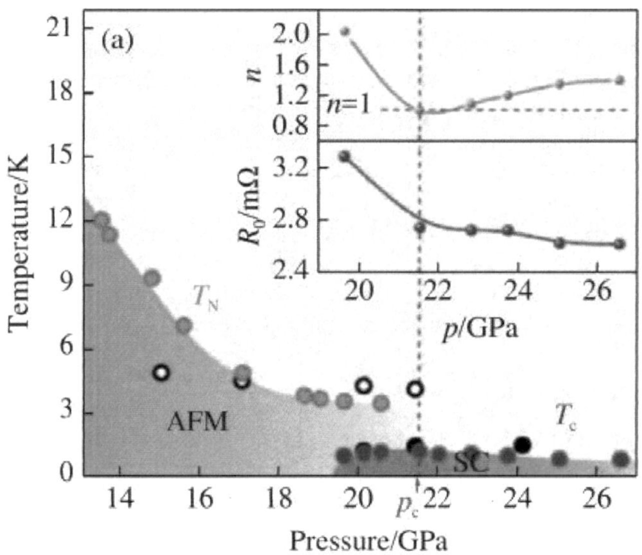

Figure 3. Phase diagram of high pressure

2.2. Copper oxide superconductor

The superconducting transition temperature of this content is higher than that of liquid nitrogen, which has a broad prospect in practical technology research and application, and is also the main content of condensed matter physics. After nearly 30 years of deep research, researchers from all over the world regard the research on high temperature superconducting materials and unconventional superconducting mechanism as the core content. Copper oxide, as a representative strongly correlated electron system, can produce HTS based on the interaction between character, orbit and spin, and can also form cooperative or competitive relations with multiple ordered directions, so as to obtain HTS phase diagrams with various physical properties. Understanding the 2D-3D superconducting state transition process of copper oxide superconductor Bi2212 under high pressure, it is found that since the ground state of Bi series copper oxide parent phase is an antiferromagnetic insulator, the actual change can be continuously controlled with the increase of flux concentration, and superconductivity can be generated after the removal of insulation. At present, researchers have not defined the relationship between pseudogap and superconductivity correctly, but put forward two viewpoints: on the one hand, the pseudogap state is the precursor phase of superconductivity, and the superconductivity phase needs to develop effectively according to the pseudogap state; On the other hand, the pseudogap is an ordered state competing with the superconducting state, and has no direct relationship with the formation of the superconducting phase in practice. For example, Guo et al. regard the Bi2Sr2Ca2o8+δ superconductor as the main research content in their research, and rationally use the high-voltage in-situ resistance synchronous measurement technology to analyze the changing trend of the resistance on the AB surface and the C direction of the sample, so as to clarify the 2D properties of the voltage-induced HTS phase and the singular metal state. The specific results are shown in Figure 4 below:

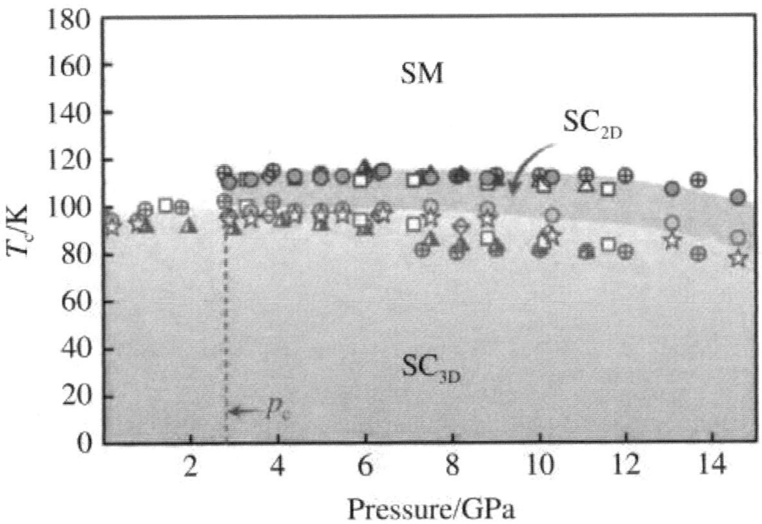

Figure 4. Results of synchronous measurement of high voltage in-situ resistance under different pressures

According to the above analysis, copper oxide has a pressure of 2.8GPa, and a 2D superconducting state similar to BKT can be found in its titanium. After continuous reduction of temperature, copper oxide will enter a 3D superconducting state. It can be seen that the superconducting state transition process of copper oxide under pressure is fully presented by the measurement results, which can not only clarify the main factors affecting the stability of superconductivity, but also provide an effective basis for the study of the relationship between pseudogap and superconductivity.

2.3. *Iron based superconductors*

Since there are many common characteristics between this content and copper oxide superconductor, both of them have basic two-dimensional superconducting structural unit and belong to strongly correlated electronic system, so doping and pressure can be used to control antiferromagnetism effectively, so as to produce superconductivity. In this paper, the double critical points of anti-ferromagnetic superconductivity in Ca0.73La0.27FeAs2 are investigated. This kind of superconductor is a new iron-arsenic based system studied by researchers at present. It has a unique monoclinic structure, and its actual crystal structure and electronic structure are quite different from that of "Type 122" iron-based superconductor. According to Zhou et al. 's high-pressure study of a single antiferromagnetic superconductor under atmospheric pressure, it is found that there is a first-order phase transition between antiferromagnetic superconductor and antiferromagnetic superconductor, and the critical pressure and temperature of the phase transition are in the double critical points. Combined with the pressure-temperature phase diagram shown in FIG. 5 below, it can be seen that the antiferromagnetic transition temperature will continue to decline under the condition that the pressure continues to rise. Under the condition of critical pressure, the antiferromagnetic transition temperature will disappear suddenly, but superconducting transition will occur. At the initial stage of the transition, the superconducting temperature and the antiferromagnetic transition disappear at the same temperature. It should be noted that although the experimental results are consistent with the predicted results, the anti-ferromagnetic transition temperature and superconducting transition temperature of ca0.73LA0.27 FeAs2 are separated during the final experimental process, which requires further investigation based on practical experience.

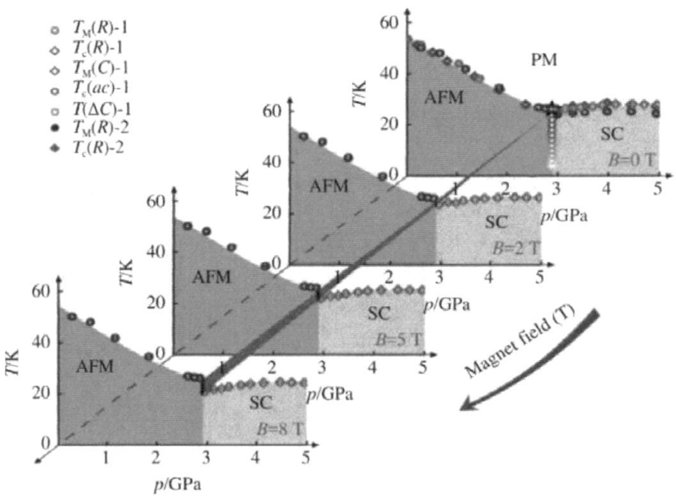

Figure 5. Pressure-temperature phase diagram

3. Result analysis

In this paper, three kinds of superconductors, namely heavy Fermions, volume and copper oxide, have been empirically analyzed by means of pressure regulation. It can be found that although the three kinds of superconductors have their own characteristics, they also have the same properties. For example, superconductivity usually occurs at the edges where traditional structures are continuously suppressed, suggesting that spin interactions have a positive effect on the formation of unconventional superconductivity. Through the system research the application of three kinds of superconductor commonness, grasp the mechanism of high temperature superconductor with practical cases, and can be found that unconventional superconductors superconducting can fine and can spin resonance is a linear relationship between, and can be based on the comprehensive control methods such as pressure, a comprehensive understanding of superconducting material characteristics and evolution law of orderly state, the superconducting mechanism analysis to provide a new train of thought.

4. Conclusion

To sum up, according to the above empirical research analysis, high-temperature superconductors have broad application prospects in the current technical field of unconventional superconductor technology research. Therefore, scholars of various countries should reasonably use high-voltage methods to conduct research on the generation of superconductivity of strongly correlated electronic systems, so as to master more technical information.

References

[1] Qi, Y. P., Ying, T. P., Wu, X. X., et al. (2021) Chinese Science Bulletin, 66(4):6.
[2] Huang, W. H., Li, M., Guo, Z., et al. (2020) Control Parameter Analysis and Optimization method for Multi-terminal Hybrid HVDC Transmission System Based on Small Disturbance Stability [J]. Power Grid Technology, 44(8):9.
[3] Cui, H. Y., (2020) Modern Physics, 10(3):18. (in Chinese)
[4] Lu, Y. F., (2022) Cause Analysis and Treatment of The Discharge Fault of Handcar Sheath of

High Voltage Circuit Breaker [J]. Journal of Anhui Water Resources and Hydropower Vocational Technical College, 22(1):3.
[5] Zhang, Y., Ren, L., Cao, Z. W., Xu, Y., Shi, J., Li, J. D., (2020) Research on influence of Flux-constrained Superconducting Current Limiting Switch on Power System [J]. Cryogenics and Superconductivity, 48(12):5.
[6] Chen, Y., Zhang, Y. J., Lu, X., et al. (2020) Progress in the study of quantum state evolution and its regulation in heavy Fermion system [J]. Basic Science in China, (2):13.
[7] Qi, Y., Zhang, L., (2022) Clinical experience of Zhang LAN in the treatment of Hashimoto's thyroiditis based on data mining [J]. Chinese Medical Forum, 37(1):4.
[8] Liu, X., Mi, W. B., (2020) Structure, magnetic properties and electric transport properties of Fe_3O_4 at Verwey Phase Transition [J]. Acta Physica Sinica, 69(4):17.
[9] Li, Y., Sheng, Y. T., Yang, Y. F., (2021) Progress in theory and materials of heavy Fermionic superconductivity [J]. Acta Physica Sinica, 70(1):31.
[10] Jiang, T., Ren, J. J., Shuai, Z. G., (2020) Advances in Renormalization Group of Spatial Density Matrix in Frequency Domain [J]. Chemical Journal of Chinese Universities, 41(12):19.

Hopping time estimation algorithm of frequency hopping signal under impulse noise

Bing Liu[1,*], Xinyong Yu[1,a], Hongguang Li[1,b], Yinsong Yang[1,c], Zhipeng Qi[1,d]

[1]Air Force Communications NCO Academy; Dalian, China
[a] Yuxinyong99@163.com; [b] toumingwings@163.com; [c] Yangyinsong131@163.com; [d]734667780@qq.com

* liubingbs@163.com

Abstract. Aiming at the hopping time estimation problem of frequency hopping (FH) signals under alpha stable distributed noise, a hopping time estimation algorithm based on time-frequency pattern gradient sequence is proposed in this paper. Firstly, the FH signal model and the alpha stable distribution noise model are established according to the received signals; Secondly the fractional lower order SPWVD (FLOSPWVD) transform is used to suppress the impulse noise in the time frequency pattern, and a clear and reliable time frequency pattern is obtained by using the morphological filtering. Finally, the hopping time estimation of the frequency hopping signal is completed according to the jump of the gradient sequence of the time frequency graph.

1. Introduction

Frequency hopping communication has been widely used in military and civil communications since last century because of its strong anti-interference and anti-fading ability, easy access to multiple access networks, easy to be compatible with narrowband communication systems and good security [1-3]. Aiming at the hopping time estimation problem of frequency hopping, in [4], the FH signal is transformed by STFT, and the hopping time value is estimated in the time frequency ridge, but the performance of this algorithm is seriously degraded in low SNR. On the basis of [4], the FH signal is transformed by SPWVD in [5] which improves the performance of the algorithm to some extent. Two combination time-frequency analysis methods are proposed in [6, 7] to estimate the hopping time. This method can retain the advantages of two time-frequency transform at the same time, which greatly improves the performance of hopping time estimation, but the algorithm is more stringent on the requirement of SNR.

However, in almost all existing hopping time estimation algorithm, the noise signals are all Gaussian noise. However, non-Gaussian noise exists in the actual complex electromagnetic environment, such as underwater acoustic clutter, radar clutter, thunderstorm noise and low frequency air noise which have a certain pulse characteristic[8-10]. Among such noises, the conventional signal processing method based on the Gaussian assumption is no longer applicable. A parameter estimation algorithm based on Myriad filtering of FH signals is proposed in [11], this algorithm has a certain inhibitory effect on alpha stable distributed noise, but when the pulse characteristics are strong, the performance degrades greatly.

Based on the above issues, aiming at the hopping time estimation problem of FH signals under alpha stable distributed noise, in this paper, we propose a pulse noise reduction algorithm based on fractional low order SPWVD. Based on it, a clear and reliable time-frequency graph is obtained by using morphological filtering method, and finally the hopping time is estimated based on the gradient sequence of time-frequency pattern.

2. FH signal model under impulse noise

Suppose that the time of initial hop is Δt_{0n}, there are K hops during the observation period. The time of the last non-complete jump in the observation period is Δt_{ns}, the center frequency is Δt_{ns}. The $\boldsymbol{s}_n(t)$ can be expressed as[11-14]:

$$\mathrm{s}_n(t)=v_n(t)\begin{pmatrix}\exp\left[j\left(2\pi f_{n0}t+\varphi_{n0}\right)\right]rect\left(\dfrac{t}{\Delta t_{n0}}\right)\\+\sum_{k=1}^{K}\exp\left[j\left(2\pi f_{nk}t+\varphi_{nk}\right)\right]rect\left(\dfrac{t-(k-1)T_n-\Delta t_{n0}}{T_n}\right)\\+\exp\left[j\left(2\pi f_{ns}t+\varphi_{ns}\right)\right]rect\left(\dfrac{t-KT_n-\Delta t_{n0}}{\Delta t_{ns}}\right)\end{pmatrix} \tag{1}$$

Assuming that there are N FH signals in the observation time T, the received FH mixed signal $x(t)$ can be defined as:

$$x(t)=\sum_{n=1}^{N}s_n(t)+v(t) \tag{2}$$

Where, $v(t)$ is an additive stable distributed pulse noise. So the formula (2) is the model of the FH signal under the condition of impulse noise.

The discrete Cohen time-frequency distribution of FH signals $x_i(\mathrm{t})$ and $x_j(t)$ can be defined as :

$$\boldsymbol{D}_{x_ix_j}(t,f)=\sum_{\tau=-\infty}^{\infty}\sum_{l=-\infty}^{\infty}\begin{pmatrix}\varphi(l,\tau)x_i(t+l+\tau)\\x_j^*(t+l-\tau)e^{-j4\pi f\tau}\end{pmatrix} \tag{3}$$

The space-time-frequency distribution of FH signal $x(t)$ can be expressed as:

$$\boldsymbol{D}_{XX}(t,f)=\sum_{\tau=-\infty}^{\infty}\sum_{l=-\infty}^{\infty}\begin{pmatrix}\varphi(l,\tau)X(t+l+\tau)\\X^H(t+l-\tau)e^{-j4\pi f\tau}\end{pmatrix} \tag{4}$$

Where $[\boldsymbol{D}_{XX}(t,f)]_{ij}=\boldsymbol{D}_{x_ix_j}(t,f)$ denotes the mutual time frequency distribution between the output signals of each array.

Therefore, the covariance matrix in the time-frequency domain of the array can be expressed as:

$$\begin{aligned}E[\boldsymbol{D}_{XX}(t,f)]&=E[\boldsymbol{D}_{YY}(t,f)]+E[\boldsymbol{D}_{NN}(t,f)]\\&=\boldsymbol{A}\boldsymbol{D}_{SS}(t,f)\boldsymbol{A}^H+E[\boldsymbol{D}_{NN}(t,f)].\end{aligned} \tag{5}$$

$E[\boldsymbol{D}_{XX}(t,f)]$ and $\boldsymbol{A}$ have the same subspace characteristics, the joint diagonalization or average processing is used to estimate the multiple self-term time-frequency points of the FH signal.

3. Impulse noise removal in FH signals

3.1. Alpha stable distributed impulse noise

The unified characteristic function of alpha stable distribution can be expressed as[15]:

$$\varphi(t)=\exp\left\{jat-\gamma|t|^\alpha\left[1+j\beta\operatorname{sgn}(t)w(t,\alpha)\right]\right\} \tag{6}$$

Where

$$w(t,\alpha)=\begin{cases}\dfrac{2}{\pi}\log|t| & (\alpha=1)\\ \tan\dfrac{\alpha\pi}{2} & (\alpha\neq1)\end{cases} \tag{7}$$

$$\operatorname{sgn}(u)=\begin{cases}1 & (\mu>0)\\ 0 & (\mu=0)\\ -1 & (\mu<0)\end{cases} \tag{8}$$

α stable distribution has two important properties:
(1) stability: For any positive numbers A and B, if there is a positive number C and a real number D:

$$AX_1+BX_2\overset{d}{=}CX+D \tag{9}$$

Where X_1 and X_2 are independent samples of the random variable X; The symbol $\overset{d}{=}$ denotes the equation has the same distribution on both sides. If $D=0$, Eq. (9) still holds, indicating that X is strictly stable.
If $X_1, X_2, X_3, \ldots, X_n$ is independent of each other, and α stable distribution of random variables. And with the same L parameters, the linear combination $\sum_{i=1}^{n} A_i X_i$ obeys the α stable distribution and also has the same α, β parameter.
(2) Generalized Central Limit Theorem: If and only if the random variable X obeys a stable distribution, and $X_1, X_2, X_3, \ldots, X_n$ are independent random variables with α stable distribution. When $n\to\infty$, X obeys the limit distribution of normalized sum:

$$S_n=\frac{X_1+X_2+\ldots+X_n}{a_n}-b_n \tag{10}$$

Where a_n is an arbitrary positive real number and b_n is an arbitrary real number. Special, if X_i is independent and identically distributed and has a finite variance, the Gaussian distribution is its limit distribution, and X_i satisfies the general central limit theorem. Regardless of whether the finite variance of the random component X_i exists, as the number of variables increases.
X is a non-Gaussian a stable distributed random variable with an unknown parameter of zero, then:

$$\lim_{t\to\infty} t_{\alpha} P(|X|>t)=\gamma C(\alpha) \tag{11}$$

Where $C(\alpha)$ is a positive constant that depends on α. Thus, the α stable distribution has algebraic smearing. In contrast, Gaussian distributions have exponential tailing.
The finite second order moment of the α stable distribution does not exist, which makes the variance of noise meaningless, so the GSNR needs to be defined as:

$$\text{GSNR}=10\lg\left(\sigma_s^2/\gamma_n\right) \tag{12}$$

When $\alpha=1.5$, GSNR=-5dB, the α stable distribution is shown in Fig.1(a). When $\alpha=0.8$, GSNR=-5dB, the α stable distribution is shown in Fig.1(b)

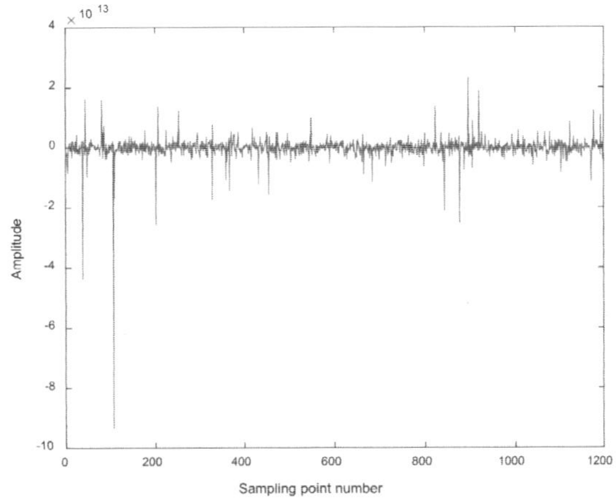

(a) α=1.5, GSNR = −5dB

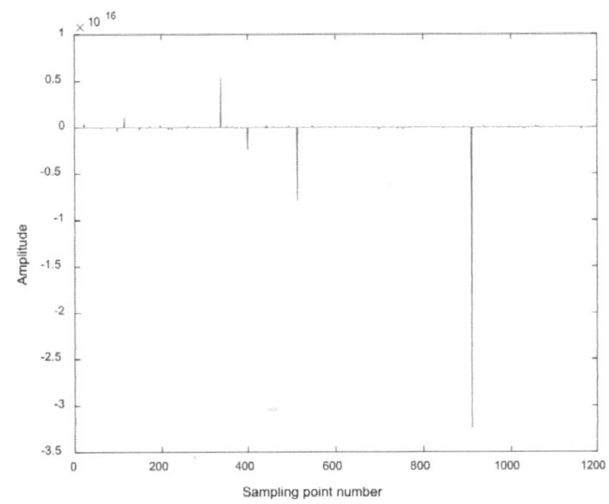

(b) α=0.8, GSNR = −5dB

Figure 1. The α stable distribution

It can be seen that as the value of α becomes smaller, the impulse characteristics of the signal become stronger. In fact, these high-amplitude outliers in the graph correspond to the polar probability of the two ends of the Gaussian density function.

3.2. Pulse noise removal based on FLOSPWVD

α stable distributed class of important non-Gaussian random distributions, with significant impulsiveness and thicker tailing compared to Gaussian distribution. The pulse amplitude is strong, and the duration of the pulse is very short. In the limit case, it can be regarded as an impulse function $\delta(t)$ with an infinitesimal duration and an infinite magnitude. Under this assumption, α stable distributed noise per burst can be expressed as $x = a_i\delta(t)$ in the time domain, where a_i denotes the impulse intensity of the burst. And after FFT transform. It appears as a straight line parallel to the frequency axis. α stable distribution of noise after time-frequency transformed distribution can be seen and time-frequency characteristics, while Gaussian noise has no such feature relative to α stable distribution of noise.

In the time-frequency transform, the crosstalk term of the FH signal can be largely eliminated. The SPWVD transform can be defined as:

$$\mathrm{SPWVD}_x^{(p)}(t,f)=\int_{-\infty}^{+\infty}\int_{-\infty}^{+\infty}\begin{pmatrix} x\left(t-v+\frac{\tau}{2}\right)x^*\left(t-v-\frac{\tau}{2}\right) \\ \cdot h(\tau)g(v)e^{-j2\pi f\tau}\end{pmatrix}d\tau dv \tag{13}$$

Where $x(t)$ is the FH source signal; $h(\tau)$ is the time domain window function; $g(v)$ is the frequency domain window function.

Although the alpha stable distribution is not a two-order moment process, it has a finite moment of p order. So the FH signal can be calculated by the fractional lower order operator of the p order to obtain the transformed FH signal as the second-order moment process. The FLOSPWVD is defined as:

$$\mathrm{FLOSPWVD}_x^{(p)}(t,f)=\int_{-\infty}^{+\infty}\int_{-\infty}^{+\infty}\begin{pmatrix} x^{\langle p\rangle}\left(t-v+\frac{\tau}{2}\right)x^{-\langle p\rangle}\left(t-v-\frac{\tau}{2}\right) \\ \cdot h(\tau)g(v)e^{-j2\pi f\tau}\end{pmatrix}d\tau dv \tag{14}$$

Where $0<p<2$, operator $\langle\cdot\rangle$ is defined as $Z^{\langle p\rangle}=|Z|^{p-1}Z^*$, $Z^{\langle -p\rangle}=\left(Z^*\right)^{\langle p\rangle}=\left(Z^{\langle p\rangle}\right)^*$. If Z is a real variable, then $Z^{\langle p\rangle}=|Z|^{p-1}Z^*\operatorname{sgn}(Z)$. From (7), we can see that the difference between FLOSPWVD and traditional SPWVD is that the signal term is the p order statistics.

4. Hopping time estimation of FH signal

4.1. Time frequency pattern correction based on morphological filtering

Although the pulse noise reduction of fractional low order SPWVD has been done for FH signals, but the interference-free noise points may still occur at low signal-to-noise ratios. In order to find and eliminate these points, we use morphological filtering to modify the time frequency pattern.

Suppose that the frequency modulus value of the FH signal at any time point (t,f) on the time-frequency patter after the FLOSPWVD transform is $\omega(t,f)$. The time frequency pattern matrix after the threshold segmentation can be expressed as:

$$\boldsymbol{R}=\begin{cases}1 & \omega(t,f)\geq\varepsilon \\ 0 & \omega(t,f)<\varepsilon\end{cases} \tag{15}$$

Where ε is the threshold. Because each of the binary images has only a number of nonzero values, the probability of a continuous occurrence of a larger fluctuating point is smaller. Therefore, the matrix [1 1 1] is used as the structural element $\boldsymbol{c}$ to calculate the corrosion and expansion of the binary images $f(x,y)$. i.e.

$$\begin{cases} f\Theta\mathbf{C}=\min\begin{Bmatrix} f(x-x',y-y') \\ -\mathbf{c}(x',y')\,|\,(x',y')\in D_c\end{Bmatrix} \\ f\oplus\mathbf{C}=\max\begin{Bmatrix} f(x-x',y-y') \\ +\mathbf{c}(x',y')\,|\,(x',y')\in D_c\end{Bmatrix}\end{cases} \tag{16}$$

Where $\mathbf{c}(x',y')$ is the structural element; D_c is the domain of $\mathbf{c}(x',y')$. Corrosion can eliminate the scattered points that affect a jump, and it can shrink the boundary as well; expansion can fill the void in the signal and expand the boundary.

Therefore, the corrupted and swept points in the time-frequency pattern can be eliminated by the corrosion and expansion processing without affecting the other points in the pattern, so as to obtain a clear and stable time-frequency pattern.

4.2. Hopping time estimation based on gradient sequence of the time frequency pattern

Suppose that the time frequency matrix corresponding to the corrected time frequency pattern is $\boldsymbol{R}$; the element at t_i time in the time frequency pattern corresponding to the $\boldsymbol{R}$ is $\%_i$; the element at t_{i+1} time is $\%_{i+1}$. So the gradient sequence Δ_i of hopping time estimation can be expressed as:

$$\Delta_i = \sum_{i=1}^{P} \left| \%_{i+1} - \%_i \right| \tag{17}$$

Where P is the number of columns in matrix $\boldsymbol{R}$. From a matrix point of view, Δ_i is the sum of the absolute values of the element values in $(i+1)$-th column of the matrix minus the element values in the i-th column. By comparing the value of the gradient sequence Δ_i , the moment of the mutation point is the hopping time of the FH signal.

5. Algorithm steps

According to the above deduction and discussion, the specific steps of the hopping time estimation algorithm for the FH signal under the alpha stable distribution noise is summarized as follows:
Step1: Establish the communication model of frequency hopping signal under alpha stable distribution pulse noise by (2);
Step2: Using FLOSPWVD transform to remove the impulse noise of FH signal by (8);
Step3: Binaryzation the time frequency pattern obtained by short time Fourier transform by (9), and using morphological filtering to modify the binary image by (10);
Step4: Estimate the hopping time parameter of FH signal by (11).

6. Simulation and analysis

Suppose that there are two asynchronous FH signals FH1 and FH2 in the space, the sampling rate is 100MHz, the hop rate is 1000hop/s, the hopping time is 0.3ms, the observation time is 1200 sampling points, and the number of the observations hop is 7. The frequency set of the FH1 signal is {100, 200, 75, 175, 103, 25, 225} MHz; the frequency set of the FH2 signal is {25, 225, 275, 10, 165, 175, 50} MHz. The estimated success rate of the hopping time is defined as:

$$\eta = N_1 / N \tag{18}$$

Where N_1 denotes the number of accurate estimate experiments, and N denotes the total number of experiments.

6.1. Pulse noise removal experiment

Suppose that the time domain window of the SPWVD transform is a Hamming window with the length of 201 and the frequency domain window is a Hamming window with the length of 257; the fractional low order operator p = 0.2. When $\alpha = 1.5$ and GSNR=-5dB, the time frequency spectrum and the gradient sequence of FH signal obtained by the SPWVD transform and the FLOSPWVD transform is shown in Fig.2.

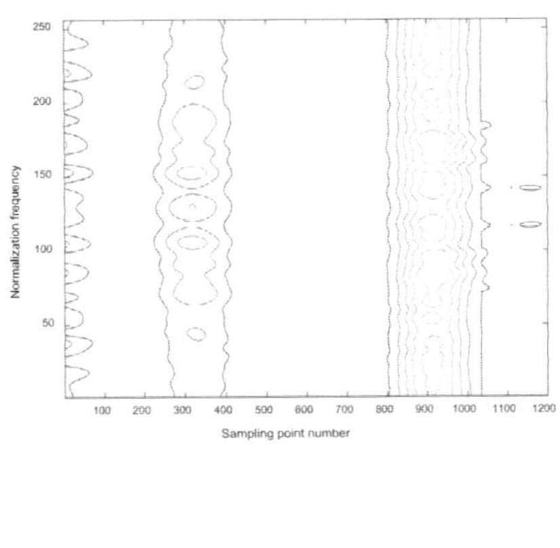

(a1) Time frequency spectrum of the SPWVD

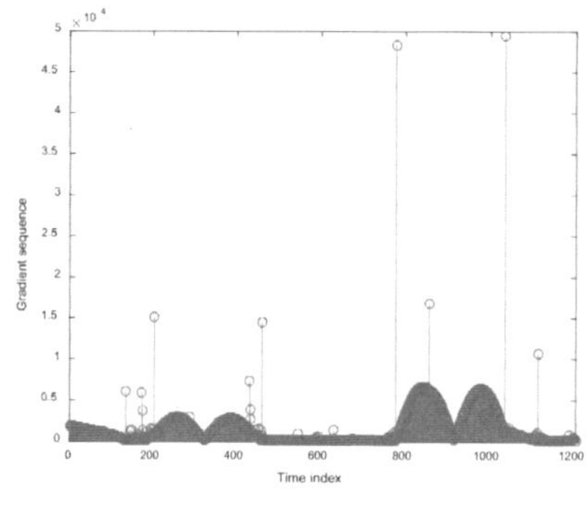

(a2) Gradient sequence of the SPWVD

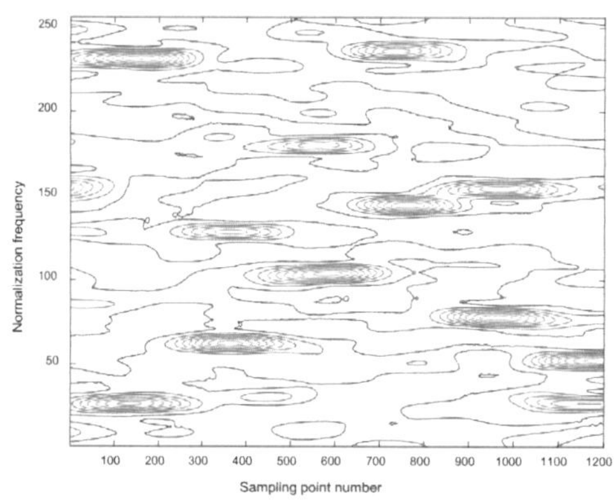

(b1) Time frequency spectrum of the FLOSPWVD

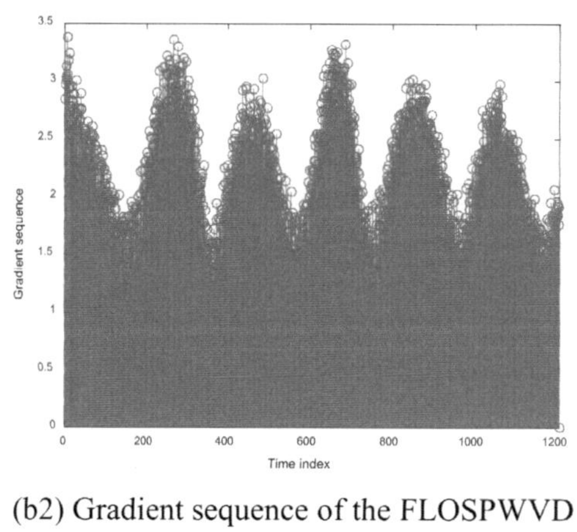

(b2) Gradient sequence of the FLOSPWVD

Figure 2. Comparison of SPWVD and FLOSPWVD

It can be seen from Fig.2(a) that alpha stable distribution impulse noise has great interference to FH signal, the FH signal is completely submerged by the impulse noise signal in the time frequency spectrum by the SPWVD transformation, and there are only two highlight points in the gradient sequence, the hopping time of the FH signal can not be estimated. From Fig.2(b), we can know that after FLOSPWVD transform suppresses impulse noise processing, most of the impulse noise is suppressed but the time frequency characteristics of the FH signal still can not be seen. It can be seen from the gradient sequence that there are five hopping time points in the frequency hopping signal.

6.2. Time frequency spectrum correction experiment

The hopping time estimation of the FH signal is directly related to the performance of the time frequency spectrum. The time frequency spectrum and the gradient sequence of FH signal after the morphological filtering is shown in Fig.3.

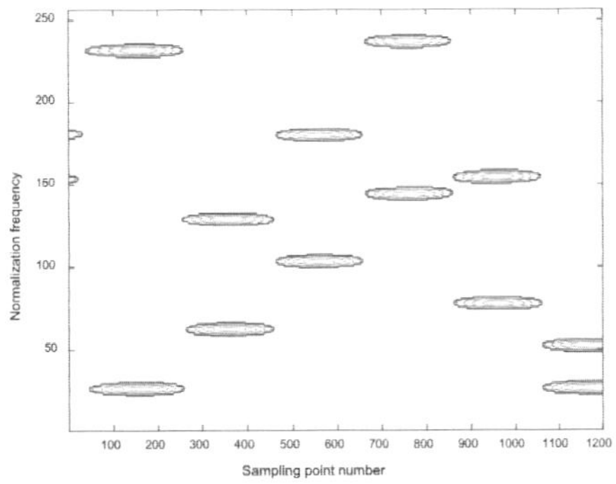

(a) Time frequency spectrum after morphological filtering

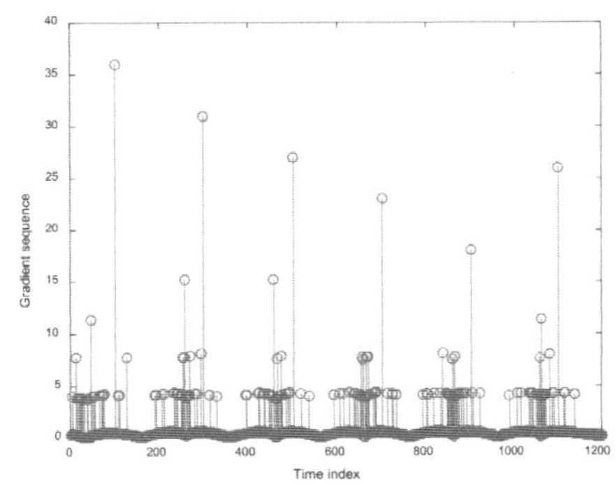

(b) Gradient sequence after morphological filtering

Figure 3. Time frequency spectrum and the gradient sequence after morphological filtering

It can be seen from Fig.3 that after the morphological filtering, the residual noise in time frequency spectrum can be removed completely, which has a good spectrum for subsequent hopping time estimation. The five hopping times of the FH signal can be obtained accurately by the gradient sequence processed by morphological filtering.

6.3. Hopping time estimation experiment

Fig.4 show the performance comparison of hopping time estimation in FLOSPWVD algorithm and after morphological filtering processing algorithm when SNR is between -15dB and 10dB.

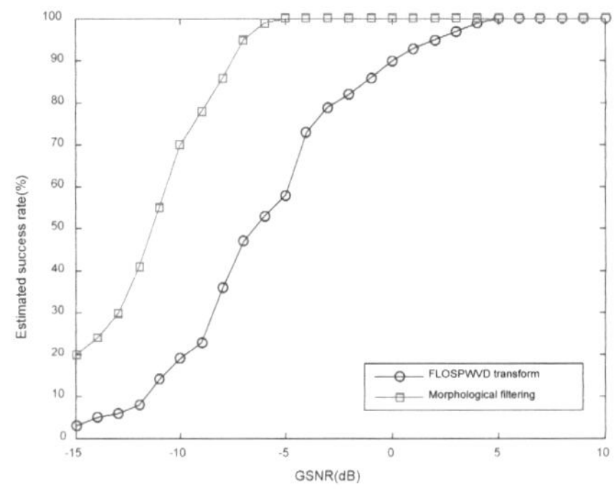

Figure 4. Comparison of hopping time estimation

From Fig.4, we can get that the hopping time estimation accuracy of the two algorithms increases gradually with the increase of signal-to-noise ratio. After the morphological filtering process, the estimated success rate can reach 100% when the GSNR reaches about -5dB; without morphological filter process, the GSNR needs to be about 5dB when estimated success rate can reach 100%.

7. Conclusion

Under the environment of intensified electronic warfare, especially in wartime, it is of great significance to estimate the hopping time of multiple FH signals under the complex electromagnetic environment. In this paper, the hopping time estimation method of frequency hopping signal based on the gradient sequence of time-frequency pattern in alpha stable distribution noise is elaborated and deduced in detail. The simulation results show that the proposed algorithm can effectively suppress the impulse noise and realize the hopping time estimation of the FH signal in low SNR.

References

[1] Torrieri, D. J. (2000) Mobile frequency-hopping CDMA systems[J]. IEEE Transactions on Communications. 48(8):1318-1327.
[2] Zhao, L., Wang, L., Bi, G., Zhang, L., Zhang, H. (2015) Robust frequency-hopping spectrum estimation based on sparse Bayesian method[J]. IEEE Transactions on Wireless Communications. 14(2):781-793.
[3] Feng, T., Yuan, C. W. (2010) A method of time-frequency analysis for frequency-hopping signals[J]. Journal of Beijing University of Posts and Telecommunications, 33(3):10-14.
[4] Zheng, X. W., Zhao, Q. G., Luo, Y. J. (2006) Hop rate estimation for frequency hopping signals[J]. Systems Engineering and Electronics, 28(10): 1500-1501.
[5] Feng, T., Yuan, C. W. (2011) Blind parameter estimation of frequency-hopping signals based on the time-frequency distribution maxima[J]. Acta Electronic Sinica, 39(12): 2921-2925.
[6] Stankovic, L. (1994) A method for time-frequency analysis[J]. Signal Processing IEEE Transactions on, 42(1): 225-229.
[7] Lei, Y. K., Wu, Y. H. (2008) A new hop rate estimation method for high-speed frequency-hopping signals[C]. 11th IEEE Singapore International Conference on Communication Systems, Guangzhou, China,19-21 Nov, 1330-1333.
[8] He, J., Liu, Z. (2006) Robust adaptive beam forming method in implusive noise[J]. Acta Electronica Sinica, 34(3):464-468.
[9] Wang, X. Y., Yang, Z. C., Huang, J. J., et al. (2017) Robust beamforming method with weigthed sparse constraint in impluse noise[J]. Journal of signal processing, 33(3):288-293.
[10] Pander, T., Przybyła, T. (2012) Impulsive noise cancelation with simplified Cauchy-based p -norm filter[J]. Signal Processing, 92(9):2187-2198.

[11] Liu, S., Zhang, Y. D., Tao, S., et al. (2018) Structure-Aware Bayesian Compressive Sensing for Frequency-Hopping Spectrum Estimation with Missing Observations[J]. IEEE Transactions on Signal Processing, 66(8):2153-2166.
[12] Manouchehri, S., Zeinalpour-Yazdi, Z. (2018) Expected distortion of simultaneous transmission of multiple source symbols with multicarrier frequency hopping modulation[J]. Wireless Networks, 33(10):1-11.
[13] Xing, L., Liang, Z. (2018) New Bound on Partial Hamming Correlation of Low-Hit-Zone Frequency Hopping Sequences and Optimal Constructions[J]. IEEE Communications Letters, 33(9):1-9.
[14] Zhang, C., Wang, Y., Jing, F. (2018) Underdetermined Blind Source Separation of Synchronous Orthogonal Frequency-Hopping Signals Based on Tensor Decomposition Method[J]. IEEE Access, 6(99):69407-69414.
[15] Georgiou, P. G., Tsakalides, P., Kyriakakis, C. (2002) Alpha-stable modeling of noise and robust time-delay estimation in the presence of impulsive noise[J]. IEEE Transactions on Multimedia, 1(3):291-301.

Design and application of new power load management system

Gang Liu[1,2,3,*],Xubing Xiao [1,2,3],Zhicheng Zhou [1,2,3],Gang Sun[1,2,3],Yunqiang Si[1,2,3]

[1]NARI Group Corporation (State Grid Electirc Power Research Institute), Nanjing, China
[2]NARI-TECH Nanjing Control System Ltd. , Nanjing, China
[3]NARI Technology Co.,Ltd. , Nanjing, China

*Corresponding author email: liugang2@sgepri.sgcc.com.cn

Abstract. In order to improve the load management ability of the new power system, this paper proposes a design scheme of the new power load management system. The design scheme is based on the cloud platform and adopts the micro service and micro application architecture to build applications such as load control customer management, equipment file management, load monitoring, load characteristic analysis, load control scheme formulation, load regulation and implementation, load implementation monitoring, implementation effect analysis, operation management, power safety, early warning monitoring, command and drill, and realize the real-time monitoring and accurate control of load resources on the customer side. The practical application proves the effectiveness of this scheme in improving the management of customer side load resources.

1. Overview

With the deepening of industrialization and urbanization at this stage, the demand for electricity will inevitably continue to grow, and the supply shortage is the biggest energy insecurity. At the same time, the output of new energy has randomness and volatility, which brings great challenges to the safe and stable operation of power system; Power supply security is related to the national economy and the people's livelihood, so it is necessary to make full use of demand side management means, and adhere to the "demand response priority, orderly power utilization to ensure the minimum, and power conservation assistance"; At the same time, the marketing function of State Grid of China has changed greatly. Under the situation of building a new power system, the marketing function of State Grid Corporation of China has also changed[1].

Therefore, it is necessary to build a new power load management system to improve the flexibility of power grid regulation, make every effort to ensure the smooth and orderly power supply, protect the electricity consumption of residents, public services and important customers, so as to support the normal operation of the demand side management system, and complete the "promotion of the construction of technical means of load control system to achieve accurate control of power side load".

2. System architecture

2.1. Overall architecture

The new power load management system is designed based on the cloud platform. It responds to the load control requirements during peak summer, peak winter and load gap, and issues control instructions to the load control customers to guide customers to cut peak and fill valleys and use electricity reasonably[2]. The system is composed of load regulation module, load application module and unified collection service components. The load regulation module realizes the decomposition of load regulation tasks and the issuance of instructions; The load application module realizes scheme management and effect evaluation. According to the requirements of the safety protection scheme of the power consumption information acquisition system, the load regulation module is deployed in the marketing and production control domain, the load application module is deployed in the marketing and management information domain, and the load regulation module is connected with the load application module through the isolation device to form a complete load management system, which responds and executes the load command at the second level, realizes the rapid control of customer load, and meets the safety protection requirements at the same time. The overall architecture of the new power load management system is shown in Figure 1.

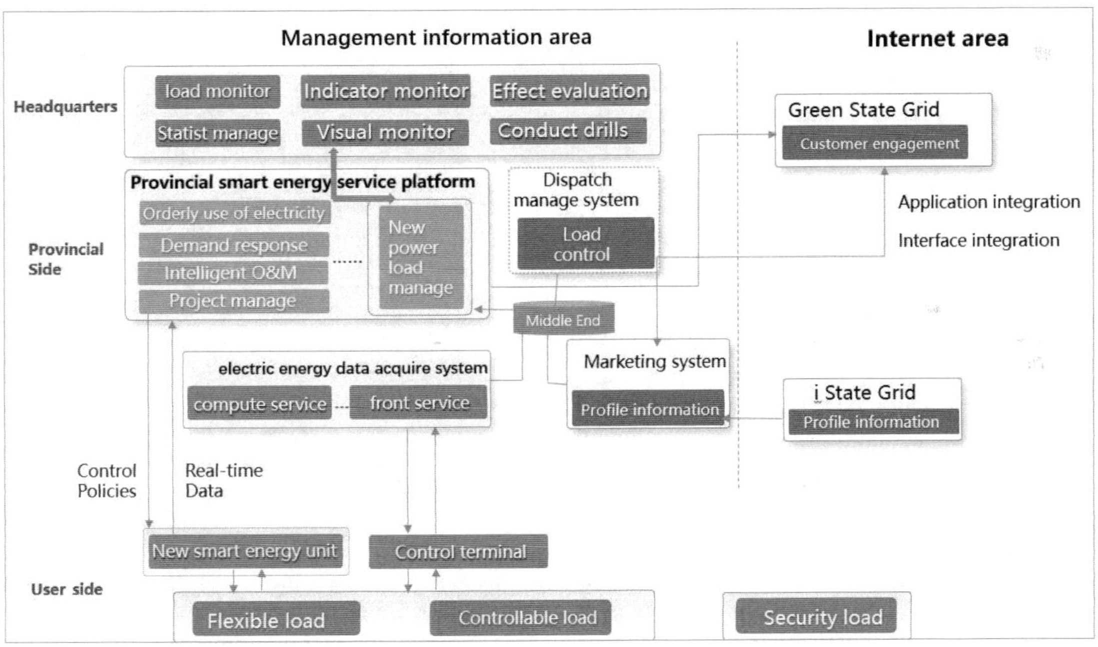

Figure 1. Overall architecture of new power load management system

The load regulation module is responsible for real-time data exchange and intermediate process control, real-time collection of customer load data, and dynamic perception of customer switching action. Realize the quasi real-time monitoring of customer operation load and status information, the process research, judgment, analysis and alarm of real-time data, the cross domain scheduling and instruction issuance of the negative control management function in the execution process of execution strategy, interact with the negative control application module upward, realize the through access between the marketing production domain and the marketing management information domain, and complete the reception of control tasks and the feedback of execution results; Connect with the power consumption system downward, call the pre service of the power consumption information acquisition system, and the power consumption information acquisition system completes the message assembly and real-time communication with the terminal[3].

The load application module is responsible for realizing load application management such as scheme management and effect evaluation, and realizing the upper application display and provincial personalized application construction of load management. It includes the functions of adding and batch importing load customers, the maintenance function of accessing switch circuit information, the personalized display function of load execution effect, and the monitoring and evaluation function of command drill. The load application module interacts with the load regulation module in real time to realize the horizontal connection between load management and load control, ensure the dynamic real-time visual display of load management tasks, and support the load management needs of different posts and positions.

The unified collection of support components is located in the collection of general resources and services in the new power load management system, with the principle of "horizontal consistency, vertical connection and elastic adaptation", the goal of global unification of core data and the construction of common business standards, and the focus of data management and utilization. Through the construction and implementation of unified collection support components, on the one hand, through the component support based on business service components, ensure the consistency, availability, security and reliability of the core data involved in the new power load management business within the business scope of the national network, and reduce the damage to the unity of the core data model caused by the self construction of provincial power companies, Reduce the complexity of the core data interaction between the State Grid headquarters and the provincial power companies, and improve the credibility of the data involved in the business control of the provincial power companies by the State Grid headquarters; On the other hand, through the construction of business application components, we can provide standardized common business support applications for provincial power companies, effectively reduce the waste of resources caused by the repeated construction of common business and basic business by provincial power companies, so that provincial power companies can further strengthen the resource inclination and investment in localization and process business, And then realize the improvement of load control reliability and process adaptability of provincial power companies.

2.2. Application architecture

Based on the system deployment requirements and platform business requirements, the business application of the new power load management system includes the following three parts: the business application of the State Grid headquarters side, the business application of the provincial power company side, and the support service application.

The business application of the State Grid headquarters covers load monitoring, index management, load execution effect evaluation, command drill and other applications, which are used to support the development of data extraction, strategy distribution and unified command and control of provincial power companies by the State Grid headquarters. Realize two-level connection, panoramic statistics, accurate regulation and intelligent analysis of power load regulation business.

Based on the business requirements of system monitoring and control, the new power load management system carries out three kinds of applications in the side of provincial power companies, including load management, load control, operation management, specifically including load control customer management, equipment file management, load monitoring, load characteristic analysis, load control scheme formulation, load regulation and implementation, load implementation monitoring, implementation effect analysis, operation management, power safety, early warning monitoring 11 applications such as command drill are used to support the professional operators of load control to carry out daily management work.

Support service application mainly refers to the unified collection of support components, including basic service components and business service components, which are used to unify and standardize the management of core basic data, provide unified services, realize the import and export control of the above-mentioned data, ensure the consistency, availability, security and reliability of core basic data throughout the network, and reduce the damage of self construction to the unity of core data model, Reduce the complexity of the core data interaction between the State Grid headquarters and the provincial power companies, and improve the credibility of the data involved in the business control of the provincial power companies by the State Grid headquarters[4]. The application architecture of the new power load management system is shown in Figure 2.

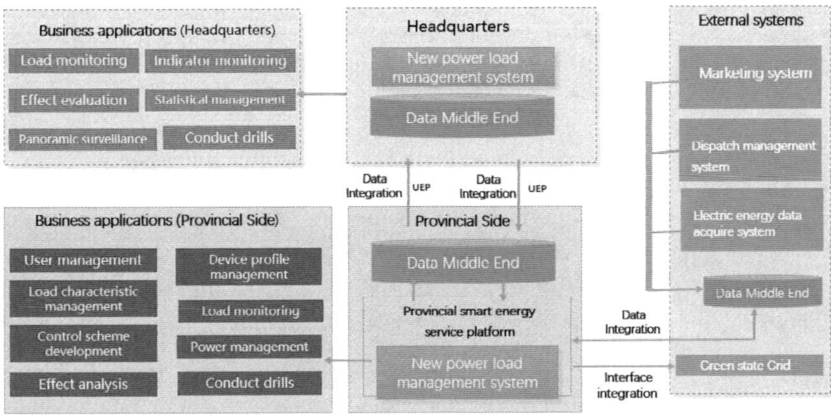

Figure 2. Application architecture of new power load management system

2.3. *Technical architecture*

The new power load management system follows the design ideas of smart energy service platform, marketing business application system 2.0 and power consumption information collection system 2.0, and adopts SOA architecture, Cloud Architecture and micro application and micro service development specifications. The provincial smart energy service platform controls the application module, receives the execution plan of the orderly power consumption module and issues control instructions. The load control and regulation module, up and down with the load control application module, realizes the cross domain through access between the marketing and production domain and the information region, down and docking with the power consumption information collection system, and invokes the collection pre service of the power consumption information collection system. The overall technical architecture of the new power load management system is shown in Figure 3.

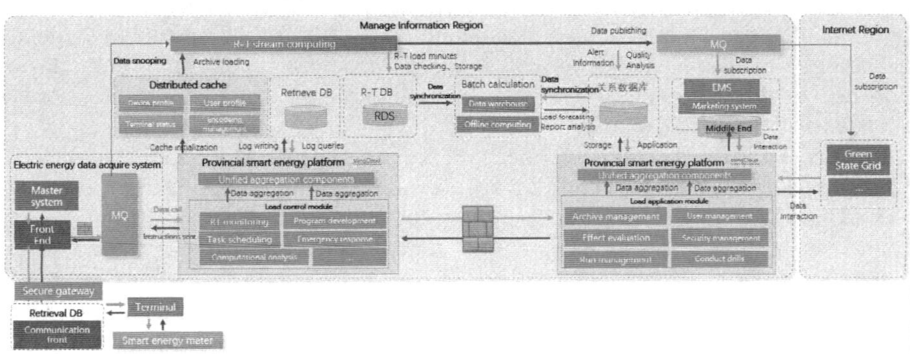

Figure 3. Technical framework of new power load management system

Data Storage

Data storage layer, including relational database, big data platform storage and cache hot data storage[5]. Compared with the technical guidelines of the State Grid, the description is as follows:

(1) Message Oriented Middleware: provide a shared channel for real-time load data, meet the use of multi-party application data and improve the timeliness of data;

(2) Distributed big data storage clusters ： HBase+Hive+Kudu+HDFS Realize big data storage and build data warehouse;

(3) Upgrading the relational database architecture and reusing the existing Oracle server;

(4) Build a cache architecture for hot data, use distributed cache redis to proxy Oracle, and reduce Oracle IO consumption; Use the efficient row key design in HBase to realize the rapid query and retrieval of a single big data record;

(5) Build multi data storage synchronization services, and use secondary Kafka to ensure the final consistency of data between the big data storage platform and Oracle production library.

Calculation and Analysis

Calculation and analysis include distributed data offline calculation and streaming real-time calculation[6].

(1) Build a distributed big data offline computing platform, which includes MapReduce and spark, with spark as the main, taking into account performance and data scale;

(2) Build a distributed real-time computing platform, use storm and Flink to realize streaming computing engine, realize data cleaning, completion and exception handling, and use Phoenix to realize JDBC type fast query operation.

Data Sharing

The data sharing layer includes OGG, WebService, dblink and other methods. In addition, when facing the data sharing requirements with high real-time requirements, Kafka and rocketmq queue methods are used to realize the rapid transmission of data to improve the performance of data sharing[7].

(1) Collect data in real time and share the channel concurrently, use Kafka to realize massive data sharing and transmission, and use rocketmq to realize complex message function, reliable message transmission and multi tenant management function;

(2) Build heterogeneous platform data synchronization, and use sqoop, dataX and OGG to realize data extraction and sharing between heterogeneous data sources.

Data Bus

Data bus is used for data transmission between layers. The data bus technology selects Kafka, MQ and other technologies to realize the rapid and reliable transmission of data, and uses ETL tools such as sqoop, dataX, Ogg to realize the data synchronization and extraction between data sources.

Operation Maintenance and Monitoring

For the increasingly complex distributed work architecture, it is necessary to realize all-round operation and maintenance monitoring for all kinds of collection services and basic functions. The selected technologies include ZABBIX to realize system parameter monitoring, flume to realize log import, es to realize search engine to facilitate fast and complex query, and spring+vue and other technologies to realize front-end pages.

Data interaction includes data midrange, service API and other methods. In addition, when facing the demand of high real-time requirements and large amount of data, Kafka queue method is used to realize the rapid transmission of data, so as to improve the performance of data sharing.

2.4. Deployment architecture

The new power load management system can be deployed as an independent system or as a business module of the provincial smart energy service platform, reusing the security protection framework of the provincial smart energy platform, and deploying on the provincial smart energy service platform using resource virtualization technology; Its external service module is deployed in the Internet region

 2387 (2022) 012025 doi:10.1088/1742-6596/2387/1/012025

using virtual machine servers, and provides unified interactive services to all kinds of users through the green national network[8].

Other systems such as marketing business application system and power consumption information acquisition system realize data integration with the new power load management system through the data console. The overall deployment architecture of the new power load management system is shown in Figure 4.

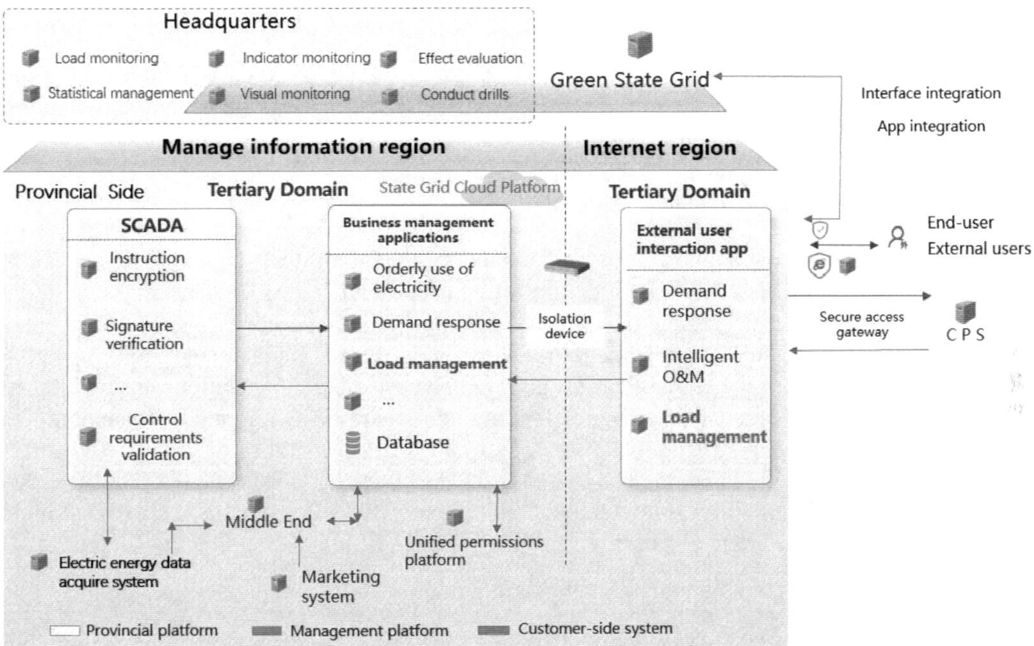

Figure 4. Deployment framework of new power load management system

3. Demonstration application

The new power load management system designed in this paper constructs a variety of business application modules, such as load control customer management, equipment file management, load monitoring, load characteristic analysis, load control scheme formulation, load regulation and implementation, load implementation monitoring, implementation effect analysis, operation management, power safety, early warning monitoring, command drill, etc. As a demonstration application in Ningxia and other provinces, this system is used to manage all kinds of loads on the user side, support demand response and be orderly and useful. The system has begun to take shape. Since the system was put into use, on the one hand, energy consumers can formulate power consumption plans timely, accurately and effectively according to the management plan of the new power load management system; On the other hand, energy service providers can timely monitor the power consumption of energy consumers during demand response and orderly power consumption implementation, and restrict their unplanned power consumption behavior. After the system went online, it continued to be stable and achieved good benefits.

4. Conclusion

Since the trial operation of the new power load management system designed in this paper, it has continued to operate stably, meeting the business needs of energy consumers and energy service providers, and proving the feasibility of the design. In future practice, we plan to introduce machine learning and artificial intelligence technology to better improve the ability of system services.

Acknowledgments
This research was supported by the project of Research and application of smart energy cloud of NARI-TECH Nanjing Control System Ltd. Grant No. 524609210054

References

[1] Zhou, Z. C., Gong, G. J., Cheng, W., Si, Y. Q and Zhou, G. (2020) Design of Intelligent Energy Service System Based on Industrial Internet ,IOP Conference Series Earth and Environmental Science 2020 International Conference on New energy, Power and Environmental Engineering, vol 696, pp 79-83.
[2] Fan, Y. J. (2020) Design of Load Management System Based on Smart Grid , Electric Power System Equipment , (20), pp144-147.
[3] Xu, S. S. (2019) Application of power load management system in electricity marketing management, China New Technologies and New Products ,(17) ,pp133-134.
[4] Wang, R. Z., Sun, Y., Feng, X. A., Qi, B. (2008) Analysis of data security algorithm for electric load management system, Power System Protection and Control, 36(2), pp54-57,63.
[5] Zhang, J., Zhao, S. B. (2022) Application of intelligent substation operation and maintenance management system based on inter-net of things technology , Information Technology, (2), pp167-172.
[6] Cai, X. W., Chen, M. X., Chen, G., Fang, Y. C., Zhang, S. X., Cheng, H. Z. (2021) Calculation Method for Maximum Allowable Access Capacity of Distributed PV in Active Distribution Network Considering Active Management and Demand-Side Management, Electric Power Construction, 42(5), pp38-47.
[7] Ding, Y., Wang, B. Y., Yuan, F., Wang, Y. J., Zhang, K., Tian, L. (2020) Secure Data Aggregation Scheme in Smart Grid with Third-Party Arbitration, Acta Electronica Sinica, 48(2), pp350-358.
[8] Bai, X. L., Zhang, K., Feng, J., Zhang, L. L., Chen, Ke., Gong, F. X. (2021) Design of demand-side management platform for fine-tuning electrical load control, Power Demand Side Management, 23(5), pp86-90.

Explore the verification and improvement method of insulation of 10~35kV high voltage switchgear

Chuansheng Luo

Nanning Power Supply Bureau,Guangxi Power Grid Corporation, Guangxi, China.

*13597112939@139.com

Abstract. In the rapid development of China's power system, the number of applications of high-voltage switchgear is increasing. Every year, tens of thousands of switchgear are put into the power industry. Since the performance and quality of high-voltage switchgear directly determines the operation efficiency of the system and has a positive impact on the safety production and effective power supply of the power grid system, scholars from all over the world have gradually strengthened the research on the insulation failure of switchgear and conducted verification and analysis based on practical cases. In this paper, based on the understanding of the development status and fault research technology of 10-35kV high voltage switchgear, according to representative insulation accident cases, taking the high voltage switchgear in a certain area as an example to verify and analyze, and finally put forward effective improvement methods.

1. Introduction

Through systematic investigation of switchgear accidents in power system of our country in recent years, it can be found that insulation accidents account for a high proportion of the total accidents, and are very easy to cause serious consequences such as equipment damage and bus voltage disappearance, which bring irreparable losses to the country. Specific typical insulation accident cases are shown in Table 1 below[1]:

Table 1. Typical insulation accidents

93.10.25	Guangxi XX Power Plant	Due to the low insulation and three-phase short circuit of 6kV section 1 bus, the fire was caused, and the secondary protection problem amplified the accident, causing the accident to affect the 220kV and 500kV systems in Guangxi at the same time, which caused the disconnection between parts of Guangxi and the main network, and the disconnection between 500kV system and Guangdong. 34 6kV switchgear devices and 1632 meters of severely laminated cables were burned down. The hV bushing of the hV standby transformer is broken, and the hV and LV coils are damaged due to long-term short-circuit current.

94.1.30	Shandong XX Power Plant	The bus of 6kV section iii B is short circuited by three-phase arc light due to the condensate single-phase discharge to ground, resulting in the disappearance of bus voltage, resulting in the shutdown of two machines
96.9.21	Shandong XX Power Plant	The 6kV bus of unit 3 is overvoltage caused by porcelain bottle grounding, resulting in arc grounding, three-phase short circuit and burning of switch cabinet
96.12.21	Shandong XX Power Plant	The 6kV section ii A busbar insulator of station A broke down and flasmowed to the ground, resulting in three-phase short circuit. Due to protection design problems, the accident was enlarged and 10 sides of switch cabinet, several cables and two large transformers were burned down.
97.10.20	Shandong XX Power Plant	Overvoltage was caused by pump operation in section A of 6kV, resulting in discharge between switch and switch cabinet, three-phase short-circuit of 6kV bus and burning of switch cabinet, resulting in power loss of Unit 1 and shutdown of unit
98.7.13	Shandong XX Power Plant	In Shizhong station, 10kV CT caused short circuit between phases, trip of main transformer and burnt out switch cabinet due to overvoltage operation
98.7.15	Shandong XX Power Plant	No. 1 high power plant 10kV side three-phase support porcelain bottle surface climbing point flashover, causing the switch short circuit burned, 6kV,35Kv, 110kV bus voltage disappeared, two machines trip and stop.

Deep exploration of common switchgear insulation accidents shows that domestic switchgear often works under extreme conditions such as pollution, condensation and high altitude, so it is easy to find that there are many weak links in insulation. From the perspective of power system operation, switch cabinet is an important distribution equipment on the power line, its internal collection of isolation switch, circuit breaker and related protection devices, mainly used to disconnect and close the normal power line, transmission and switching load. It can be in the power system in the operation of equipment and lines suddenly failure, timely and effective disconnect lines, so as to ensure the safety of important equipment, at the same time has been out of operation of equipment and lines for reliable grounding, so as to ensure the safety of the operation of staff [2].

Regular maintenance of switchgear is an effective way to ensure its safe operation. However, there are many switchgear in the distribution network and the maintenance task is heavy. Blindly disassembling or replacing the important equipment in the switchgear may not only lead to artificial destruction of the original normal equipment but also cause unnecessary waste. Therefore, how to effectively detect and evaluate the insulation condition of switchgear without affecting its normal operation is the focus of the electric power operation department.[3-4]

Partial discharge and partial overheating will lead to decomposition of organic insulating materials and gas insulating media in the switch cabinet, and a series of complex chemical reactions with trace oxygen and water in the equipment. Especially for the fully enclosed degraded gas insulated switchgear, partial discharge may lead to the generation of H2S, SO2, HF, CO and other derived gases, which will not only lead to the reduction of the insulation performance of the equipment, but also corrode the internal metal parts of the equipment, and further lead to the decline of the insulation performance of the equipment. The detection process of partial discharge is shown in Figure 1 below:

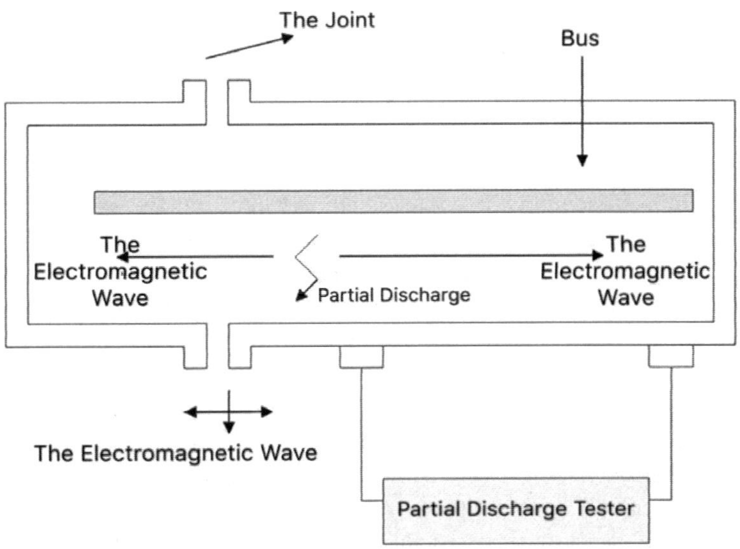

Figure 1. Detection diagram of partial discharge

Therefore, researchers in various countries have proposed effective detection methods for these two problems. For example, high-frequency current detection technology. This Technology has been studied in related PD research institutions and companies in various countries to a certain extent. Currently, TechImp (Technology Improvements) in Italy is the most popular research report and field application. TechImp was founded in 1999. Based at the University of Bologna, TechImp develops online and offline insulation diagnostics for power system equipment. UHF detection technology refers to the PD detection system based on ultra-high Frequency (UHF) principle, which was first proposed by Boggs and Stone in 1982. [5-6] This method uses electromagnetic wave signals radiated by PD in the switch cabinet for detection. Because electromagnetic wave has rich frequency components, it covers information from low frequency band to microwave band. The advantages of UHF technology are mainly reflected in its high sensitivity and excellent anti-interference ability, which makes it very suitable for on-site electric detection of equipment in complex electromagnetic environment. It should be noted that although UHF technology has great potential and a lot of research achievements have been made, it cannot fully meet the needs of engineering applications. Ultrasonic PD detection method is one of the early PD detection methods. As a non-invasive detection method, it has been the main PD detection method of power equipment before the uHF detection method was proposed. For transformer and switch cabinet, ultrasound, as a detection and auxiliary positioning technology, is the key to achieve accurate positioning of PD source of equipment. The research on ultrasonic PD detection is more in-depth in the Norwegian Electric Power Research Institute. Based on the research results of the institute, the former Norwegian TransiNor Company (now part of the American Doble Company) developed AIA PD ultrasonic detector. Based on the understanding of insulation failure detection technology of 10-35kV high voltage switchgear, this paper verifies and analyzes the insulation failure of high voltage switchgear in a certain area according to the current accumulated experience, and puts forward appropriate improvement countermeasures.

2. Method

Based on the application of power supply facilities in China, it can be seen that high-voltage switchgear has been widely used in the power system, and the actual improvement involves the following two points[7-8]:

2.1. Study the components of all-working high-pressure switchgear

The insulation technology along the surface is regarded as the basic condition of practical research, and the latest insulating elements are selected for 10~35kV high voltage switchgear through establishing good cooperation and exchange relations with relevant manufacturers, which specifically involves insulating connectors, pillar insulators, insulating bushings, transformers, contact boxes and other contents. Among these components, the staff should optimize the design of the specific structure and component installation size to improve the insulation creeping distance, in which the specific creeping distance of porcelain along the surface should not be less than 1.8cm/kV, and the specific distance of organic insulation along the surface creeping pad should not be less than 2.0cm/kV, so as to further improve the insulation performance of the internal components of 10-35kV high voltage switchgear. According to the flow chart analysis of intelligent high-voltage switchgear shown in Figure 2 below, impact condensation experiment, pollution insulation experiment and power frequency insulation experiment should be conducted on these components.

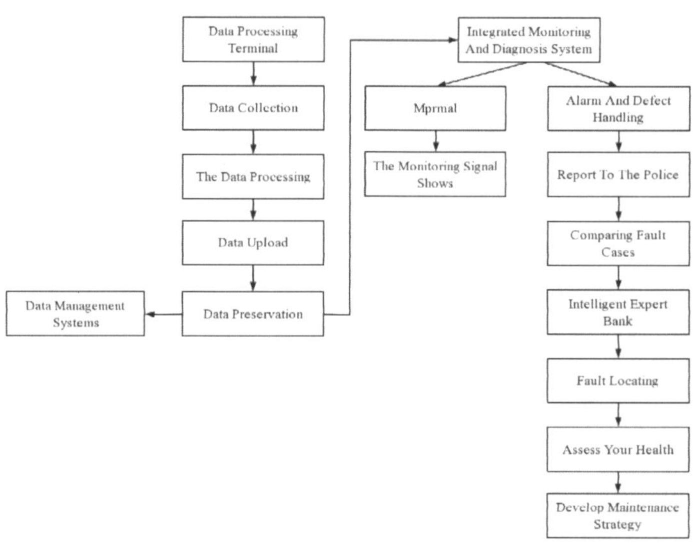

Figure 2. Flow chart

The system understands the final experimental results and finds that these components meet the insulation verification requirements after modification. For example, the 10kV KYN28 switch cabinet can use the new contact box, because the thickness of the insulation layer has been optimized design, can ensure that the outer insulation modeling becomes more scientific and reasonable. The busbar and static contacts can be fixed with five bolts to reduce contact resistance and heat. At the same time, to add air vents on the back, so as to facilitate the ventilation and heat dissipation of components. Because of the larger creepage distance of the contact box, the insulation performance of the high voltage switchgear can be improved in polluted environment or condensation environment. In addition, composite insulation mainly using SMC insulating version optimized design, the content is the use of unsaturated polyester glass fiber reinforced plastic sheet membrane after the compression molding, so the application materials and related products have high mechanical strength and flame retardant, in the practical application of the leakage resistance and good mechanical properties, It has positive influence on improving insulation performance of high voltage switchgear.

2.2. Optimize insulation of high voltage switchgear

In view of the current market development, there are obvious deficiencies in the insulation parts selected in the design and application of high-voltage switchgear in China, which leads to the insulation performance of the high-voltage switchgear applied in power system can not achieve the desired effect, or a large number of insulation failures occur in the working state, resulting in insulation flasticity accidents. Therefore, on the basis of clarifying the operation structure of intelligent power system, the staff should put forward effective improvement countermeasures according to practical problems. According to the operation structure diagram of intelligent power system as shown in FIG. 3 below, the following improvement measures are proposed in this paper: Firstly, the insulation busbar should be utilized to continuously optimize the insulation performance of wireless in high-voltage switchgear. For example, in the 10kV KYN61 switch cabinet, the internal selection is fully insulated bus, which can not only guarantee the insulation of the system design on the basis, but also reduce unnecessary safety problems; Secondly, priority should be given to selecting insulating components to improve the insulation performance of system design. At the same time, the new material and new process should be optimized to accurately grasp the structural parameters to provide effective basis for subsequent inspection and analysis; Again, in conjunction with the composite insulation, ensure that the charged body and in the 10 kv high-voltage switchgear insulation version of the air gap between not less than 30 mm, and the charged body in the 35 kv high-voltage switchgear and insulation between the air gap no less than 60 mm, only in this way can ensure the high voltage switch cabinet in power system, give full play to its insulation performance; Finally, the minimum air spies in the high voltage switchgear are strictly controlled to avoid discharge accidents, which directly harm the physical and mental health of surrounding residents[9-10].

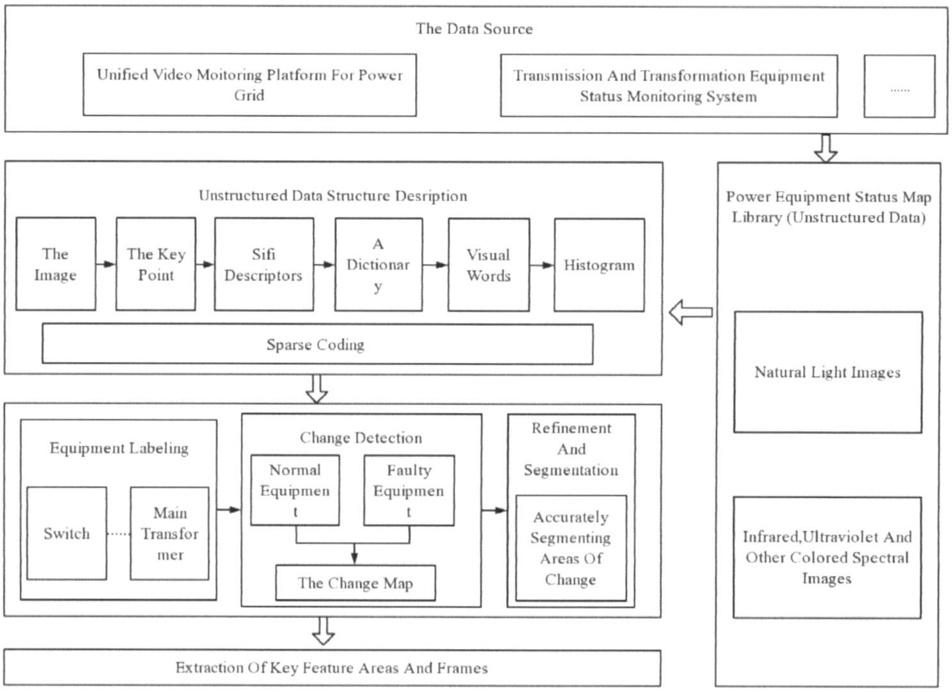

Figure 3. Operation structure of intelligent power system

3. Result analysis

In this paper, on the basis of clarifying the current high voltage switchgear insulation improvement measures, 10kV high voltage switchgear as an example, using uHF method and voltage method live test and other methods for analysis. It should be noted that because there are complex interference sources in the substation, the actual insulation verification analysis is difficult. The staff can combine transient to learn from voltage method and UHF method to improve the effectiveness of the actual verification analysis.

On the one hand, it can be seen that a power supply enterprise has carried out insulation verification on 10kV high voltage switchgear in the area through the preliminary judgment of local leakage scheme by voltage method live test, thus finding hidden fault signals. The operation structure of the substation is shown in Figure 4 below. After studying the partial discharge phenomenon in detail, the staff finally found abnormal signals when detecting the switch cabinet of no. 3 high voltage chamber. The high voltage chamber contains 23 cabinets, divided into two rows. The final results show that the signal amplitude of the high-voltage switchgear is larger than the background value, which proves that the signal originated from the switchgear. The signal amplitude of the switchgear on the left is significantly higher than that on the right, while the signal amplitude variation of the two rows is basically the same. It can be seen that the signal amplitude of the high voltage switchgear on the left is higher than that of other switchgear, showing a trend of unilateral attenuation as a whole. It is proved that partial discharge fault occurs in 532BCT high voltage switch cabinet.

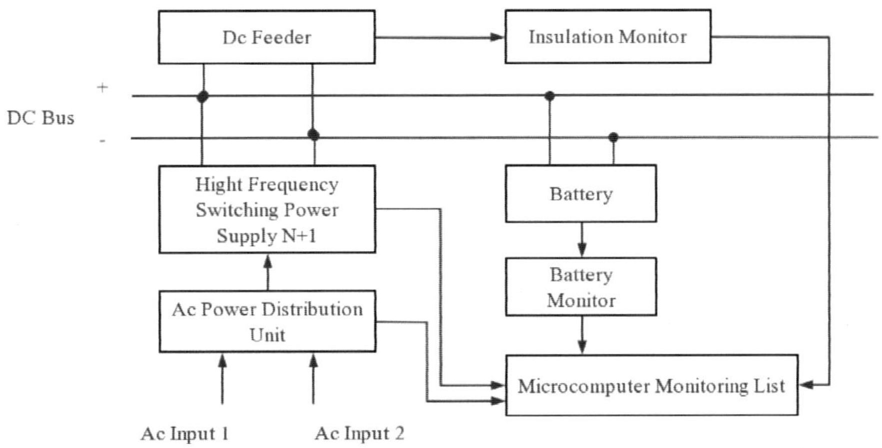

Figure 4. Operation diagram of substation

On the other hand, uHF method is used to judge and analyze local housing cases. An enterprise has carried out insulation verification for the local 10kV high voltage switchgear with partial discharge charged. According to the analysis of oscilloscope waveform change, the discharge is basically controlled within the peak range of 90 degrees to 180 degrees, and the discharge pulses of two probes are basically the same, so it can be judged that the discharge power supply belongs to the same.

4. Conclusion

To sum up, as an important part of modern power grid construction and operation, high-voltage switchgear has certain defects in its working state, so the staff should accumulate experience according to practice and think about solutions based on specific problems. According to this article research high voltage switchgear insulation validation and improvement measures are put forward by the analysis, the insulation fault is 10 kv ~ 35 kv high-voltage switchgear application of the main factors of the quality and efficiency, thus grid staff to strengthen the daily supervision at the same time, actively introduce advanced technology for testing analysis, and proven effective improvement measures are put forward. At the same time, it is necessary to do a good job in the technical training of professionals, encourage and support department employees to participate in the study and application of new knowledge and technology, so as to improve the insulation performance of high-voltage switchgear, guarantee the safety and stability of power grid system operation, and provide better power services for social residents.

Reference

[1] Tang, Q., Yang, Y., Wei, M., et al. (2021) Multi-physical Field Analysis of KYN61-40.5kV High Voltage Switchgear and Experimental Research on Insulation Characteristics of Components [J]. High Voltage Apparatus, 57(12):9.
[2] Wu, C. Q., Chen, J., Liu, F., Xia, X., Li, J. B., Wang, T. (2020) Design and application of performance Test platform for Temperature Sensor with Contact of High voltage Switchgear [J]. High Voltage Apparatus, 56(11):6.
[3] Leng, Z. Y., Chen, P., Zhang, S. H., Gu, Y., Miao, J., Li, D., Zhou, X. Y. (2020) Research on key

Technologies of Installation and Application of 40.5 kV Switchgear wall Bushing [J]. High Voltage Apparatus, 56(11):6.
[4] Guo, L. J., Wang, G. Q., Xu, S., et al. (2021) Research on Optimization of RADIO Energy Transmission System for DC High Voltage Switchgear [J]. Electric Drive, 51(12):7.
[5] He, J. M., Liu, J. G., Yi, F. F., et al. (2022) High Voltage Appliances, 58(3):6. (in Chinese)
[6] Cao, P., Xu, P., He, J. M., et al. (2021) China Electric Power, 54(10):9.
[7] Wu, H., Li, S. Y., Qi, Y. X., et al. (2020) Research on State Evaluation Method of High Voltage Switchgear Based on Comprehensive weight assignment and WRSR [J]. High Voltage Appliances, (2):6.
[8] Wang, C. X., Li, Y. H., Peng, Z. M. (2021) New Technology of Electrical Engineering and Energy, 40(11):8. (in Chinese)
[9] Lin, X. Y., Nie, Y. X., Zhu, T. C., et al. (2020) Research on three-dimensional Transient electric Field Calculation of High Voltage Switchgear [J]. High Voltage Apparatus, 56(10):7.
[10] Ren, S. Z., Zeng, X. M., Wu, J. F., et al. (2020) Development of Portable Detector for Partial Discharge in Switch Cabinet Based on TEV and Ultrasonic Wave [J]. Electrical Measurement & Instrumentation, 57(8):5.

Journal of Physics: Conference Series **2387** (2022) 012027 doi:10.1088/1742-6596/2387/1/012027

Judgment of Low-voltage leakage fault type and fault point finding method in power distribution platform area

Gang Hong[1]* ,Song Zhu, Hengying Huang, Xiangxiang Wei
Laibin Power Supply Bureau of Guangxi Power Grid Co. Ltd.

*hong_g.lbg@gx.csg.cn

Abstract. In the development of modern social economy and technological innovation, the technical application of the power industry is more and more demanding, and researchers have gradually strengthened the judgment and search technology research of low-voltage leakage fault type in distribution platform area. From the perspective of practical application, when the leakage fault occurs in the low-voltage line, it is necessary to find the area as soon as possible and put forward an effective solution, so as to avoid further expansion of the fault problem. In this paper, on the basis of understanding the causes of low-voltage leakage faults in distribution platform area, according to the various judgment schemes proposed by scholars at home and abroad, an intelligent fault identification and positioning system is deeply discussed, so as to clarify the types of faults and find methods. Finally, the practical results show that the causes of leakage faults are different, and the practical judgment and search methods should be targeted and diversified.

1. Introduction

Under the background of new era, because people the growing demand for electricity in the development of modernization in our country, gradually enhance the security and stability of power system operation, and the development practice to face the opportunities and challenges, has carried on the empirical analysis to distribution area of low pressure leakage fault, both to reduce unnecessary safe hidden trouble, It can further improve the economic benefits of enterprise development and ensure the normal operation of social production and life. [1] From the perspective of practical operation, leakage fault, as the most common type of low-voltage distribution radio operation, if it appears in the working state, is likely to cause excessive loss of power energy and other problems, directly threatening the life and property health of social residents. Therefore, effective solutions should be proposed for specific problems in practical exploration. And rational use of advanced technology to quickly find the fault area, so as to enhance the reliability of power system operation. According to the accumulated experience of low-voltage leakage in distribution platform area, there are two main reasons for leakage faults: on the one hand, leakage faults exist at the user end. Client problems are the main factors leading to leakage faults in low-voltage distribution stations, such as insulation damage of pumps, refrigerators and other equipment, not strictly in accordance with the standard installation of indoor lines, unauthorized wiring or stealing electricity behavior. According to the practical investigation, the low voltage leakage fault is usually caused by the residual current action protection fault. Because the leakage fault of the client side has certain concealment, it is more and more difficult to find and judge the actual fault. On the other hand, low voltage line leakage fault. In the construction of power distribution area lines, if there are crossing points or redundant trees, it will also affect the line transportation to a certain extent, which is another factor for the leakage of low-voltage lines. If the operation is not carried out in strict accordance with the requirements during connection, it will also increase the difficulty of fault diagnosis and search. At present, some scholars have proposed a 400V distribution network portable non-contact leakage detection device in practical exploration, which includes magnetic field signal detection module, signal amplification module, bandpass filter module, data processor module and LED display module. In practice, the use of portable leakage detection device can not only achieve non-contact detection, high security, good mobility, easy to carry, but also greatly improve the location and troubleshooting of leakage faults. On the one hand, the hand-held signal detection device has comprehensive signal detection, processing and communication functions, receive the specific signal sent by the signal source, can select the fault line according to the detection, convenient and fast correct choice of fault point; On the other hand, handheld data receiving device is easy to carry, operation and maintenance personnel hold mobile device, electronic screen can display measured data, the system has voice function. This paper mainly discusses a power distribution system and fault identification and location system, and studies its practical operation and existing fault search methods, so as to ensure the safety of power system operation on the basis.[2-3]

2. Method

2.1. System features

According to the analysis of the system structure diagram shown in Figure 1 below, the characteristics of the overall system design involve the following points: First, the wireless relay terminal. This module is mainly designed in the incoming line side of the user table box, which is used to collect the user's electrical information and upload the relevant content as relay information to the wireless centralized terminal corresponding to the incoming line side of the line where the wireless relay terminal is located. Secondly, wireless centralized terminal. This module is mainly designed on the incoming side of the branch line, which is used to collect the electrical information on all branch lines. Finally, the collected information, received information and terminal content are uploaded to the wireless centralized terminal or fault identification and location terminal on the branch line and the line. Finally, fault identification locates the terminal. This module is mainly designed in the outgoing side of the intelligent distribution terminal, which is used to collect the centralized information of the wireless centralized terminal on the outgoing side of all branch boxes, so as to complete the fault identification and location analysis[4-5].

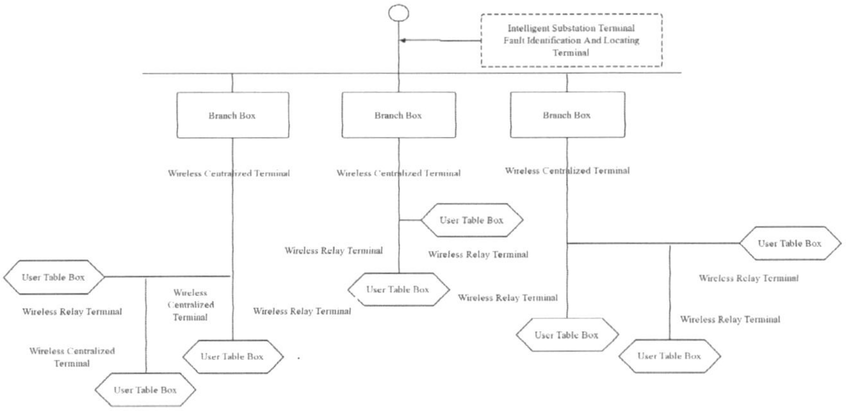

Figure 1. System structure diagram

2.2. System technology

This paper studies distribution area of low pressure system is mainly used to solve the current problem of leakage fault location is more difficult, in practical application, the system includes wireless relay terminal and wireless terminal, fault identification terminal this three part, can help staff faster access to the user table box and branch lines in all kinds of information. The fault identification and location terminal can quickly identify and locate faults based on the mapping between user table boxes and lines. When the current change exceeds the set value, it is proved that there is a fault. The electrical information of the set time before and after the fault should be extracted. The specific communication diagram of the system is shown in Figure 2 below[6]:

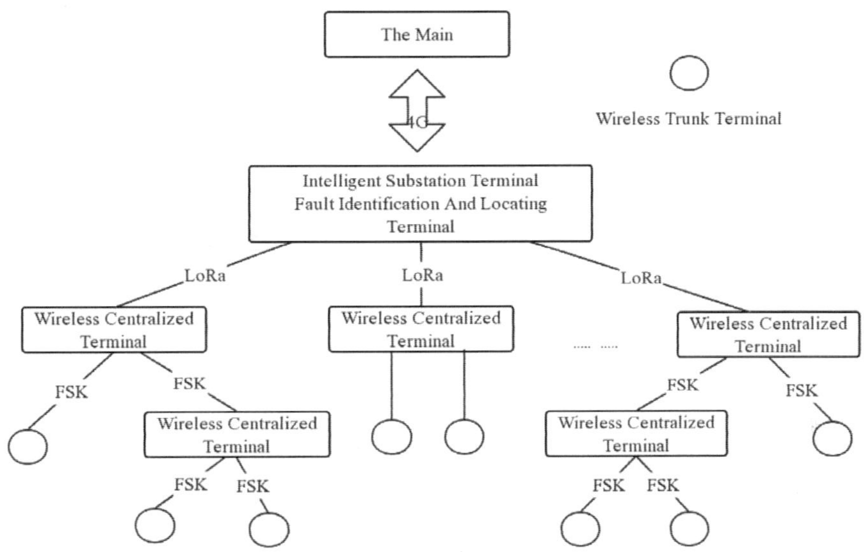

Figure 2. System communication diagram

Shown in figure 3, according to the following system information transmission schematic analysis shows that the users table box to install on the branch box outlet side branch lines, branch box to install in the smart match the outlet side of the terminal, it can ensure the user's information appliances can be collected by collection module, and as the ultimate information uploaded to the specified area. Among them, the location information of the wireless relay terminal and the wireless centralized terminal will be obtained through the GPS module, and FSK will be used to communicate with each other, while the centralized terminal and the fault identification and location terminal will be used to communicate with LoRa[7-9].

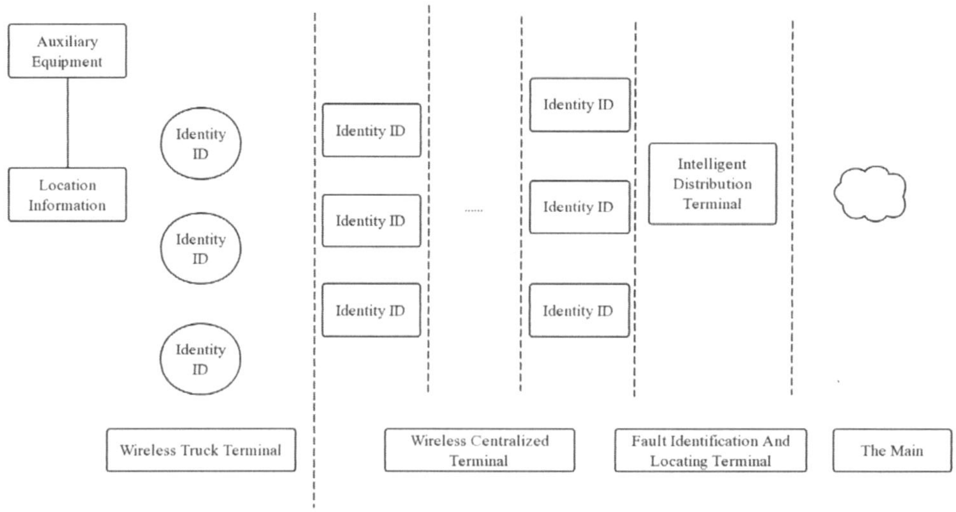

Figure 3. Schematic diagram of system information transmission

The fault diagram of the low-voltage leakage fault location and identification system in the distribution platform area studied in this paper is as follows:

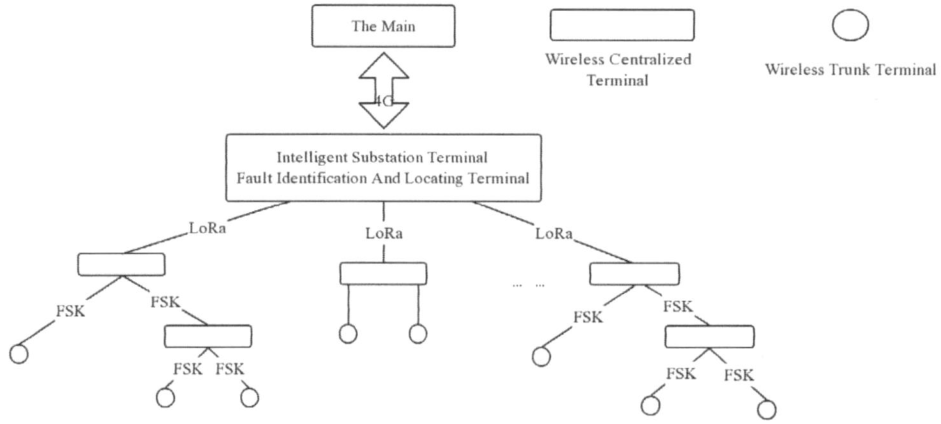

Figure 4. Schematic diagram of system failure

From the perspective of practical application, the relay terminal of the system does not need to enter the user load, nor need to deal with the power outage, so it can obtain the electrical information of the line in real time and upload it to the wireless centralized terminal. The wireless centralized terminal is installed on the incoming side of the line branch and the outgoing side of the branch box, which is relatively speaking. It can obtain the electricity information on the line at the same time, receive the regular data from the subordinate wireless relay terminal or centralized terminal, and send it up to the superior wireless centralized terminal or fault identification and location terminal. The fault identification and location terminal will be installed at the outgoing side of the intelligent distribution and transformation terminal, and the fault identification and location will be completed according to the information obtained by the wireless centralized terminal at the outgoing side of each branch box of the distribution and transformation terminal. These three module designs have unique identifiers, and prior to installation, all terminals store the identifiers of their parent terminals according to the initial topology saved by the master station. At the same time, when installing the terminal, use the auxiliary scanning device to obtain their location information and store it effectively.

3. Result analysis

3.1. System results

According to the following figure 5 shows the flow chart of system analysis, this article studies the intelligent fault identification system applied to the distribution area of low pressure leakage fault detection system, can improve the efficiency of practical work, and can quickly find the fault location, accurately judge the fault types, puts forward the effective solution.

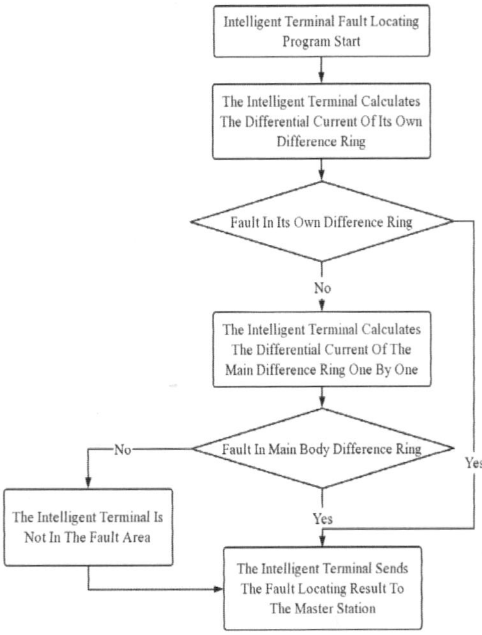

Figure 5. System flow chart

Firstly, the information will be verified by security encryption chip when uploading, so as to ensure the perfection and security of data information transmission. Secondly, the fault identification and location terminal uses LoRa communication technology to obtain the topology of the wireless centralized terminal line at the outlet side of the branch box, and supports access to real-time information of multiple lower terminals or sensors. According to the location information and electrical fault information of the wireless relay terminal and the wireless centralized terminal, the terminal position in the topology diagram of the distribution platform area is gradually determined by edge calculation. Finally, in the use of the wireless relay terminal, the installation of line current monthly failure, according to fault interception function, for 20 seconds before and after the fault occurred during the fault information, upload it to the superior centralized wireless terminal are studied, finally the results of the study and identification information to pass to intelligent distribution terminal, Therefore, the updated topology is provided to the master station to realize the topology update of the distribution platform area.[10]

3.2. Search technology
On the one hand, clamp ammeter. This method means that when the circuit is working normally, the synthetic current in the three phase four wire usually meets this condition. If four wires are placed in a clamp ammeter at the same time, the values of the primary side and the secondary side can be displayed inside the ammeter amplification system. When detecting the leakage fault, if the switch of the leakage fault equipment cannot appear on the river, it is necessary to detect whether the specific location of the fault is the line or the switch itself. If the line leakage fault occurs, and there is a trip phenomenon, then only need to close the leakage switch, let it stop working.

Megohm meter, on the other hand. If the leakage fault switch can be closed, but the trip phenomenon occurs again after a period of power supply, the difficulty of finding the actual leakage fault will be greatly increased. Integration of accumulated experience shows that more than 90% of the leakage accidents are due to improper use of electrical appliances by users. Although there has been a degree of innovation and innovation, the final leakage switch has been widely installed, but the standard leakage current is up to 30 ma. If the leakage current is only 20 mA, it will be difficult to use this switch. There are multiple faults in the line, and the leakage switch does not meet the rated value, so the leakage switch is difficult to maximize the role. In this case, the staff should carefully check the main cause of tripping, and analyze whether it is caused by the user's electricity or overhead line. If it is the former, the staff will use a 500-volt megohmmeter to telemetry all the households of the faulty line, and turn off the user switch to test the insulation resistance remotely. Generally speaking, in sunny weather, insulation resistance should reach more than 0.5M ω; But in rainy and snowy weather, the insulation resistance is more than 0.1M ω. If this standard is not met, a megohmmeter is used for rapid location.

4. Conclusion
To sum up, the system to know distribution area in low voltage main influence factors of leakage fault, in the domestic and foreign scholars on the basis of research results, failure to make use of existing technical methods reasonable judgment, to help repair people quickly find effective countermeasures, so that can speed up the pace of technological innovation of science and technology, and the future economic development lay the foundation for the nation. Therefore, based on the low-voltage leakage fault of distribution platform area, China has put forward a variety of positioning system and search technology, and gradually strengthen the training of professional and technical personnel, create a positive and enterprising convenient enterprise operation environment, in order to provide high-quality convenient services for social residents. At the same time, we should actively learn from the advanced technical research results at home and abroad, pay attention to the basic situation of China's power system operation, build a targeted and timely fault location detection system, reduce the probability of low-voltage leakage fault in the distribution platform area.

Reference
[1] Guo, S., Lin, J. Y., Wang, P., et al. (2020) Topology Verification of low-voltage distribution Network based on ROF Outgroup Detection [J]. Control Engineering, 27(1):6.
[2] Zhang, C. Y., Dou, X. B., He, G. X., et al. (2020) Power Automation Equipment, 40(9):8. (in Chinese)
[3] Shen, P. F., Wang, X. Y., Zhang, H. L., et al. (2021) Risk assessment of low-voltage interconnected distribution platform area considering large-scale fast charging load [J]. China Electric Power, 54(5):9.
[4] Li, K. M., Jiang, Y. Q., Huang, S. F., et al. (2021) Identification method of Low-voltage Topological Structure in Distribution Platform Area Based on DTW Distance and Cluster Analysis [J]. Power System Protection and Control, 49(14):8.
[5] Xu, W. T., Wang, X., Wu, Z. X., et al. (2022) Modern Electronics Technique, 45(4):6. (in Chinese)

[6] Zhang, M. M., Qin, P., Chen, Y. J, et al. (2020) Power Capacitors and Reactive Power Compensation, 41(5):7. (in Chinese)
[7] Chen, Z. A., Huang, C., Zhang, Z. D., et al. (2021) Topology Identification method of Low-voltage Distribution Network Based on T-type Grey Correlation degree and KNN Algorithm [J]. Power System Protection and Control, 49(1):7.
[8] Luo, J. T., Zhang, J. M., Yao, L., et al. (2021) Optimization modeling and application of low voltage user phase identification based on voltage and power data [J]. Automation of Electric Power Systems, 45(7):9.
[9] Wan, L. Y., Ying, D. Z., Xin, J. Q. (2021) Electrical Measurement & Instrumentation, 58(10):7. (in Chinese)
[10] Guo, P. Z., Cai, T. T., Deng, Q. T., et al. (2022) Electronic Technology and Software Engineering, (4):4. (in Chinese)

Monitoring and prediction of thermal failure of high-voltage switchgear insulation materials by studying the thermodynamic properties of FR-4 epoxy resin and the characteristic gas of thermal decomposition

Mengzhu Hu[1], Bin Tang[1,*], Qinqin Liang[1], Zongchang Luo[1], Chuansheng Luo[2], Jialin Wang[3], Longfei Zhang[1], Liping Zhu[1]

[1]Guangxi Key Laboratory of Intelligent Control and Maintenance of Power Equipment, Electric Power Research Institute of Guangxi Power Grid Co., Ltd.
[2]Nanning Power Supply Bureau of Guangxi Power Grid Co., Ltd.
[3]Guangxi Power Grid Co., Ltd.

*Corresponding author e-mail: tang_b.sy@gx.csg.cn

Abstract. FR-4 epoxy resin is a common material used as internal parts in high voltage switchgear. In this paper, we adopted the accelerated thermal aging method to explore the thermal aging behavior of FR-4 epoxy resin, simulating the practical application of this material in high voltage switch. The effects of thermal aging on the thermodynamic properties and thermal decomposition characteristic gases of FR-4 epoxy resin were characterized by FT-IR, DSC, TG and GC-MS analyses. The experiment results showed that thermal aging could destroy the internal crystal structure of FR-4 epoxy resin, which could lead to this material easier to be decomposed after heating. We also detected that the thermal decomposition rate of FR-4 epoxy resin aging was greatly accelerated compared with the materials that had not been aged. In addition, unaged FR-4 epoxy resin could release 9 main volatile gases under 120 °C, and irreversible defects or structural damages occurred in the materials along with the volatilization of small organic molecules. Furthermore, after continuous thermal aging at 120 °C for 672 h, the molecules of FR-4 epoxy resin were more cracked and volatilized, we observed a more obvious phenomenon of thermal decomposition in this materials. In total, we compared the types of thermal decomposition gases and the decomposion mechanism of aged and unaged FR-4 epoxy resin. In application, our results can provide reference for condition monitoring and fault prediction for this material in high voltage switchgear in the future.

1. Preface

FR-4 epoxy resin is one of the commonly used insulating materials for high-voltage switchgear. During long-term operation, it is prone to aging due to various stresses such as electricity, heat, and machinery, resulting in irreversible defects or structural defects. destroy. At present, domestic and foreign scholars have carried out some researches on the thermal decomposition process of different epoxy resin insulating materials [1-6]. The previous literature research mainly focused on the thermal decomposition characteristics of epoxy composites at different heating rates, but few related reports were based on thermal aging experiments on the thermodynamic characteristics of commercial epoxy resin insulation materials and the influence of thermal decomposition characteristic gases. During the actual operation of the high-voltage switchgear, the aging of epoxy resin is inevitable, and the thermal effect caused by electrical aging aggravates the thermal aging, which eventually leads to the decline of insulation performance and mechanical performance [7]. Therefore, it is of great significance to analyze the thermal aging behavior of FR-4 epoxy resin materials, and then to perform state monitoring and fault prediction for high-voltage switchgear.

In this paper, accelerated thermal aging is used to simulate the thermal aging behavior of FR-4 epoxy resin commonly used in high-voltage switchgear in actual operation. To explore the thermodynamic properties and thermal decomposition characteristic gases of FR-4 epoxy resin, Predict the aging state of FR-4 epoxy resin material under working temperature conditions [8], The components of the thermally decomposed gases at different temperatures were analyzed by GC-MS [9], To determine the type of characteristic gas and summarize its thermal decomposition law, The effect of temperature on the composition and content of thermally decomposed gases was systematically explored. It can provide an important reference for predicting the state of epoxy resin used in high-voltage switchgear and fault detection.

2. Experimental part

2.1. Raw materials and instruments

FR-4 epoxy resin; 400P Shixin cutting prototype, Korea Shixin Co., Ltd.; Nicolet 6700 Fourier transform infrared spectrometer, Thermo Fisher Co., Ltd., USA; STA409 PC/PG synchronous thermal analyzer, NETZSCH, Germany Instrument Manufacturing Co., Ltd.; Mettler Differential Scanning Calorimeter, Mettler-Toledo, Switzerland; DHG-9030A Aging Test Box, Beijing Luxi Technology Co., Ltd.; OTF-1200X Tube Furnace, Hefei Kejing Company; 0.1 L gas collection bag, Ningbo Hongpu Instrument Technology Co., Ltd.; QP2010 PLUS gas chromatography-mass spectrometer, Shimadzu Corporation, Japan.

2.2. Experimental process

(1) After the preparation of the sample, the thermal aging test was carried out in the air aging test box, the temperature was set to 120 °C, and the aging was performed for 672 h. (2) The samples before and after thermal aging were cut into small pieces of 1 cm × 1 cm × 5 mm, with a mass of about 1.5 g; (3) The samples were subjected to thermal decomposition experiments at 120 °C, 200 °C and 270 °C, respectively, and the The gas collecting device collects the thermally decomposed gas; (4) The gas generated by the thermal decomposition of the sample is analyzed by GC-MS, and the decomposition law of the characteristic gas is studied.

2.3. Testing and characterization

The following tests were performed on the samples before and after heat aging:

(1) Fourier infrared scanning range was 4000-500 cm-1, scanning 32 times, spectral resolution was 4 cm-1, using KBr tablet; (2) DSC analysis of the melting and crystallization of the sample, indium For temperature calibration, sample mass is 3-5 mg, nitrogen atmosphere, gas flow rate is 50 mL/min, heating rate is 10 °C/min, and temperature range is 60-300 °C; (3) Thermal decomposition of sample for TG analysis Weight loss characteristics, sample mass 2mg~4 mg, nitrogen atmosphere, gas flow 50 mL/min, heating rate 10 °C /min, temperature range 30 °C -800 °C ;(4) GC-MS gas chromatography-mass spectrometer, using SH-Rtx-Wax (30 m × 0.25 mm × 0.25 μm) elastic quartz capillary column, the column temperature was 45 °C, the constant temperature was 1 min, and the temperature was programmed at 10 °C/min to The temperature was kept at 250 °C for 10 min, the carrier gas was He, the pre-column pressure was 40 kPa, and the split ratio was 30:1. MS detector

adopts EI ion source, electron energy 70 eV, scan 29-450 u, quadrupole temperature 150 ℃, ion source temperature 230 ℃, electron multiplier voltage 2300 V, GC-MS interface temperature 280 ℃, each chromatographic peak The corresponding mass spectra were searched by computer spectral library (standard mass spectrum library NIST14s) and determined by manual analysis.

3. Results and Discussion

3.1. *Infrared (FT-IR) spectral analysis*

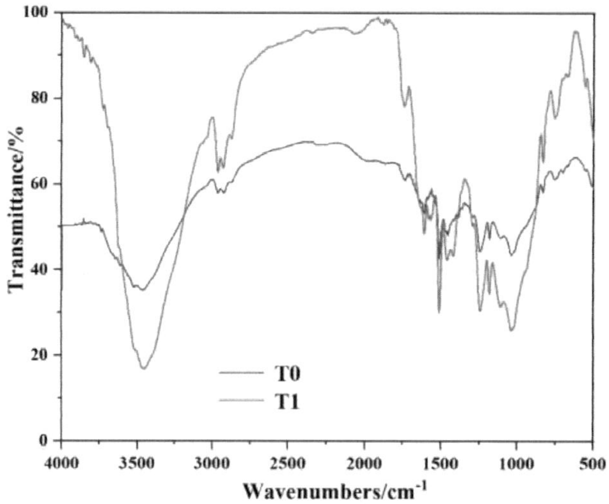

Figure 1. Infrared spectrum of FR-4 epoxy resin before and after aging

From the analysis of the results in Figure 1, it can be seen that the absorption peaks near 1460 cm-1 and 1509 cm-1 correspond to the characteristic absorption peaks of the benzene ring [10], indicating that the epoxy resin material contains a benzene ring structure; 2960 cm-1- The absorption peak at 2850 cm-1 is attributed to the contraction vibration peak of CH bond; the absorption peak at 1718 cm-1 is attributed to the characteristic absorption peak of carbonyl C=O; the absorption peak at wavelength 3435.49 cm-1 corresponds to The hydroxyl-OH absorption peak of epoxy resin; because the epoxy group in the epoxy resin reacts with the amine group in the curing agent, resulting in a large amount of hydroxyl-OH and amino-NH2 after its ring opening, so at 3356 cm- The stretching vibration absorption peaks corresponding to -OH and -NH2 appeared in the range of 1-3480 cm-1 [11].

From the comparative analysis of the infrared spectra of FR-4 epoxy resin before and after thermal aging in Figure 1, it can be seen that the benzene ring structure of the epoxy resin has changed after thermal aging. Compared with the sample before thermal aging, the sample after thermal aging is at 1413. New characteristic absorption peaks appeared at both cm-1 and 1560 cm-1, indicating the formation of carbonyl C=O or carboxyl-COOH[12].

3.2. *Differential Scanning Calorimetry (DSC) spectral analysis*

Figure 2 shows the DSC curves of the FR-4 epoxy resin material before thermal aging (T0) and after thermal aging (T1). Integrate the endothermic peak of the DSC curve corresponding to Figure 2, and obtain the corresponding results of the endothermic enthalpy according to the peak temperature of the endothermic peak and the integrated area of the endothermic peak, as shown in Table 1. It can be seen from Figure 2 that the DSC curves of FR-4 epoxy resin before and after thermal aging show a single hump-shaped endothermic peak, and the peak temperature is around 266 ℃ and 273 ℃. It should be noted that FR-4 epoxy resin is a thermosetting polymer material, and the matrix needs to undergo a curing reaction during the molding process, so that it has the characteristics of high cross-linking. But at high temperature, the material usually transforms from glass state to high elastic state, and the DSC curves of FR-4 epoxy resin before and after thermal aging show a long melting range endothermic peak. It can be seen that thermal aging does not directly affect the heat resistance of the material, but thermal aging will destroy the internal crystal structure of the material.

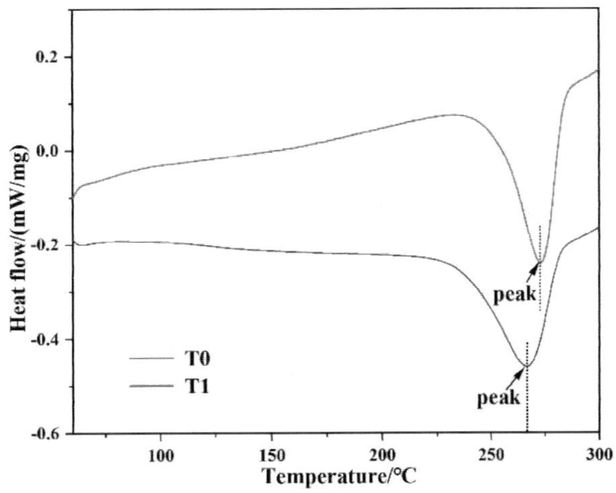

Figure 2. DSC diagram of FR-4 epoxy resin before and after aging

Table 1. DSC endothermic peak data of FR-4 epoxy resin before and after thermal aging

different cycles	peak	
	Peak temperature/℃	Endothermic enthalpy/mJ
T0（Not heat aged）	266.4	-174.32
T1（672 h heat aging）	273.1	-206.42

3.3. Thermogravimetric (TG) result analysis

The thermal weight loss curves of FR-4 epoxy resin before and after thermal aging are shown in Figure 3. From Figure 3, the weight loss characteristics of the sample can be divided into three stages: 30 °C-300 °C, 300 °C-450 °C and 450 °C-800 °C, as shown in Table 2.

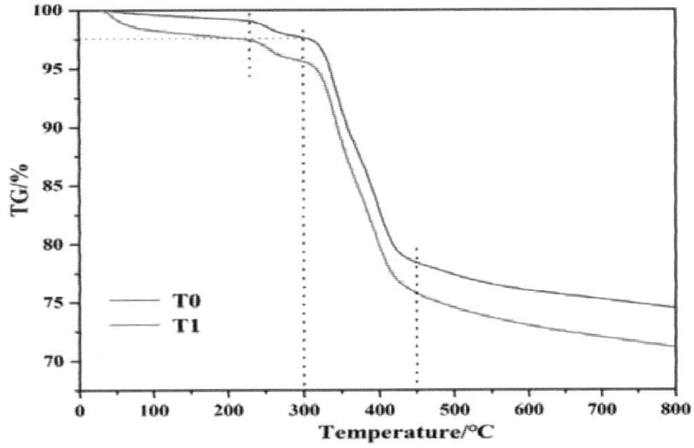

Figure 3. TG curve of FR-4 epoxy resin before and after aging

Table 2. Analysis of decomposition characteristics of FR-4 epoxy resin

Temperature/℃	Weight loss stage analysis	Analysis of decomposition characteristics
30-300	slight weightlessness	The small molecules and other substances in the sample are decomposed and volatilized;
300-450	rapid weightlessness	In the early stage of decomposition, random decomposition mainly occurs, and the macromolecular chain of FR-4 epoxy resin is broken from weak bonds, but the quality remains basically unchanged; in the later stage of decomposition, intramolecular and intermolecular transfer reactions mainly occur, and small molecules are exposed to high temperature. volatilization, causing a rapid decline in the quality of the material;
450-800	Quality tends to be stable	The residual rate remained basically unchanged, and the remaining part was mainly carbonization products.

It can be seen from Figure 3 and Table 2 that before and after thermal aging of FR-4 epoxy resin, the trend of T0 and T1 thermal weight loss curves are basically the same, but the thermal weight loss rate of epoxy resin after thermal aging is significantly higher than that of unaged epoxy resin. The thermal weight loss rate is lower, and the residual mass of the sample is lower.
According to the results of thermal weight loss, the thermal weight loss characteristics of FR-4 epoxy resin materials can be divided into three stages: (1) When the temperature is lower than 300 ℃, the material undergoes slow thermal decomposition before and after thermal aging, which may be due to the The internal small molecule additives are gradually decomposed and released, so that the mass fraction begins to decrease; (2) In the temperature range of 300-450 ℃, the epoxy resin material undergoes rapid thermal weight loss decomposition, and the corresponding weight loss rate increases sharply, which may be due to the internal The macromolecular network cross-linked substance undergoes chain scission and decomposes into small molecular substances, and the rapid volatilization leads to a rapid increase in the weight loss rate; (3) In the temperature range of 450-800 ℃, its decomposition is slow and tends to be gentle, and the residual rate basically remains unchanged. Change. The remainder is mainly carbonized or graphitized products.

3.4. Comparative study of characteristic gases before and after thermal aging

Based on the test results of DSC and TG, combined with the actual situation that the high-voltage switchgear usually operates at a lower temperature, the thermal decomposition and calcination temperature and holding time of the epoxy resin sample are determined, and the samples generated at this temperature are collected simultaneously. Thermal decomposition gas. Among them, the gas collection temperature of the samples before heat aging was set to 120 °C, 200 °C, and 270 °C; the gas collection temperature of the samples after heat aging was set to 200 °C, 270 °C, and the FR-4 epoxy resin was thermally decomposed. All gaseous products were tested by GC-MS. The GC spectra of the thermally decomposed gases of the samples before and after thermal aging are shown in Figures 4 to 6.

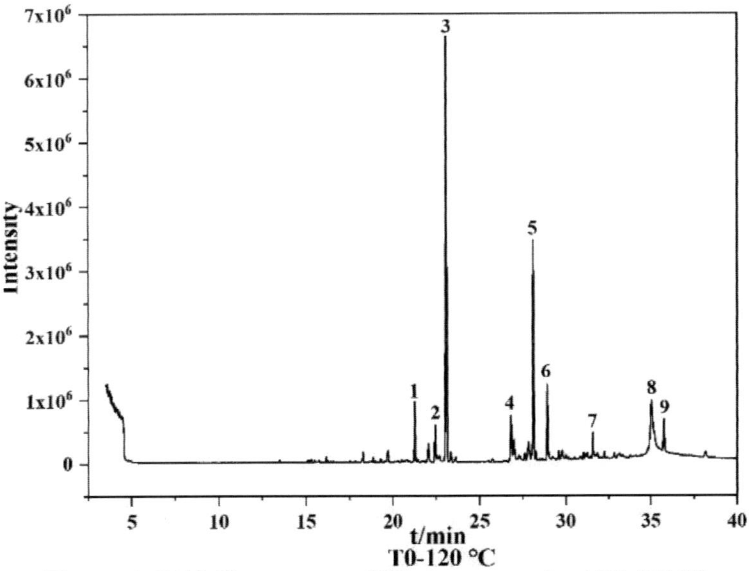

Figure 4. GC-MS spectrum of FR-4 epoxy resin at T0-120 °C

It can be observed from the experimental results in Figure 4 that the GC spectrum of the gas decomposition products of the unaged FR-4 epoxy resin material under the pyrolysis condition of 120 °C contains 9 main characteristic peaks (numbers: 1-9), according to The NIST14S.LIB database was searched and compared, and the gas species collected by the unaged FR-4 epoxy resin samples under the condition of thermal decomposition at 120 °C were analyzed in detail, including macromolecular hydrocarbons, alcohols, phenols, etc. The detailed gas types and molecular formula results are shown in Table 3.

Table 3. Main characteristic gases of FR-4 epoxy resin samples at T0-120 ℃

serial number	keep time	molecular weight	molecular formula	Molecular Structure
1	21.260	194	$C_{13}H_{22}O$	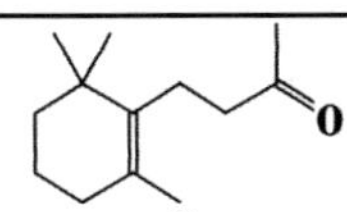
2	22.440	222	$C_{15}H_{26}O$	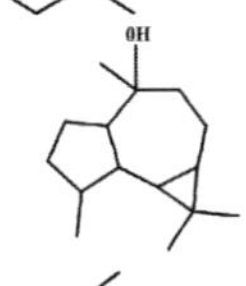
3	23.080	208	$C_{11}H_{16}N_2O_2$	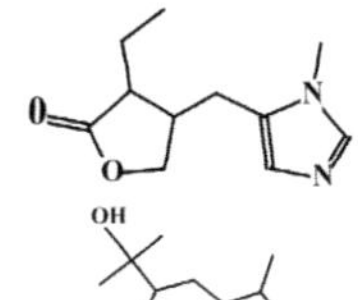
4	26.785	222	$C_{15}H_{26}O$	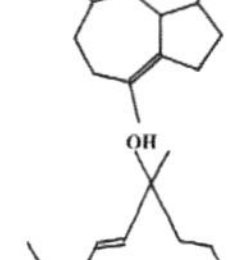
5	28.095	290	$C_{20}H_{34}O$	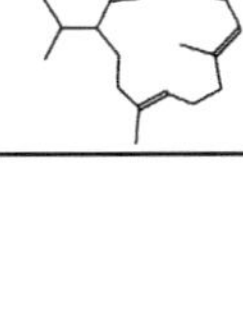

6	28.920	194	$C_{13}H_{22}O$	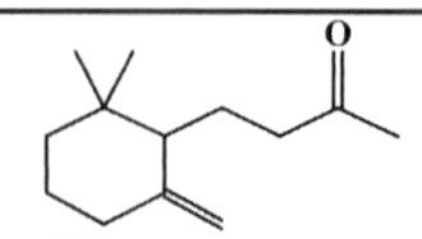
7	31.575	222	$C_{15}H_{26}O$	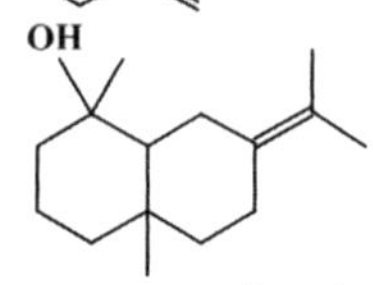
8	35.040	410	$C_{30}H_{50}$	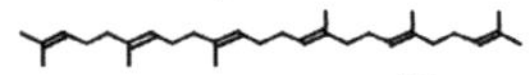
9	35.755	302	$C_{20}H_{30}O_2$	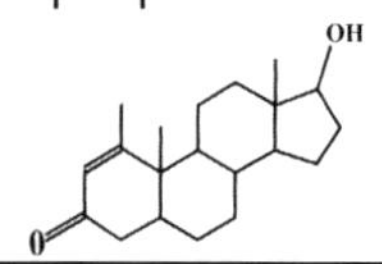

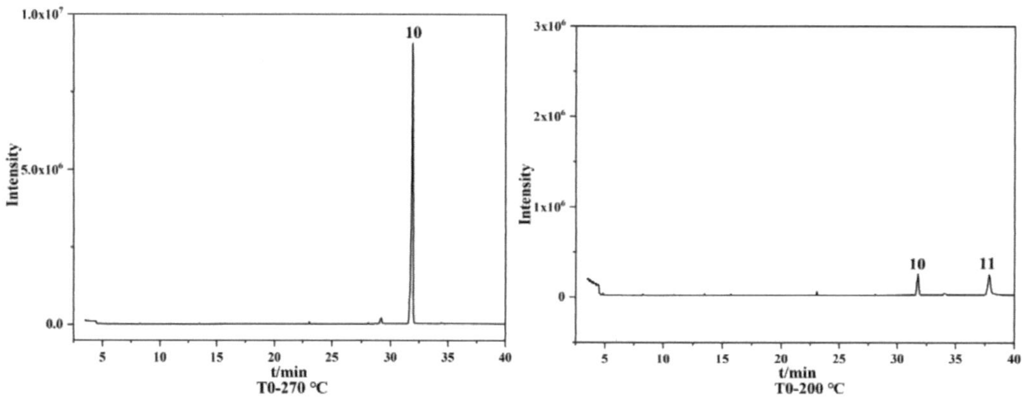

Figure 5. GC spectrum of FR-4 epoxy resin at T0-200 °C and T0-270 °C

From the GC spectrum results of the FR-4 epoxy resin sample at T0-200 ℃ in Fig. 5, it can be seen that the GC spectrum of the gas decomposition products of the FR-4 epoxy resin material with the T0 period under the pyrolysis condition of 200 ℃ contains two The main characteristic peak (No.: 10-11), under the condition of T0-270 ℃ pyrolysis, in the GC spectrum corresponding to the obtained gas decomposition products, there is only one main characteristic peak and the GC spectrum at T0-200 ℃. The retention time of characteristic peak No. 10 is the same. The main characteristic gases under the decomposition conditions of T0-200 ℃ and T0-270 ℃ are shown in Table 4. Compared with the GC spectrum under the condition of T0-120 ℃, the GC spectrum under the condition of T0-200 ℃ has significantly reduced 7 characteristic peaks, and the GC spectrum under the condition of T0-270 ℃ has reduced 8 characteristic peaks, indicating that with the With the increase of thermal decomposition temperature, the low-boiling point easily decomposable substances in FR-4 type epoxy resin have gradually volatilized completely, and the high-boiling point refractory substances begin to decompose. However, the two characteristic gases generated by the decomposition of the unaged FR-4 epoxy resin sample at 200 ℃ are inconsistent with the gas generated at 120 ℃, and the gas corresponding to the No. 10 peak at 200 ℃ is a polymer. The plasticizer commonly used in materials, dioctyl phthalate (DOP), the escape of this substance indicates that the internal structure of the epoxy resin material has undergone irreversible structural damage, resulting in a decline in material properties. The gas corresponding to the No. 11 peak is oleic acid amide, which is often used as a dispersant for polyolefins, antistatic agents, mold release agents, pigments, slip agents for PVC plastics, dyes, etc. It can also be used as polyethylene, polypropylene , Synthetic fiber and other kinds of concentrated color masterbatch and cable (insulation) material lubricant, release agent, etc.

Table 4. Main characteristic gases of FR-4 epoxy resin samples at T0-270 °C

serial number	keep time	molecular weight	molecular formula	Molecular Structure
10	31.795	390	$C_{24}H_{38}O_4$	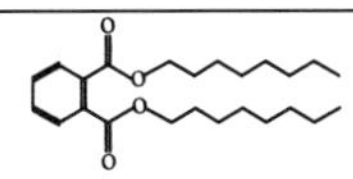
11	37.850	281	$C_{18}H_{35}NO$	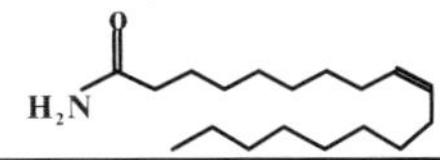

It can be seen from Figure 6 that there are no obvious gas characteristic peaks in the GC spectrum of the gas decomposition products of the FR-4 epoxy resin material after thermal aging at 120 °C for 672 h under the pyrolysis conditions of 200 °C and 270 °C . , this is due to the continuous decomposition of small-molecule volatile substances, macromolecular DOP, oleic acid amide, etc. during the aging process. After 672 h of thermal aging, the content of DOP and oleic acid amide in the material decreased significantly. With the volatilization and release of the additives in the FR-4 epoxy resin material, the material properties deteriorate gradually, so no obvious gas decomposition products were detected after aging.

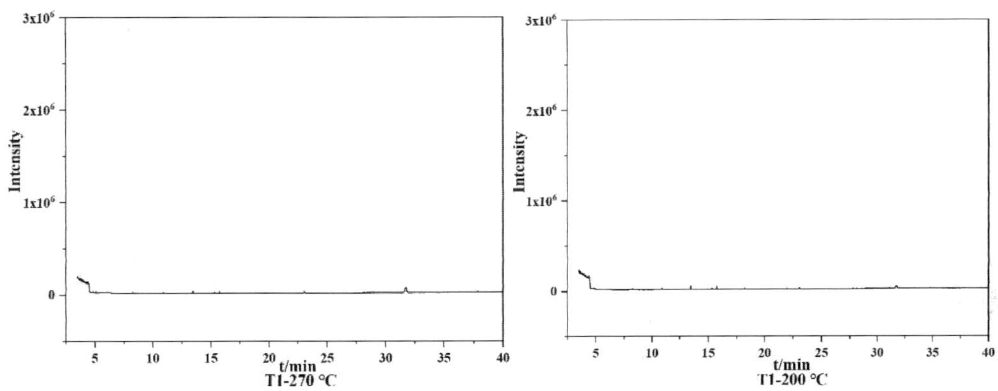

Figure 6. GC spectrum of FR-4 epoxy resin at T1-200 °C and T1-270 °C

4. Conclusions

1) The FT-IR results show that, compared with the samples before thermal aging, the aged samples have new characteristic absorption peaks near the wavelengths of 1413 cm-1 and 1560 cm-1, indicating that there are carbonyl C=O or carboxyl-COOH As a result, the molecular chains of oxygen-containing groups inside the FR-4 epoxy resin material increase, and the unsaturated hydrocarbons gradually increase.

2) TG and DSC show that the FR-4 epoxy resin material exhibits good heat resistance, and it can be used for a long time at a high temperature of 120 °C, but thermal aging will accelerate its thermal decomposition rate.

3) GC-MS results show that at 120 °C, the decomposition of the unaged epoxy resin material produces 9 characteristic gases, at 200 °C, 2 characteristic gases are produced, and at 270 °C, only one characteristic gas is produced. No obvious characteristic gas peaks were detected in the GC-MS spectra of aged samples at 120 °C, T1-200 °C and T1-270 °C.

4) The fault temperature and fault type of the high-voltage switchgear can be preliminarily judged by monitoring the thermal decomposition gas of the epoxy resin, so that the fault type of the high-voltage

switchgear can be monitored and predicted, which is convenient for maintenance personnel to overhaul.

References

[1] Cao, J. Y., Fu, D. H., Guo, M., et al. (2001) Study on the structure of bisphenol A epoxy resin by pyrolysis gas chromatography/mass spectrometry [J]. Modern Instruments, (04): 28-30.
[2] Qiu, K. Q., Wu, Q., Zhan, Z. H. (2009) Vacuum pyrolysis and product analysis of epoxy resin for waste circuit boards [J]. Journal of Central South University (Natural Science Edition), 40(05): 1209-1215.
[3] Yang, L., Li, H. Y., Wang, J. (2010) Study on epoxy resin for wind turbine blades by gas chromatography-mass spectrometry [J]. Fiber Composite Materials, 27(02): 3-5.
[4] Sun, L. L., Wang, J., Cai, H. P. (2007) Effect of acid anhydride dosage on thermal properties of cured epoxy resin [J]. Foreign Building Materials Science and Technology, (05): 34-36.
[5] Li, Z. X., Ding, K., Hao, L. C., et al. (2020) Non-isothermal curing kinetics and creep properties of Al2O3/epoxy resin composites for UHV gas-insulated metal-enclosed switchgear [J]. Journal of Composite Materials, 2020, 37(03): 562-572.
[6] Ji, Z., Xu, Y. Y., Li, J. D., et al. (2020) Pyrolysis characteristics and kinetics of carbon fiber/epoxy composites [J]. Fire Science and Technology, 301(07): 106-109.
[7] Xia, L., Xiong, J., Shentu, B., et al. (2014) Thermal-oxidative degradation and accelerated aging behavior of polyamide epoxy resin-modified montmorillonite nanocomposites[J]. Journal of Applied Polymer Science, 131(19): 40825.
[8] Li, X., Yang, D. G., Shang, Y., et al. (2020) Research on the influence of different aging conditions on the insulation performance of switch cabinet insulation [J]. High Voltage Electrical Appliances, 56(10): 48-55.
[9] Chen, K., Yuan, S., Gong, Z. F., et al. (2018) Ultra-high-sensitivity photoacoustic spectroscopy for trace gas detection based on fiber-optic acoustic sensing [J]. Journal of Optics, 38(03): 185-190.
[10] Li, J. (2014) Preparation and research of waterborne epoxy resin system [D]. Beijing University of Chemical Technology.
[11] Xie, M. X., Liu, Y. (2000) The effect of chloroform on the infrared absorption peaks of aromatic amine amino groups [J]. Journal of Beijing Normal University (Natural Science Edition), (06): 791-794.
[12] Shang, X. Y., Liu, X. Y, Zhang, L. J., et al. (2019) Effects of lignin filling modification on thermal aging properties of epoxy resins [J]. Chemistry and Adhesion, 41(04): 235-239.

Power transformers fault diagnosis using graph neural networks based on dissolved gas data

Weiyun Mao[1,*], Bengang Wei[2], Xiangyi Xu[1], Lu Chen[1], Tianyi Wu[1], Zhengrui Peng[1], Chen Ren[1]

[1]State Grid Shanghai Electric Power Research Institute, China.
[2]State Grid Shanghai Municipal Electric Power Company, China.

E-mail: wzx@hdu.edu.cn

Abstract. The Dissolved Gas Analysis (DGA) is an important way and is now extensively used to detect incipient faults in oil-immersed transformers. Some conventional methods for DGA, such as the Three-Ratio Method, usually misunderstood the gas data and failed to accurately detect multiple fault combinations. Recently, most researchers investigating DGA have utilized machine learning algorithms to improve the accuracy of fault diagnosis. The dissolved gas data are usually treated as tabular data by using machine learning algorithms to achieve better results compared to conventional methods. However, those machine learning methods ignore the internal connection of the dissolved gas data in the time dimension. Therefore, this paper adopted a new DGA method for transformer fault diagnosis based on Graph Neural Network (GNN) and Multivariate Time Series (MTS) model. The results finally showed that the model outperforms the existing machine learning baseline methods and conventional methods on the dataset.

1. Introduction

Under the action of electricity and heat, insulating materials such as oil and paper inside the transformer in operation will gradually decompose and produce hydrogen gas. The decomposition gases such as H_2, CH_4, C_2H_6, C_2H_4, C_2H_2, CO_2, CO and N_2 dissolved in oil will reflect faults such as ageing, electrical discharge, and heat generation. By analyzing the density of the dissolved gas in the oil to evaluate the operating status of the transformer, it is not affected by the interference of external environmental noise and can be carried out without power failure. It is one of the most effective measures to find potential faults in oil-immersed transformers. It is widely used in online monitoring and periodic testing of transformers.

2. Background

2.1. Dissolved gas analysis

The Dissolved Gas Analysis (DGA) [1] is the most important technique used for online fault diagnosis. The main reason is that most transformers in electric power systems cannot be de-energized. This technology combined the modern science and technology of routine oil sampling and online gas monitoring theoretically. One of the key concepts of the DGA is to detect the abnormal outlier inside the generated gas data. The composition of gases is based on the type of fault. The detection of gases generated that exceed or below a certain threshold level in an oil-immersed transformer in service usually indicates a malfunction inside or outside a transformer.

2.2. Conventional methods

Various DGA methods are usually used by organizations and institutions to detect the transformer status. These methods are generally based on defined fixed and predefined rules such as key gases, key gas ratios, gas concentrations and graphical representations. Based on the previous work, we summarized the common methods used in the DGA field [2].

When the internal temperature of the transformer increases, the fault can be detected through the Key Gas method. Because the released gases which are corresponding to the faults can disrupt the chemical structure of the insulating oil.

The key to the method is to measure the individual dissolved gases rather than the gas ratios.

The Rogers Ratio method is the most common gas ratio method, which diagnoses faults by analyzing four gas ratios: CH_4/H_2, C_2H_6/CH_4, C_2H_4/C_2H_6 and C_2H_2/C_2H_4. This method is only beneficial when it is applied to a data set with a vast of positive instances corresponding to transformer failures. However, the imbalanced data set problem which appears as the scarcity of fault samples is normal, intractable, and vital, for fault diagnosis of transformers. So, it can only perform well in some cases and may fail in other cases.

2.3. Multivariate time series forecasting

A multivariate time series consists of several time-dependent variables. Those variables depend on both their past values and other variables. Time series forecasting has developed for decades, the most popular methods are statistics approaches Autoregressive (AR), Moving Average (MA), Autoregressive Moving Average (ARMA), and Autoregressive Integrated Moving Average (ARIMA)[3]. The Vector Autoregressive model (VAR) captures the relationship between multiple quantities over time. The multivariate version based on the ARMA model [4] is the Vector Autoregressive Moving Average model (VARMA) as we know. Statistics methods are all based on strong assumptions such as stationary processes, which perform poorly on real-world data. Meanwhile, multivariate time series data are not suitable for those assumptions. Compared with traditional statistics methods, deep learning methods are free from assumptions and could capture linearity efficiently without prior knowledge.

2.4. Graph neural networks

In past decades, Deep Neural Networks have strongly promoted the development of pattern recognition and data mining. Most real-world data are generated from non-Euclidean spaces, which makes the deep learning method fail to work on some of the real-world scenarios. Recently, Spatial-temporal Graph Neural Networks give us a new perspective. At the same time, we find out that we can transform the DGA data to the same format as multivariate time series data. We observe the dissolved gas density data through repeated measurements over time. With this approach, we could apply an improved spatial-temporal graph neural networks framework called Multivariate Time Series Graph Neural Network (MTGNN) [5] to the dissolved gas density data. We made some improvements to the MTGNN to make it more efficiently on the classification tasks. With the strong interpretability and visualization ability of graph neural networks, we can easily capture the relationships between the variables in multivariate time series and apply them for transformer fault diagnosis.

3. Proposed model

3.1. Model architecture

The improved MTGNN consists of m graph convolution modules and temporal convolution modules with a graph learning layer and an output module based on SoftMax. The dimensions of the input layer are $m * n$ where m is the length of the time series sequence and n is the number of gases.

Table 1. Dissolved gas data example

Time	H_2	CH_4	C_2H_6	C_2H_4	C_2H_2	CO	CO_2
2020/10/21 1:00	22.66	82.74	17.19	65.36	0.73	196.47	2969.26
2020/10/21 7:00	23.35	82.25	17.5	64.44	0.46	195.56	2944.66
2020/10/21 13:00	23.54	81.57	18.52	64.81	0.53	196.22	2974.93
2020/10/21 19:00	22.99	80.78	20.02	65.21	0.48	196.25	2973.37
2020/10/22 1:00	23.28	80.35	18.52	64.72	0.04	196.15	2964.73
2020/10/22 7:00	23.56	80.79	19.56	65.94	0.48	196.63	2962.71
2020/10/22 13:00	22.68	81.32	18.54	65.43	0.6	196.66	2968.88
2020/10/22 19:00	22.24	81.77	19.06	65.96	0.75	196.75	2973.39

The size of the output layer is one: The fault category of the transformer. With the input of 7 gases time series data, the algorithm output fault categories with their probability.

3.2. Graph learning layer

The graph learning layer captures the hidden relationships between the time series data by adaptively learning the graph adjacency matrix. The original MTGNN model adopts a sampling approach to address the limitation of computation and memory cost caused by huge graphs, which allows the algorithm to capture information inside the subgraph instead of the whole graph. However, in DGA we don't need to do that, since we have only 7 nodes in a graph. Besides the symmetry and bidirectional of the existing distance metrics, the adjacency matrix in DGA is supposed to be unsymmetric under the fact that the change of a gas condition may cause the change of other gases. Therefore, the graph learning layer in MTGNN uses a combination of ReLU and Tanh functions to extract uni-directional relationships.

3.3. Temporal & graph convolution module

The role of the temporal and graph convolution modules is to extract the space and time information inside the data and adjacency matrix. Specifically, the graph convolution layer employs two mix-hop propagation layers to process inflow and outflow information for every node, and finally extracts the adjacent relation among the nodes.

To solve the spatial information flow in the topographies, the improved MTGNN put forward the mix-hop propagation layer which includes information propagation and information selection steps. The mathematical form of these two steps is defined as follows:

$$H^k = \beta * H_{in} + (1 - \beta) * \widetilde{A} * H^{k-1} \tag{1}$$

Note that β is a hyperparameter that controls and preserves the ratio of root node information. The information filter follows the equation below

$$H_{out} = \sum_{i=1}^{K} H^{(k)} W^{(k)} \tag{2}$$

where K is the propagation depth, H_{in} and H_{out} represent the input and output hidden states. While the temporal convolution layer is to capture the time signal inside the time series data of each node.
This module extracts temporal features in high dimensional space through a set of standard dilated 1D [6], and [7] convolution filters. One dilated inception layer follows an activation function of tangent hyperbolic. The other dilated inception layer follows a sigmoid activation function.
The dilated inception layer follows the form below

$$z = \mathrm{concat}(z \star f_{1\times 2}, z \star f_{1\times 3}, z \star f_{1\times 6}, z \star f_{1\times 7}) \quad (3)$$

where the four outputs of these filters are clips with the same size according to the largest filter and concatenate in the channel dimension. The dilated convolution [8] denoted by $z \star f_1$ is defined as

$$z \star f_1(t) = \sum_{s=0}^{k-1} f_1(s) z(t - d \times s) \quad (4)$$

where d is the dilation factor.

3.4. Skip connection layer & output module

The skip connection layers consist of $1 \times L$ standard convolutions where L is corresponding to the sequence length of the inputs to the ith skip connections layers. In this case, the layer standardizes information to the same sequence as the input of the SoftMax module. The output module integrates one SoftMax layer and two 1×1 standard convolution layers to transform information to the desired output dimension. The SoftMax activation function would assign decimal probabilities to each fault category in terms of previous information flow.

4. Case study

We evaluated the network performance by using cross entropy loss function. The lower value corresponding to the better performance. In this section, several experiments confirmed that the proposed model has good performance on prediction and classification. All experiments are coded in PYTHON 3.7 and implemented on Intel Core i7-12600K CPU, RTX 3090 GPU computer with a 16 GB memory.

4.1. Data overview

In Table 2, we summarize the statistics of the example dataset we used. There are 1408 samples in the example dataset. The majority is Carbon dioxide (CO_2), which is usually over 2000. Other gas concentrations such as Carbon monoxide (CO) are much lower than Carbon dioxide (CO_2), which is around 200. Considering the unit of measurement is the same for all 7 gases, we do not scale the data in the data processing progress.

Table 2. Data Statistics

	H_2	CH_4	C_2H_6	C_2H_4	C_2H_2	CO	CO_2
COUNT	1408.0	1408.0	1408.0	1408.0	1408.0	1408.0	1408.0
MEAN	64.24	136.13	30.44	111.50	0.61	218.04	2667.99
STD	137.40	183.00	39.78	157.53	1.17	47.92	967.45
MIN	0.11	0.0	0.0	0.0	0.0	0.0	340.23
25%	19.54	86.38	19.18	68.29	0.0	197.38	2695.24

50%	21.74	87.54	19.97	69.3	0.18	206.47	2790.46
75%	23.51	90.94	21.46	70.67	0.47	240.55	3183.14
MAX	665.86	868.48	191.35	744.84	5.7	314.7	3836.7

4.2. Baseline methods for comparison

The improved MGTNN are our model to be evaluated. Baseline methods are summarized in the following:

- Key Gas: A method based on the chemical structure of insulating materials inside the transformer of temperature to measure the released fault gases in the oil.
- Rogers Ratio: A method that diagnoses faults by analyzing four gas ratios: CH_4/H_2, C_2H_6/CH_4, C_2H_2/C_2H_4 and C_2H_4/C_2H_6.
- Logistic Regression: A basic form of the statistical model is to use the logistic function to model the binary dependent variable and with many more sophisticated extensions, Logistic/logit regression is estimated by the parameters of the logistic model which is a form of binary regression in regression analysis [9].
- Support Vector Machines: A Supervised Learning model of correlation learning algorithms is used to analyze data for classification and regression analysis [10].

4.3. Main results

In our experiment, the actual data composed of normal and abnormal states of transformers from China national grid and China southern power grid are used. The dataset has a total of 5680 samples and is randomly divided into a training set: 4544 samples, accounting for 80%, and a test set: 1136 samples, accounting for 20%.

To illustrate the difference between normal and abnormal states, we compared the adjacency matrix learned from time sequences. We chose Hydrogen (H_2) as the standard gas in this case. According to guidance for power transformer fault Diagnosis in China, the threshold for Hydrogen (H_2) is 150. When the density of Hydrogen (H_2) is below 150, the power transformer state is recognized as normal. On the contrary, the power transformer state is recognized as abnormal with a density of Hydrogen (H_2) over 150.

The normal state graph is shown below:

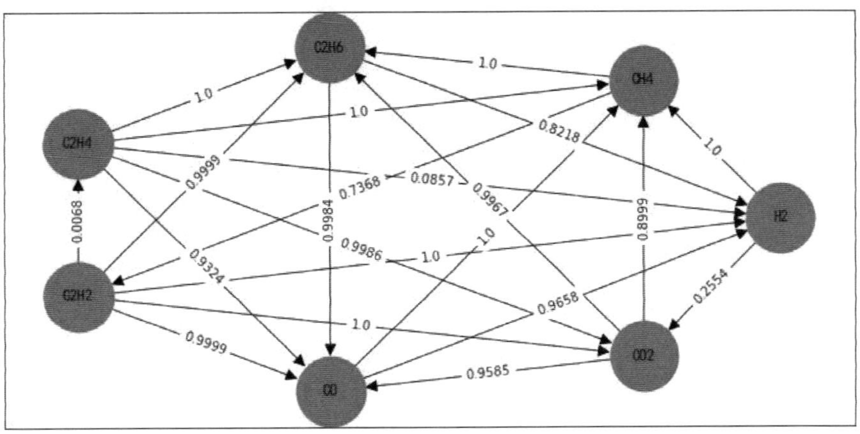

Figure 1. Normal State Graph of Dissolved Gas

The abnormal state is shown below

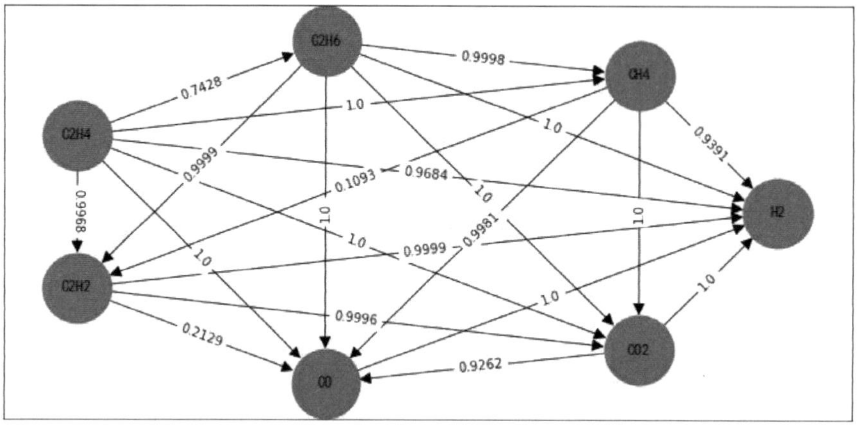

Figure 2. Abnormal State Graph of Dissolved Gas

In the abnormal states graph, the effect from Acetylene (C_2H_2) to Ethylene (C_2H_4) disappeared and the effect direction turned the opposite from Ethylene (C_2H_4) to Acetylene (C_2H_2). Meanwhile, the effect factor (weight of edges) become larger than before from 0.0068 to 0.9968. Moreover, the effect factor from Methane (CH_4) to Acetylene (C_2H_2) decreases a lot - from 0.7368 to 0.1093. The effect factor from Ethylene (C_2H_4) to Ethane (C_2H_6) also decreases slightly from 1.0 to 0.7428. The effect of Acetylene (C_2H_2) on Carbon monoxide (CO) decreased rapidly from 0.9999 to 0.2129.

The above graph shows that there is a certain difference between a normal state and an abnormal state. Every gas record at a time point corresponds to a state graph. Any slight change in the environment or the power transformers would be reflected on the state graph. In this perspective, the edges' weight could be used for guiding the practical transformers' troubleshooting, which could significantly reduce the time-consuming troubleshooting.

We conduct a comparative experiment for oil-immersed transformers from different brands to validate the performances of the algorithm. Three brands are chosen in this experiment and the algorithm performs well with high ac- curacy for all three brands. Besides, the state graph of those transformers is similar, which indicates the algorithm extracts the right relationship of gas.

The results of this investigation shown below in Table 3 indicate that the improved MTGNN model performs well in faults classification, compared with the traditional method such as the Rogers ratio method.

Table 3. Partial discharge (PD) fault correct prediction percentages

	11kV	22kV	66kV
Key Gas	10%	22%	13%
Rogers Ratio	57%	22%	36%
Logistic Regression	70%	64%	55%
Support Vector Machines	74%	79%	69%
Improved MTGNM	78%	92%	83%

We validate the improved MTGNN on the task of partial discharge fault prediction. At first, the improved MTGNN is compared with other conventional methods for predicting PD faults.

Furthermore, compared with some machine learning methods to show the significant performance of the improved MTGNN.

5. Conclusion

In this paper, we apply an improved multivariable model, namely, MTGNN to the power transformer fault diagnosis. As far as we know, it is the first application of a Multivariable Time Series model to the DGA problem. The Graph Neural Network framework like MTGNN achieves excellent performance in the oil-immersed transformer fault prediction task. Its accuracy is much higher than traditional industrial detection methods. Our paper demonstrates a new road in Dissolved Gas Analysis by using the Multivariate Time Series model and Graph Neural Network. In addition, the generalizability of these results is subject to certain limitations. For instance, the experiment illustrates that there is still a lot of room for improvement in the Intelligent Manufacturing field.

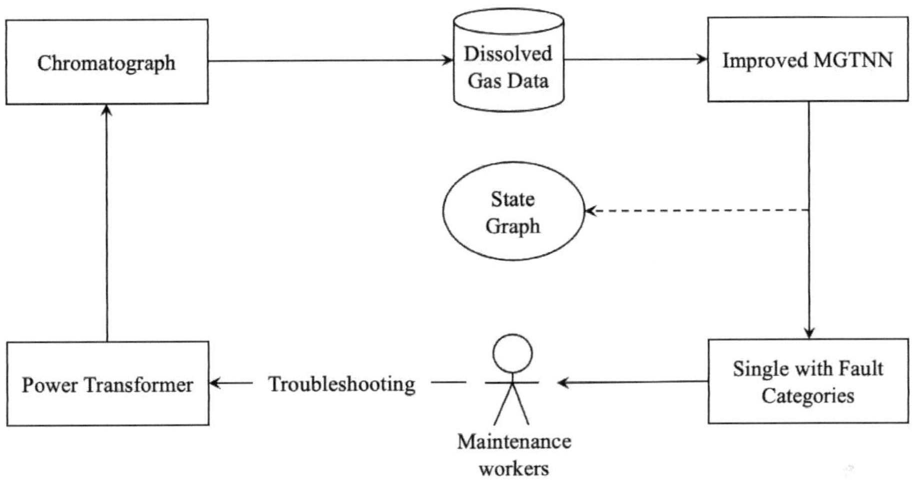

Figure 3. Warning System Workflow

The whole system framework is shown in Figure 3. In real-world applications, the algorithm analyses the data of the dissolved gases generated inside the transformer and continuously prints graphs consisting of those gases, which shows the relationship between the relationships and weights of relationships among the gases. The transformer state graph allows the maintenance staff to know the real-time status of the oil-immersed transformer. When the weight of the edges in the graph exceeds or falls below a certain threshold, it indicates an abnormal state of the transformer. Afterwards, the maintenance staff will get a signal of the abnormal state and then fix the faults according to the output fault category, which greatly improves the real-time efficiency of maintenance.
Finally, we still need more research on the interpretability of the improved MTGNN. The directed graph is printed during the training process, and the stability and how to guide the actual troubleshooting needs further research.

Acknowledgements
This work was supported in part by China's State Grid Shanghai Electric Power Company under Grant 520940200067.

References

[1] Committee, T., et al. (1991) "Guide for the interpretation of gases generated in oil-immersed transformers; ieee std c57. 104," IEEE: New York, NY, USA.
[2] Sun, H. C., Huang, Y. C and Huang, C. M. (2012) "A review of dissolved gas analysis in power

transformers." Energy Procedia, vol. 14, DOI https://doi.org/10.1016/j.egypro.2011.12.1079, pp. 1220-1225, 2011 2nd International Conference on Advances in Energy Engineering (ICAEE).

[3] Box, G. E., Jenkins, G. M., Reinsel, G. C and Ljung, G. M. (2015) Time series analysis: forecasting and control. John Wiley & Sons.

[4] Lütkepohl, H. (2006) "Forecasting with varma models," Handbook of economic forecasting, vol. 1, pp. 287-325

[5] Wu, Z., Pan, S., Long, G., Jiang, J., Chang, X and Zhang, C. (2020) "Con- necting the dots: Multivariate time series forecasting with graph neural networks," in Proceedings of the 26th ACM SIGKDD International Conference on Knowledge Discovery & Data Mining, pp. 753-763.

[6] Yan, S., Xiong, Y and Lin, D. (2018) "Spatial temporal graph convolutional networks for skeleton-based action recognition," in Thirty-second AAAI conference on artificial intelligence.

[7] Yu, B., Yin, H and Zhu, Z. (2017) "Spatio-temporal graph convolutional networks: A deep learning framework for traffic forecasting," arXiv preprint arXiv:1709.04875.

[8] Yu, F and Koltun, V. (2015) "Multi-scale context aggregation by dilated convolutions," arXiv preprint arXiv:1511.07122.

[9] Kleinbaum, D. G., Dietz, K., Gail, M., Klein, M and Klein, M. (2002) Logistic regression. Springer.

[10] Kecman, V. (2005) "Support vector machines–an introduction," in Support vector machines: theory and applications, pp. 1-47. Springer.

Molecular dynamics simulation of inhomogeneous density distribution mechanism of Ultra-Thin perfluoro sulfonic acid ionomer films on Triple-phase interface of PEM fuel cells

Yue Wan*[a], Diankai Qiu[a], Peiyun Yi[a], Linfa Peng[a], Xinmin Lai[a,b]

[a] State Key Laboratory of Mechanical System and Vibration, Shanghai Jiao Tong University, Shanghai, 200240, PR China
[b] Shanghai Key Laboratory of Digital Manufacture for Thin-walled Structures, Shanghai Jiao Tong University, Shanghai, 200240, PR China

*wanyue2018@sjtu.edu.cn

Abstract. The proton exchange membrane (PEM) fuel cell is a clean power generation component with high energy conversion rate and high energy density. The triple-phase interfaces of catalytic layer are where the electrochemical reaction occurs in proton exchange membrane (PEM) fuel cell, and its distribution directly affect the reactivity. The triple-phase interface is a nano-thick perfluorosul-fonic acid ionomer (PFSA) film that covers the catalyst particles. When the density of the ultra-thin PFSA film is not uniform, the gas transport resistance increases significantly, which is not conducive to the progress of reaction. However, ultra-thin PFSA film will have inhomogeneous density distribution due to the influence of forming process. Therefore, how to design the process to improve density distribution of ultra-thin films is an urgent problem. Molecular dynamics simulation model is built to study the influence of manufacturing process on inhomogeneous distribution of ultra-thin PFSA film, and the microscopic mechanism is revealed.

1. Introduction

The proton exchange membrane (PEM) fuel cell is a clean power generation component with high energy conversion rate and high energy density [1, 2]. In PEM fuel cells, the reactions take place at the triple-phase interface of the catalytic layer (CL). The main components of triple-phase interface are carbon-supported Pt particles, ultrathin ionomer films, and gas transport cavities [3-6]. Among them, the thickness of ultra-thin ionomer films is only 4-10 nm [7,8]. When the density of the ultra-thin PFSA film is not uniform, the gas transport resistance increases significantly, which is not conducive to the progress of reaction [9,10].

Process has a significant effect on the density distribution of ultrathin PFSA films. In general, PFSA ionomer mixture is coated on the surface of PEM or GDL by electrostatic spraying, then the solvent is removed by heating and evaporation. Finally, PFSA ionomer condense into ultra-thin films and CL. In this process, the electrostatic field will always exist, and the heating evaporation process has two modes: annealing and non-annealing. All these processes may have an impact on the density distribution of ultrathin films. Therefore, how to design the process to improve density distribution of ultra-thin films is an urgent problem.

In this study, the molecular dynamics simulation method is used to establish an evaporation process model of ultra-thin PFSA film, and the influence of non-annealing, annealing and applied electric field is studied to reveal the mechanism of ultra-thin film's density distribution.

2. Simulation details

The geometric models of platinum crystal and PFSA ionomer to form triple-phase interface are established, as shown in Figure 1. The size of molecular dynamics simulation box is 35.3 Å × 35.5 Å × 292.6 Å as shown in Figure 1(a), which consists of 3 parts: Pt catalyst, PFSA ionomer, and vacuum region. During the simulation, the size of the vacuum region along the Z axis is 150.0 Å, which reserves enough relaxation space and motion freedom for the PFSA molecular chain. In this model, the Z-direction refers to the direction along the normal line of the Pt crystal plane and pointing to the outside of the Pt crystal plane. The shape of the PFSA molecular chain is shown in Figure 1 (b), which includes a main chain composed of -(CF2)- and a side chain containing a sulfonic acid group, and the molecular formula is $C_9HF_{17}O_5S$. In Figure 1 (c), the Pt catalyst is a single crystal face of the nano-Pt crystal, which keeps its position unchanged and has zero momentum during the forming process. The lattice constant of platinum is 3.9 Å, and 10 single crystals are periodically tiled along the X and Y directions. Since the top three layers of Pt atoms in the Z direction accounts for 99.7% of forces on the molecular chain of PFSA, 4 layers of single crystals **were** tiled along Z direction to ensure a complete evaluation of the force of CL on the PFSA molecular chains.

In the actual forming process, the PFSA molecular chains are wrapped on the surface of the catalyst particles in the form of dispersion liquid, and gradually agglomerate with the evaporation of the solvent, and finally form the triple-phase interface. This model proposes a distributed molecular dynamics model for simulating evaporation.

First, a dispersion model of the PFSA molecular chain was established, and Langevin implicit solvent was applied to the system to simulate the force of the actual solvent on the motion characteristics of PFSA. An evaporation surface is added on the top surface of ionomer, and this evaporation surface is pressed down at a constant speed along the -Z direction. This step simulates the passive condensation process of the PFSA molecules in the system by the evaporation, and under the NVE system, the internal system energy of the page is kept constant to simulate the solvent evaporation's loss. Finally, using the density of solid PFSA film as the evaporation cut-off condition, maintain the system temperature to release the stress in the PFSA molecular chain. The simulated annealing process reduces the energy of the system to the local minimum energy. This step simulates the condensation process of PFSA molecular chains at the triple-phase interface. After dispersion, evaporation and condensation, the PFSA molecular chain is formed into a solid ultra-thin PFSA film attached to the Pt surface, that is, a three-phase interface is formed.

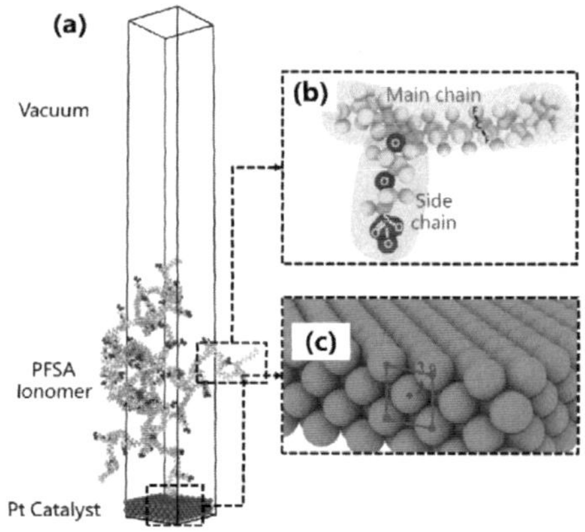

Figure 1. Forming simulation model of ultra-thin PFSA film at triple-phase interface
(a) Forming simulation box and initial system, (b) PFSA molecular chain, and (c) grain type of Pt catalyst particles

3. Results and discussion

To verify the feasibility of the model, one of these three processes is selected and the motion state of PFSA molecular chain under this process is extracted. The parameter conditions of the temperature of 360K, no annealing process, and no electric field strength are selected to investigate the state of the PFSA molecular chain at different time. In Figure 2, as the evaporation surface moves down, the molecular chains close to the surface are compressed and formed an atom-dense layer. As the evaporation progresses, the evaporation surface gradually approaches the bottom Pt, and the surface layer exerts force on the molecular chain near the substrate, finally forming a ultra-thin PFSA film. In the process of ultra-thin film formation, the compression state of each position of PFSA molecular chain is different, which can easily lead to inhomogeneous distribution of density.

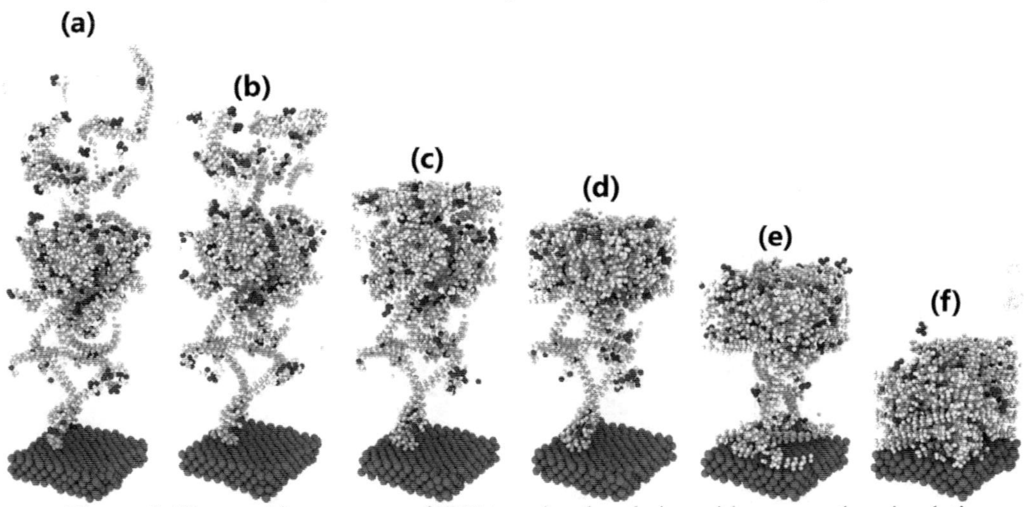

Figure 2. The stacking process of PFSA molecular chains with evaporation simulation (a) 0ps, (b) 50ps (c) 100ps (d) 150ps (e) 200ps (f) 250ps

The atomic density of the ultra-thin PFSA film was sliced along the Z direction, and the effects of different processes on the density distribution are compared, as shown in Figure 3. The Z-direction atomic density distribution is an index to characterize the uniformity of the spatial distribution of the triple-phase interface.

The molecular motion process shown in Figure 2 applies to all the three processes. In fact, the density distribution of ultra-thin PFSA films has a similar law regardless of any process. Atoms of PFSA pack significantly close to the Pt surface, forming a red "dense layer" as shown in Figure 3(a). At the same time, a relatively uniform density is maintained in most positions, and a loose area is formed on the outermost side of the ultra-thin film. This situation is mainly caused by the uneven stacking of PFSA molecular chains due to the evaporation surface.

Further, the parameters of the dense layer were quantified, and the differences in the influence of different forming processes on the thickness and maximum density of the dense layer are compared, as shown in Figure 3 (b), (c).

As shown in Figure 3(b), for the non-annealing process, the dense layer thickness of the ultra-thin films is about 3.7Å, which is like that of the annealed ultra-thin PFSA films (4Å). This indicates that the Z-direction density distribution at the triple interface is hardly affected by the annealing process. Also, considering the electrostatic spraying process of actual forming, the influence of electric field strength at 10V/cm on the spatial distribution of the triple-phase interface is simulated. The results show that the electrostatic field also does not affect the thickness of the dense layer (3.95Å). The maximum density of the dense layer is compared, as shown in Figure 3(c). The maximum density of the ultra-thin film dense layer obtained at non-annealing process is 151.0. Annealing reduces this value slightly to 144.9, while electric field increases it significantly to 155.1. In conclusion, the forming process does not change the thickness of the dense layer at the three-phase interface, but annealing and electric field strength change the maximum density of the triple-phase interface.

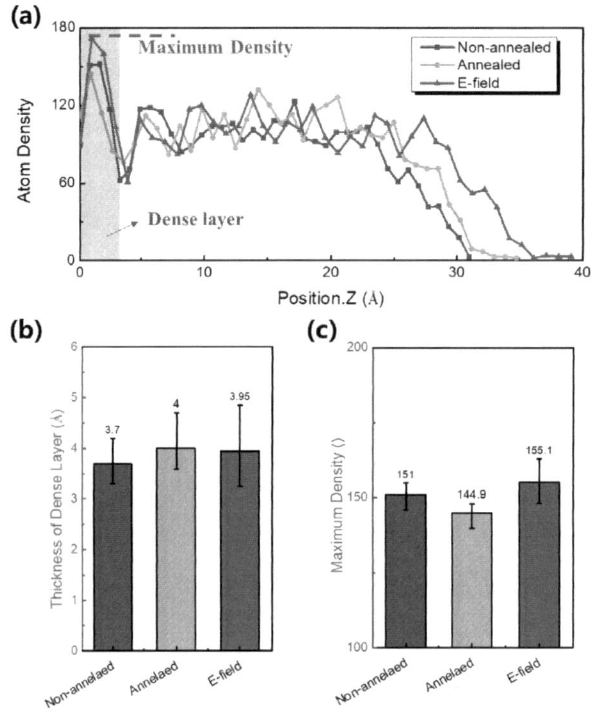

Figure 3. Influence of forming process on density distribution of ultra-thin PFSA film (a) Atomic density distribution in the Z direction, (b) thickness of dense layer, and (c) maximum density.

To explore the influence mechanism of the process on the density distribution of ultra-thin PFSA films, the radial probability density function g(r) was compared. For PFSA molecular chains, the handedness of the side chains and the curling degree of the main chain are important parameters for evaluating the molecular chain morphology. In this section, based on the ideal state molecular chain theory, the radial distribution function between any two points of the molecular chain is investigated and counted as an index to evaluate the morphological characteristics of the molecular chain. The g(r) function refers to the probability distribution function that assumes that one end of the molecular chain is fixed at the origin of the coordinate system (0, 0, 0), and the other end of the molecular chain randomly appears in the volume unit near any position (x, y, z) in the three-dimensional space.

The agglomeration process of PFSA molecules at non-annealing, annealing, and additional electric field is simulated respectively, and the obtained g(r) results are shown in Figure 4. In this figure, g(r) of different processes has four identical coordination numbers, which are r=3.8 Å, 4.8 Å, 6.2 Å and 7.3 Å respectively. The results show that g(r) will not change due to different forming processes, which means, the final curled state of PFSA molecular chains after forming an ultra-thin film will not be affected by the forming process.

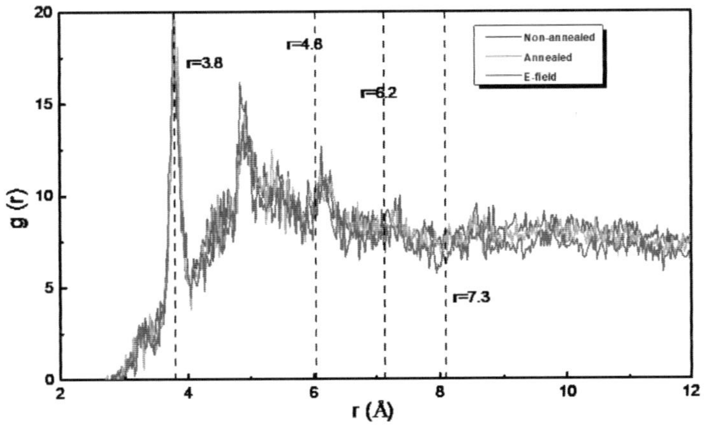

Figure 4. Radial probability density function g(r) of PFSA under different forming processes

To explore the mechanism of the forming process on PFSA molecular chain, the mean square displacement (MSD) of the PFSA molecular chain is investigated. MSD of molecular chain can characterize the total moving distance in within the time. When the MSD is larger, the movement of molecular chain is longer, which means the diffusion of molecule is more sufficiently. The MSD versus elapsed time is obtained by averaging the squares of displacements for each atom. Based on this method, the MSD value of all atoms on PFSA molecular chain is calculated, and the relationship between MSD and time under different processes is obtained, as shown in Figure 5.

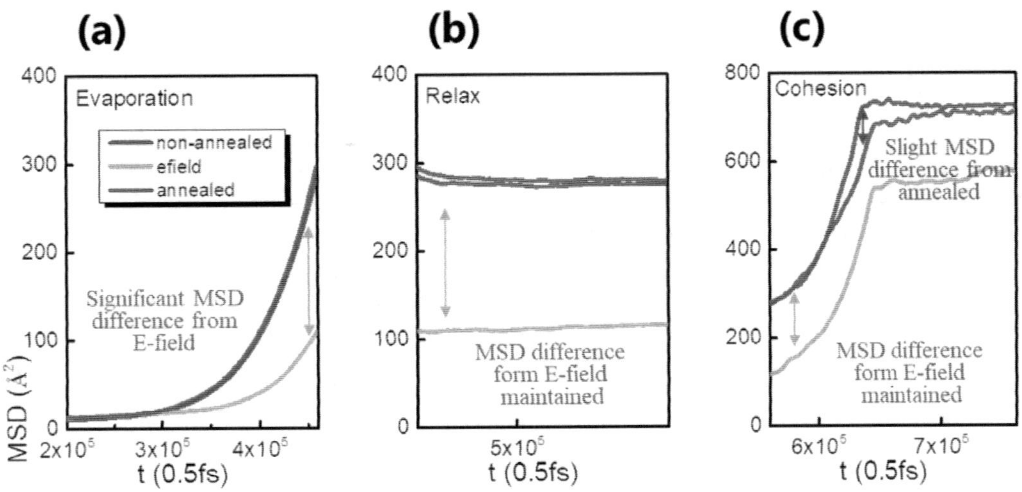

Figure 5. Influence mechanism of process on MSD
(a) Variation trend of MSD in the whole process of evaporation simulation, (b) comparison of MSD in evaporation process, and (c) comparison of MSD in condensation process.

In Figure 5(a), in evaporation stage, MSD curves of the non-annealing and annealing processes basically overlap, indicating that the MSD is not affected by these two processes; the MSD of PFSA with an electric field of 10 V/cm is significantly smaller than that of the other two processes. It indicates that applying an additional electric field leads to a reduction in the MSD of the PFSA molecules during the evaporation stage. Currently, the compression path of PFSA molecules to the Pt crystal plane is shorter.

In Figure 5(b), the MSD curves of the three processes in the relaxation stage are parallel to each other, and the MSD is unchanged at this stage, which means the relaxation stage is not affected by the forming process.

Figure 5(c) shows the MSD of the three processes in the cohesion stage. It can be found that the MSD curves of non-annealing process and the application of electric field are basically parallel, indicating that the electric field does not affect the motion process of PFSA molecules in the condensation stage. Comparing the non-annealing process and annealing processes, the MSD of PFSA molecules during annealing is significantly smaller than that of the non-annealing process, indicating that annealing weakens the stretching process of the molecular chain during the cohesion stage.

4. Conclusion

In this study, a molecular dynamics model is established to simulate the forming process of ultra-thin PFSA films, the influence of the density distribution of ultra-thin films by different forming processes is obtained, and g(r) and MSD are used to reveal the microscopic mechanism in the forming process. In general, the electric field mainly acts on the evaporation stage of the PFSA molecular chain, and the annealing process mainly acts on the cohesion stage. The electric field increases the maximum density of the ultra-thin film, which is not conducive to the uniformity of the ultra-thin film density; the annealing process is the opposite, which is beneficial to improve the density distribution of the ultra-thin PFSA film.

Acknowledgment

National Natural Science Foundation of China (Grant No. 52075322), Shanghai Rising-Star Program (Grant No.21QA1405000), Natural Science Foundation of Shanghai (Grant No.20ZR1426800).

References

[1] Jao, T. C., Jung, G. B., Chi, P. H., Ke, S. T., & Chan, S. H. (2011) Investigation of degradation behavior of membrane electrode assembly with polytetrafluoroethylene/Nafion composite membrane. Journal of Power Sources. 196(4): p. 1818-1825.

[2] Qiu, D., Yi, P., Peng, L., & Lai, X. (2015) Assembly design of proton exchange membrane fuel cell stack with stamped metallic bipolar plates. International Journal of Hydrogen Energy, 40(35): p. 11559-11568.

[3] Xia, L., Ni, M., Xu, Q., Xu, H., & Zheng, K. (2021) Optimization of catalyst layer thickness for achieving high performance and low cost of high temperature proton exchange membrane fuel cell. Applied Energy, 294.

[4] He, P., Mu, Y. T., Park, J. W., & Tao, W. Q. (2020) Modeling of the effects of cathode catalyst layer design parameters on performance of polymer electrolyte membrane fuel cell. Applied Energy, 277.

[5] Hou, Y., Deng, H., Pan, F., Chen, W., Du, Q., & Jiao, K. (2019) Pore-scale investigation of catalyst layer ingredient and structure effect in proton exchange membrane fuel cell. Applied Energy, 253.

[6] Chen, B., Cai, Y., Yu, Y., Wang, J., Tu, Z., & Chan, S. H. (2017) Gas purging effect on the degradation characteristic of a proton exchange membrane fuel cell with dead-ended mode operation II. Under different operation pressures. Energy, 131: p. 50-57.

[7] Hou, J., Yang, M., Ke, C., Wei, G., Priest, C., Qiao, Z., et al. (2020) "Platinum-group-metal catalysts for proton exchange membrane fuel cells: From catalyst design to electrode structure optimization." EnergyChem 2.1: 100023.

[8] Jiao, K., Alaefour, I. E., Karimi, G., & Li, X. (2011) "Cold start characteristics of proton exchange membrane fuel cells." international journal of hydrogen energy 36(18): 11832-

11845.

[9] Greszler, T. A., Caulk, D., & Sinha, P. (2012) "The impact of platinum loading on oxygen transport resistance." Journal of The Electrochemical Society 159(12): F831.

[10] He, Q., Suraweera, N. S., Joy, D. C., & Keffer, D. J. (2013) "Structure of the ionomer film in catalyst layers of proton exchange membrane fuel cells." The Journal of Physical Chemistry C. 117 (48): 25305-25316.

Journal of Physics: Conference Series **2387** (2022) 012031 doi:10.1088/1742-6596/2387/1/012031

Research on performance of thermal management system for lithium-ion power battery

Hongjiao Qiao[1,a],Wenfang Sun[1],Weizhong Zhang[1,*], Chunlei Song[2], Cheng Liu[2]

[1]Anhui Xinhua College, Intelligent Manufacturing College, Hei Fei, China.
[2]Hefei Guoxuan Hi-Tech Power Energy Co., Ltd., Product Research Institute, Hei Fei, China.
[a] 826051234@qq.com

*Corresponding author's e-mail: 23119522@qq.com

Abstract. According to battery thermal management system's design theory, a battery module with thermal management system is designed in this article, and the thermal management system model is simulated. By establishing thermal management system's simulation model, modular thermal management system's performance is analyzed.The simulation consequences show that the liquid cooling plate's outlet temperature is higher than the inlet temperature, so the temperature rise and outlet temperature of the cooling liquid of a single battery module that cools system should be properly controlled. Through the thermal imager experiment, the test consequences are consonant with the simulation analysis's conclusions.The reliability and accuracy of the simulation analysis is verified by the experiment, and shows the rationality of the design of the thermal management system for power battery.

1. Introduction

With new energy source technology's development, new energy vehicles gradually come into people's sight. Lithium batteries are often used as power batteries for new energy vehicles as a result of their no memory effect, high energy density and low self-discharge rate [1,2]. As a key component of new energy vehicle, lithium battery's performance is obviously affected by temperature, so it is necessary to control power battery's working temperature strictly, otherwise, power battery can't exert its performance in extreme environment. Therefore, it is of great significance to take power lithium battery pack's appropriate thermal management measures to improve electric vehicles' overall performance and ensure electric vehicles' driving safety [3].

During the operation of electric vehicles, the arrangement of battery cells in the closed battery box is very compact. When the batteries work in high-rate discharge, the internal electrochemical reaction rate is intense, and the generated heat increases. The batteries arranged in groups are easier to gather heat [4-5]. If the heat can't be discharged in time, there will be gas leakage and liquid leakage inside the battery, and more seriously, there will be explosion and other dangers, which will have a huge impact on the safety and overall performance of electric vehicles [6]. It will also affect the performance of the battery,shorten the battery's service life, and even lead to thermal runaway, leading to serious safety problems [7]. At the same time, the heat dissipation effect of single cells in different positions of the battery system is also different, Because temperature difference is easy to occur, the temperature uniformity between batteries decreases. If the battery pack's temperature is not effectively controlled and the temperature difference between the batteries is too large, the cycle life and service performance of the batteries will also be reduced [8-9].

A large number of studies show that overly high or overly low temperature will influence the battery's performance, and even cause safety problems such as thermal runaway and internal short circuit of the battery, which will impact on the battery's general use. So as to improve the thermal safety and usability of power battery and extend its cycle life, it is needed to study the performance of power battery pack's thermal management system.

2. The choice of power battery thermal management

2.1. Cooling method of power battery system

The power battery thermal management system, in a few words, is to take control of the battery system's temperature through heating or cooling, improve the working environment of the battery, and make the battery system work within a reasonable temperature range,in order to improve the dependability and life of the battery. There are two important performance indicators for power battery thermal management: (1) to maintain the temperature balance within and between batteries; (2) to control the battery's absolute temperature within a reasonable range[10].

At present, as stated by the heating and heat dissipation characteristics of lithium-ion power battery monomer, according to different heat transfer media, thermal management cooling methods can be divided into several forms such as natural heat dissipation, forced air cooling, liquid cooling, direct cooling, and phase change materials. Among them, natural cooling is a passive thermal management method, while air cooling, liquid cooling, and direct cooling are active. The important difference between the three is the difference in heat exchange media. During practical applications, it is necessary to select a reasonable cooling method according to different working conditions, battery type, battery shape, battery box structure, size and power requirements [11-12]. The influence of varying cooling ways is exhibited in Table 1.

Table 1. Comparison of cooling three cooling methods' effects

Compare items	Natural heat dissipation	forced air cooling	liquid cooling
Heat dissipation efficiency	bad	higher	high
temperature uniformity	Good when there is no external heat source, otherwise bad	Bad(especially in import and export)	good
Installation environment adaptability	bad (requires external insulation, ventilation)	Bad (air inlet and outlet structure)	good
Both high temperature and cold	Bad(contradiction between auxiliary heating and heat preservation and heat dissipation)	Bad (contradiction between auxiliary heating and heat preservation and heat dissipation)	Good (can dissipate heat and heat)
the complexity	the easiest	medium	complex
energy consumption	none	high	Low (easy to achieve thermal insulation)
cost	Low	higher	high (optimizable)

The three methods of natural heat dissipation, forced air cooling and liquid cooling all use the convective heat transfer that occurs when the cooling medium flows through the hot power surface to take the heat away, and phase change is not undergone by the cooling medium during the process. Phase-change refrigeration is a phase-change material undergoing a stage modification during cooling, and assimilates heat to remove the heat that is generated by the battery [11].Heating method of power battery system.

2.2. Heating method of power battery system

When the vehicle is in a cold winter area, the temperature inside the battery box is low, which will affect the normal use of the vehicle. The power battery must be equipped with a heating system to conform to the normal usage the vehicle's requirements. The heating system's function is to heat the cells' temperature quickly in the battery system to a specific temperature, and to make sure the consistence of the cells' temperature in the heating process[13]. Common heating methods for power batteries include heating film heating, PTC heating and liquid heating heating. The comparison of the three heating methods' characteristics is shown by Table 2.

Table 2. Comparison of three heating methods' characteristics

project	heating film	PTC heating	liquid heat
Heating Features	Constant power heating	self-controlled heating	Convective/Conductive Heating
Space Constraints (Thickness)	0.3mm～2mm	5mm～8mm	Integrated in liquid cooling system
Heating equipment dry burning temperature	60°C～130°C	60°C～80°C	25°C～60°C
Cell temperature rise rate（/min）	0.15°C～0.3°C	0.15°C～0.5°C	0.3°C～0.8°C
Uniformity (cell temperature difference)	≈8°C	≈10°C	≈5°C

As is visible from Table 1 and 2 above, the thermal management method of liquid cooling and liquid heating is the preferred solution for many electric vehicles at present. Because of the high coefficient is transferred by heat, large heat capability, and fast cooling or heating speed of the liquid medium, the effect is remarkable, and the liquid cooling liquid heat is integrated into one, and the volume is relatively small. It is generally arranged at the bottom of the battery module or between the battery modules. To ensure that the battery is in direct contact with the liquid, it has the effect of cooling or heating, and the effect is remarkable [14-15].

3. Module thermal management design and verification

3.1. Basic theory of thermal fluid simulation

Using the law of conservation of energy and Fourier's law, the mathematical expression of the temperature field in a thermally conductive object is established. The temperature field is simulated and calculated, and the thermal conduction mechanism of the system is generally described according to thebattery's inner power differential conservation equation [11,16], as exhibited in formula (1).

$$\rho c_p \frac{\partial T}{\partial t} = \lambda \left(\frac{\partial^2 T}{\partial x^2} + \frac{\partial^2 T}{\partial y^2} + \frac{\partial^2 T}{\partial z^2} \right) + \emptyset_V \quad (1)$$

In the formula, ρ is the density of the object (kg/m3); C_P is the object's particular heat capability[J/(kg·K)]; λ is the object's thermal conduction system[W/(m· K)]; T is the analysis temperature of the object (K) ; t is the time (S); x, y, z, is the three dimensions of the Cartesian coordinate system (m); $\varnothing_V$is the heat production per unit time (W/m3) of the object per unit volume.
Among them, a mass-weighted average of the specific heat ability of various materials that are contained in a single battery is performed by the calculation method of a battery cell's particular heat capability mostly[17-18], as shown in formula (2).

$$c_p = \frac{1}{m} \sum^i C_i m_i \quad (2)$$

In the formula:Cp is the battery cell's specific heat capacity ; m is the battery cell's mass ; Ci is the particular heat capability of the substance of the battery's each layer cell; mi is the mass of the stuff of the battery's each layer cell.
Among them, the average thermal conductivity of the series material is shown in formula (3).

$$\lambda_x = L_X / \sum_i \frac{L_{X_i}}{\lambda_i} \quad (3)$$

The average thermal conductivity of parallel materials is shown in formula (4).

$$\lambda_y = \lambda_z = \sum_i \frac{\lambda_i L_i}{L_x} \quad (4)$$

Where: λ_i is the thermal conductivity of the substance of the battery's each layer; Lxi is the thickness of the material of the battery's each layer ; Lx is the thickness of the battery cell. For the simulation calculation of heat generation in the battery, the Bernardi heat generation rate modelequation is commonly used in theory[19-20] , as shown in equation (5).

$$Q_V = \frac{I}{V} (E_{oc} - U - T \frac{dE_{oc}}{dT}) \quad (5)$$

In the formula: I is the current; V is the battery's volume; U is the battery's operating voltage; EOC is the battery's open circuit voltage; T is the battery's initial temperature.
While estimating the pole's heat generation rate, the calculation equation of the Joule heat that is generated by the pole is shown in equation (6).

$$Q_V = \frac{Q}{V} = \frac{{I_R}^2 R}{V} \quad (6)$$

In the formula: Q is the calorific value of the pole; V is the pole's volume; IR is the internal resistance current of the pole.
According to Newton's cooling law , the calculation equation of the boundary conditions of the power battery's thermal model is shown in equation(7).

$$-\lambda \left(\frac{\partial T}{\partial n} \right) = h \ (T_0 - T_\infty) \quad (7)$$

where: λ is the battery's average thermal conductivity surface material; n is the vector direction perpendicular to the battery surface; h is the heat transfer coefficient between the battery's surrounding fluid and face, which is composed of convection and radiation heat transfer coefficients and is determined by an iterative method; T0 is the surrounding fluid's temperature; T_∞ is the battery ambient temperature.
By means of the above mathematical model, the heat generation rate, particular heat capability, thermal conductivity, etc. of the battery can be obtained, which lays a certain theoretical foundation for simulation analysis and thermal management design.

3.2. Research on cooling performance of modules

1) Module thermal management simulation modeling

In order to study the key points of module thermal management design, it is required to build a normal module thermal management analysis model to analyze and study the module thermal management performance. When establishing the liquid cooling geometric model of the lithium-ion battery module, the geometric model is simplified, which mainly includes: 12 single cells, thermally conductive silicone pads and liquid cooling plates. Among them, accessories such as terminals and coolant pipes can be ignored here for the reason that they have got little effect on the simulation of the battery module's temperature field.

The liquid cooling of the battery module adopts the scheme of double-liquid cold plate arrangement, the size of the cold plate is 355*145*11mm, and the liquid flow channel is circular in cross section and 7mm in diameter. A total of 8 flow channels are divided into The two groups are arranged at equal intervals in the same direction, and the structure and shape of the liquid cooling plate are shown in Figure 1. The outlet and cooling liquid inlet are arranged on the cold plate's both sides, and the distance between the outlet and the inlet is 128mm. The length and width of the thermally conductive silicone pad are similar to the size of the cold plate, and the thickness is 1mm. The placement of the thermally conductive silicone pad is placed between the liquid cold plate and the battery module to facilitate heat transfer. Figure 2 is the geometric model of the lithium-ion battery module established.

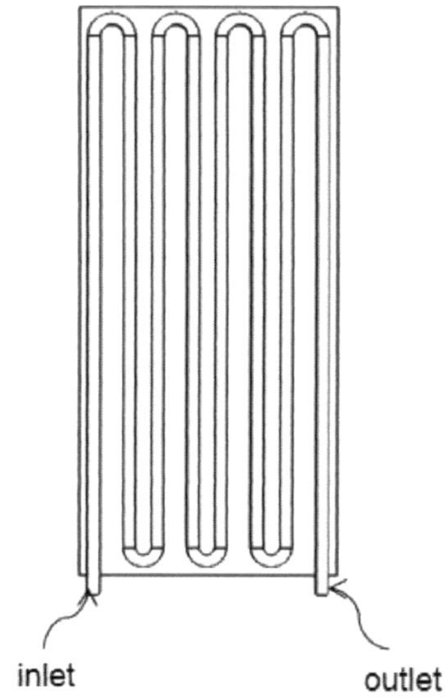

Figure 1. Schematic diagram of liquid cooling plate structure

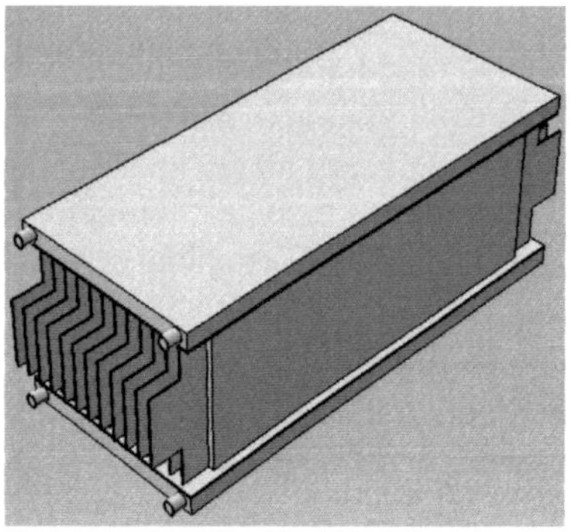

Figure 2. Geometric model of the battery module

2) Module thermal management simulation analysis

For pure electric vehicles' vehicle battery pack thermal management system, the liquid cooling system's cooling liquid temperature rise has a great influence on the energy consumption and load of the vehicle air radiator, so for the liquid cooling plate of a single battery module that cools unit It is also of wonderful importance to examine the temperature difference and temperature rise of the coolant.

In this article, it is supposed that lithium-ion battery's main physical parameters, coolant and liquid-cooling plate do not change with temperature's alteration. According to different working conditions, the liquid-cooling model of lithium-ion battery module is simulated and analyzed respectively. The first working condition is: the simulation calculation is carried out under the temperature field when the discharge rate is 1C, the ambient temperature is 25°C, the coolant flow rate is 5g/s, and the coolant inlet temperature is 20°C.

Figure 3 is a real-time residual diagram of the simulation calculation process. The figure shows that when the simulation reaches 1300 steps, each residual curve has become stable, indicating that each residual value has converged. Through the simulation calculation, it is obtainable that Figure 4 is the simulation cloud diagram of the battery module's temperature field at the simulation's end, Figure 5 is the simulation cloud figure of the XOY's temperature field and YOZ cross-sections of the battery module in the straight and horizontal directions, and Figure 6 is the surface of the liquid cold plate and the cooling liquid flow channel. Cross-section temperature field cloud map.

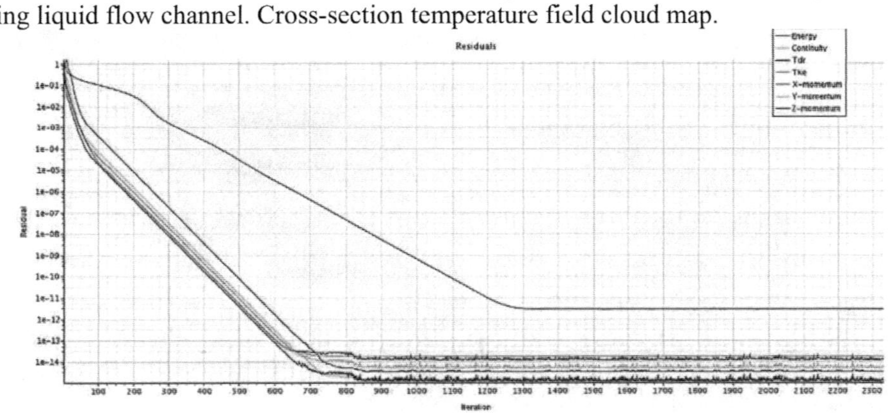

Figure 3. Real-time residual diagram of simulation calculation process

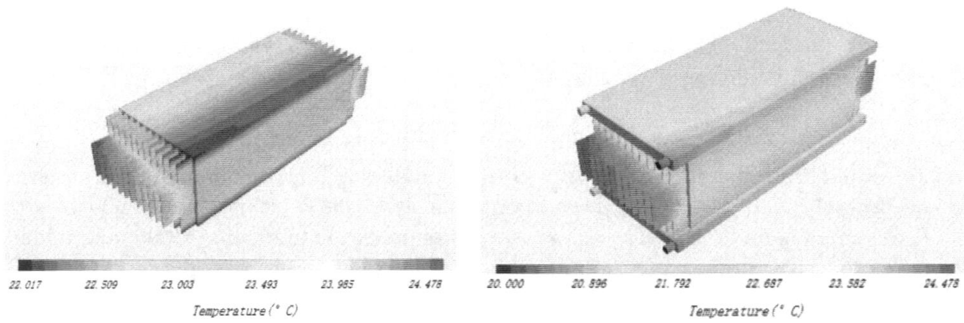

Figure 4. Overall temperature distribution of the battery module

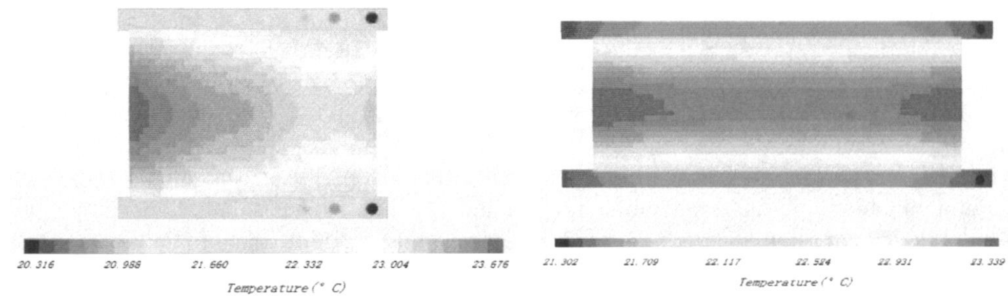

Figure 5. Cross-sectional temperature distribution of the battery module in the horizontal and vertical directions

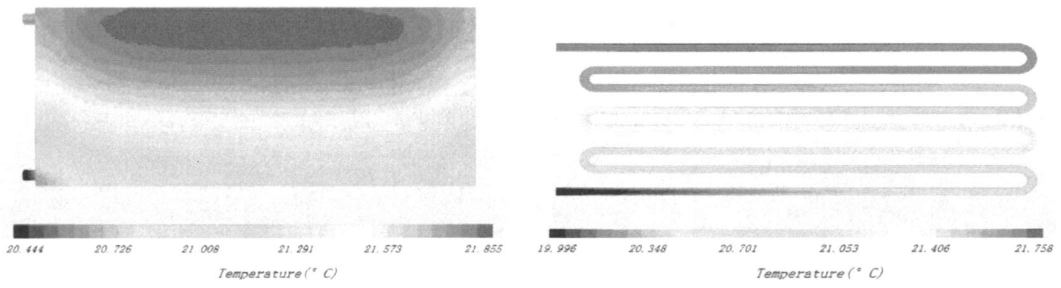

Figure 6. Temperature distribution on the liquid cooling plate's face and the cross-section of the coolant flow channel

Figure 5 and Figure 6 shows that the temperature distribution of the entire battery module shows that the cooling liquid inlet side's temperature is lower than the cooling liquid outlet side's temperature,, and due to the existence of the upper and lower double liquid cold plates, the more down and upper faces of the battery module and The temperature in the nearby area is generally more down than the temperature in the battery module's middle area. The battery module's lowest temperature appears at the single cells' tabs near the coolant inlet. With the flow of the coolant, the surface and internal temperatures of the battery module gradually increase, and the battery module's highest temperature appears on the coolant outlet side. At the tabs of the battery module, the maximumbattery module's temperature is 24.48° C, the minimum temperature is 22.02° C,and the maximum temperature difference is 2.46° C.

For this cooling system, the cooling liquid enters from the same side of the upper and lower liquid cold plates, and flows out from the other side through the serpentine flow channel. As the cooling liquid flows in the flow channel, the cooling liquid continuously absorbs the heat generated by the battery module. It increases its own temperature, so its cooling effect is gradually weakened, which also directly leads to a reduction in the battery module's heat dissipation capacity on the cooling liquid outlet side, resulting in an rise in temperature. It is known that the cooling liquid's inlet temperature is 20 °C, and the outlet temperature is 21.67 °C through simulation and calculation, indicating that the cooling liquid temperature increases by 1.67 ° C during the flow of the cooling liquid. Since the entire battery pack often contains multiple battery modules, the liquid cooling plates may be used in a common or series-parallel hybrid arrangement, so the cooling liquid outlet temperature and temperature rise of the cooling system for a single battery module should be properly controlled.

3) Visualization-based module test verification

The battery module was charged and discharged at 1C at 25° C, and the surface temperature field was recorded by an infrared thermal imager. An infrared thermal imager changes the invisible infrared power that is given out by an object into a visible thermal image. The different temperatures of the object being measured are represented by the different colors on the thermal image. The test results are shown in Figure 7 below.

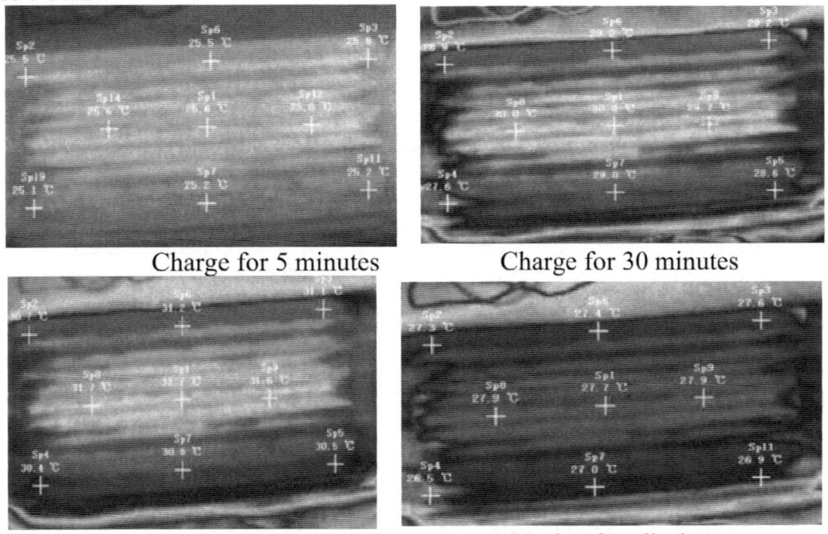

Charge for 5 minutes Charge for 30 minutes

Discharge ends 30min after discharge

Figure 7. Changes in temperature distribution on battery module's surface

As is visible from figure 7 above, the maximum battery module's temperature is 31.7° C, and the heat dissipation performance is good. According to the x-direction section's temperature field distribution, it can be seen that the temperature on the left part is more high-pitched than that on the right part at the same time, which is caused by the cooling liquid of the liquid cooling plate that flows out from the left part and entering on the another side. This conclusion is extremely consonant with the preceding simulation consequence that the outlet temperature (21.67° C) is higher than the inlet temperature (20° C) by 1.67° C. The reliability and accuracy of the simulation calculation is also verified by this conclusion.

4. Conclusion

Through the introduction of the thermal management system, it is analyzed that the liquid that cools system takes a significant influence on the power battery's temperature control in practical applications. By establishing a lithium-ion battery thermal model, a battery thermal management system with a liquid cooling system is designed. The simulation software is used to simulate and analyze the worked out thermal management system below a certain working status, and the analysis consequences are compared with the experimental test results to make sure the reliability and accuracy of the simulation results. The rationality of the modular thermal management system.

Acknowledgment

I would like to thank my colleagues at Guoxuan Hi-Tech for providing me with a lot of help and support. I would the same want to express my heartfelt thanks to the instructors in our unit.

I would like to thank the school's scientific research team-Intelligent Robot Team Project (kytd201903) for its support.

Thanks to all the scholars that be involved in this paper, several scholars' research literature is cited by this paper, without the inspiration and help of the scholars' research results.

By reason of my limited academic level, the articles I have produced will take shortcomings inevitably.

References

[1] Saw, L. H., Ye, Y., Tay, A. A. O. (2014) Electro-thermal analysis and integration issues of lithium-ion battery for electric vehicles[J]. Applied Energy, 131:97-107.
[2] Saw, L. H., Poon, H. M., San Thiam, H., Cai, Z., Chong, W. T., Pambudi, N. A., King, Y. J. (2018) Vovel thermal management system using mist cooling for lithium-ion battery packs[J]. Applied Energy, 223:146-158.
[3] Li, L. Y. (2020) Current Situation and Development Trend of Lithium Battery Industry for New Energy Vehicles in China[J].Power Technology, 44(04):628-630.
[4] Peng, Y., Huang, R., Yu, X. L., et al. (2015) Comparative study on cooling schemes of lithium-ion power batteries for electric vehicles [J]. Mechanical and Electrical Engineering, 32(4):537-543.
[5] An, Z. G., Ding, Y. Z., Liu, Q., et al. (2018) Research on the influence of liquid cooling pipes on the temperature field of power lithium battery packs [J]. Power Technology, 42(6): 795-798.
[6] Ma, J., Liu, X. D., Chen, Y. S., et al. (2018) Current situation and countermeasures of energy vehicle industry and technology development in China [J]. Journal of China Highway, 31(08).
[7] Song, B. L. (2013) Analysis and classification of advantages and disadvantages of electric vehicles [J]. Automobile Maintenance and Repair, (1):78-82.
[8] Zhu, J. G., Sun, Z. C., Wei, X. Z., et al. (2019) Research progress on low temperature characteristics and heating methods of lithium-ion batteries for vehicles [J]. Automotive Engineering, (05): 571-581.
[9] Chen, Y. S., Sun, H. L., Mao, J. (2019) Analysis of the causes and mechanism of abnormal performance degradation of power batteries [J]. Power Technology, 43(04): 654-657.
[10] Wang, F., Xia, J., et al. (2016) Safety Analysis and Design of Electric Vehicle Power Battery System [M]. Science Press, 9.
[11] Ge, R., Li, Y. Y. (2017) The key technology of thermal management system of lithium-ion power battery [J]. Power World, (12): 41-47.
[12] Wang, F., Xia, J. (2016) Safety Analysis and Design of Electric Vehicle Power Battery System [M]. Beijing: Science Press.
[13] Kizilel, R., Sabbah, R., Selman, J. R., et al. (2009) An alternative cooling system to enhance the safety of Li-ion battery packs[J]. Journal of Power Sources, 194(2):1105-1112.
[14] Qiao, H. J., Dai, D. C., Tu, J. F., Huang, J., Xu. B. N. (2020) Simulation analysis and experimental research of power battery thermal management system based on CFD [J]. Natural Science (full text), 05
[15] Wang, H., Lin, X. W., Yuan, Y. M., et al. (2016) Demand analysis and expectations for future lead-acid batteries [J]. Power Technology, 40(04): 935-937
[16] Qiao, S., Hu, M., Fu, C., et al. (2019) Experimental Study on Storage and Maitenanc1742e Method of Ni-MH Battery Modules for Hybrid Electric Vehicles [J]. Applied Sciences, 9(9).
[17] Tian, C. Z., Gao, C., Tang, X. S., et al. (2018) Industrial structure and development prospect of power lithium-ion battery [J]. Power Technology, 42(12): 1930-1932.
[18] Yang, G. S. (2016) Research on thermal management system of electric vehicle power battery pack [J]. Science and Technology Innovation Herald, (4): 178-180.
[19] Chen, T., Sun, G. H., Wang, M. Q., Cao, Y., Wang, S. F., et al. (2019) Research on the performance of liquid-based power battery thermal management system [J]. Power Technology, (4): 658-661.
[20] Bernardi, D., Pawlikowski, E., Newman, J. (1985) A general energy balance for battery systems [J]. Journal of the Electrochemical Society, 132(1): 5-12.

Optimization of silicone rubber materials for high voltage switchgear and study on characteristic gas of thermal decomposition

Bin Tang[1,*], Qinqin Liang[1], Longfei Zhang[1], Zongchang Luo[1], Chuansheng Luo[2], Mengzhu Hu[1], Jialin Wang[3], Liping Zhu[1]

[1]Guangxi Key Laboratory of Intelligent Control and Maintenance of Power Equipment, Electric Power Research Institute of Guangxi Power Grid Co., Ltd.
[2]Nanning Power Supply Bureau of Guangxi Power Grid Co., Ltd.
[3]Guangxi Power Grid Co., Ltd.

*Corresponding author e-mail: tang_b.sy@gx.csg.cn

Abstract. In this paper, three kinds of silicone rubber materials, methyl silicone rubber, phenyl silicone rubber and fluorosilicone rubber, are used as the research object, contact angle tester, thermal conductivity tester, insulation resistance tester, differential scanning calorimeter and thermogravimetric analyzer to explore the effect of side groups on the properties of silicone rubber. The results show that the thermal conductivity of three kinds of silicone rubber materials is poor, but they all have good hydrophobicity, insulation and thermal stability. Among them, phenyl silicone rubber shows better hydrophobicity and thermal stability. In addition, the type and relative content of characteristic gases of thermal decomposition of silicone rubber materials at different temperatures are analyzed in detail by gas chromatography mass spectrometer. The results show that C6H18O3Si3 and C8H24O4Si4 are the common characteristic gases of thermal decomposition of three kinds of silicone rubber.

1. Introduction

High-voltage switchgear internal contact boxes, insulators, wall bushings and other polymer materials are often overheated and decomposed due to excessive load current and poor contact, which seriously affects the safe operation of the power grid [1-3]. Silicone rubber is widely used in high-voltage switchgear due to its advantages of good insulation, high temperature resistance, ozone resistance, ultraviolet light resistance and weather resistance [4-6]. Among them, the Si atom is connected with different groups such as methyl groups, which can be divided into methyl silicone rubber (PDMS), phenyl silicone rubber (PMPS), fluorosilicone rubber (FVMQ), etc. according to the side groups [7]. Different side groups of silicone rubber have different properties. How to optimize the silicone rubber material to better meet the requirements of high-voltage switchgear for hydrophobicity, thermal conductivity, insulation, thermal stability and other properties is currently a topic of concern. Based on the above analysis, this paper selects PDMS, PMPS, FVMQ three kinds of silicone rubber materials with different side groups as the research objects, using contact angle tester, thermal conductivity tester, insulation resistance tester, differential scanning calorimeter (DSC) and thermal The gravimetric analyzer (TG) compared and analyzed the surface hydrophobicity, thermal conductivity, insulation and thermal stability of the three silicone rubber materials, and based on this, the silicone rubber material with the best performance was selected. In addition, a tube furnace was used to collect the thermal decomposition characteristic gases of the three silicone rubber materials at different temperatures, and their species and relative contents were analyzed by gas chromatography-mass spectrometry (GC-MS).

2. Experimental part

2.1. Experimental materials

Phenyl silicone rubber (PMPS), methyl silicone rubber (PDMS), fluorosilicone rubber (FVMQ), Chengdu Baoliguang Technology Co., Ltd.; 0.1 L aluminum foil air bag, Ningbo Hongpu Instrument Technology Co., Ltd.1.2

2.2. Experimental operation

2.2.1 Sample preparation

The samples were rinsed with 95% industrial ethanol and deionized water respectively to achieve the effect of surface decontamination and degreasing, and then the samples were dried in an electric heating constant temperature drying oven at 60 ℃, and the samples were cut into suitable sizes for use.

2.2.2 Collection of thermal decomposition characteristic gases

Weigh 2.500 g of the pretreated sample with an electronic balance and put it in the middle of the quartz tube in a porcelain boat, and set the heating rate of the tube furnace to 10 °C/min. Before the experiment, N2 was introduced for 30 min to exhaust the air inside the quartz tube, and the gas flow was set to 100 mL/min. During the experiment, according to the TG results of each material, a 0.1 L aluminum foil gas collection bag was selected to start gas collection at the initial decomposition temperature (Tbegin) of the sample, and end gas collection at the final decomposition temperature (Tend) of the sample.

2.2.3 Performance test

(1) Hydrophobic performance test: A801S contact angle tester, using direct measurement method; (2) Thermal performance test: TC3000E thermal conductivity tester, experimental voltage 1 V, acquisition time 5 s, sample size 50 mm × 50 mm ×3 mm; (3) Insulation performance test: TH2683A insulation resistance tester, output voltage 10 V, charging time 2.5 s, waiting time 0 s, measurement time 5 s, discharge time 2.5 s, measurement speed is slow, locked The range is 10 nA, and the sample size is 50 mm×50 mm×3 mm; (4) DSC test: (5) Mettler-Toledo differential scanning calorimeter, N2 atmosphere, heating rate 10 ℃/min, temperature range 30 ～300 ℃, the gas flow rate is 50 mL • min-1, the mass of the sample to be tested is 5 mg; (6) TG test: STA449F3 thermogravimetric analyzer, N2 atmosphere, heating rate 10 ℃/min, temperature range 30～800 ℃ , the gas flow rate is 50 mL • min-1, and the mass of the sample to be tested is 5 mg; (7) GC-MS test: QP2010PLUS gas chromatography mass spectrometer, using Rtx-5MS (30 m × 0.25 mm × 0.25 μm) chromatographic column. Chromatographic conditions: the injection port temperature was 260 °C; the temperature programmed process: the initial column temperature was kept at 50 °C for 3 min, raised to 180 °C at 10 °C/min, held for 10 min, and then raised to 220 °C at 10 °C/min, held for 10 min; mass spectrometry conditions: GC-MS interface temperature: 250 °C; ion source temperature: 200 °C; ion scan mode: Scan mode; mass-to-charge ratio monitoring range: m/z = 20-500; NIST14s mass spectral library was used for mass spectrometry retrieve.

3. Results and discussion

3.1. Hydrophobic performance analysis

Figure 1 shows the static water contact angle data for three silicone rubber materials. The results show that the surface contact angles of PDMS, PMPS and FVMQ materials are all greater than 90°, showing good hydrophobicity. This may be because the methyl and phenyl groups connected to Si atoms in PDMS and PMPS, respectively, are non-polar groups, which are opposite to the polarity of hydroxyl groups in water. The low polarizability of the F atom of the trifluoropropyl group in FVMQ reduces the polarizability of the C-F bond, resulting in a decrease in intermolecular cohesion and a decrease in surface energy, thus showing hydrophobicity [8]. In addition, the contact angle of PDMS is 100.4°, that of PMPS is 105.0°, and that of FVMQ is 97.2°. The hydrophobicity of the three silicone rubber materials is PMPS>PDMS>FVMQ. Therefore, PMPS exhibits the best hydrophobicity and is suitable for components with higher hydrophobicity requirements inside the high-voltage switchgear.

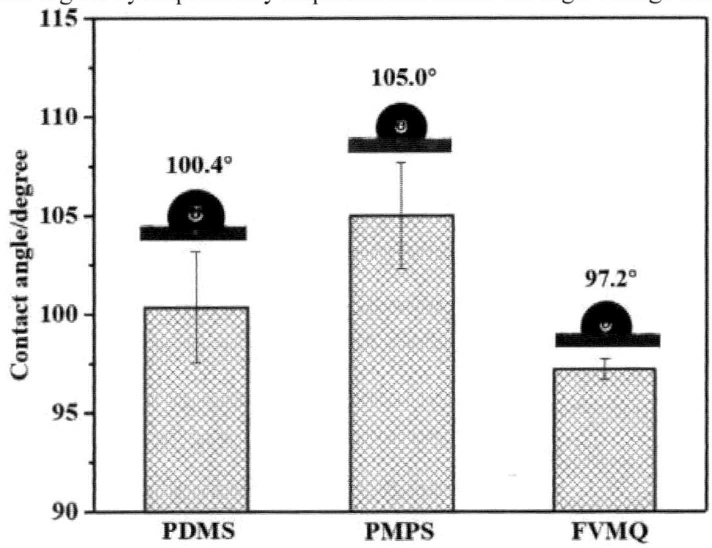

Figure 1. Static water contact angle of three kinds of silicone rubber.

3.2. Thermal conductivity analysis

Table 1. Test results of thermal conductivity of three kinds of silicone rubber.

Material	PDMS	PMPS	FVMQ
Experimental temperature/K	293.16	293.39	293.86
Thermal conductivity (W/m•K)	0.2491	0.1931	0.1846

From the results of the thermal conductivity of the three silicone rubber materials in Table 1, it can be seen that the thermal conductivity of PDMS, PMPS and FVMQ at room temperature are 0.2491,

0.1931 and 0.1846 W/m•K, respectively, and the thermal conductivity is poor. The thermal conductivity is distributed at 0.17. -0.26 W/m•K[9].

3.3. Analysis of insulation performance

Table 2. Insulation test results of three kinds of silicone rubber.

Material	PDMS	PMPS	FVMQ
Experimental temperature/K	293.16	293.39	293.86
Volume resistivity (Ω/cm)	1.92×1011	2.82×1011	3.12×1011

Table 2 shows the insulation performance results of three silicone rubber materials. The volume resistivities of PDMS, PMPS, and FVMQ at room temperature are 1.92×1011, 2.82×1011, and 3.12×1011 Ω/cm, respectively. This shows that the insulating properties of the three silicone rubber materials are in the same order of magnitude and meet the definition of insulating material (>1×109 Ω/cm). Therefore, PDMS, PMPS, and FVMQ can be used as excellent insulating materials in the internal components of high-voltage switchgear that have high requirements for insulation performance.

3.4. Differential Scanning Calorimetry (DSC) analysis

Figure 2 shows the DSC curves of three silicone rubber materials at 30-300 ℃, N2 atmosphere, and 10 ℃•min-1 heating rate. PDMS, PMPS and FVMQ have no endothermic peaks in the range of 30-300 ° C, and all show good thermal stability, which is better than that of epoxy resin and cross-linked polyethylene commonly used in high-voltage switchgear [10].

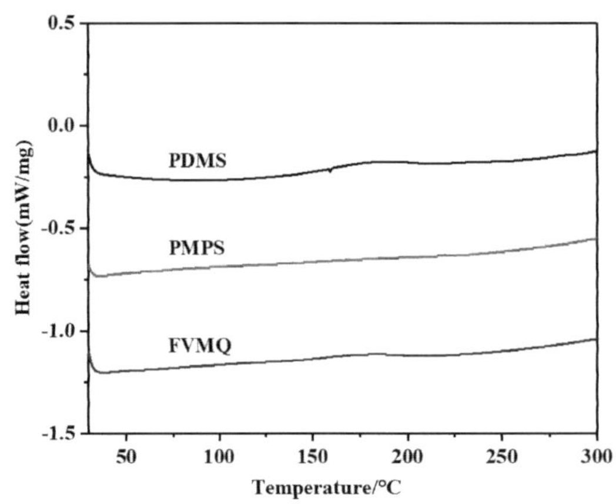

Figure 2. DSC curves of three kinds of silicone rubber

3.5. Thermogravimetric (TG) analysis

Figure 3 (a) shows the TG curves of the three silicone rubbers. It can be seen from the figure that PDMS and FVMQ have only one weight loss section during thermal decomposition, while PMPS has two weight loss sections during thermal decomposition.

Figure 3 (b)-(d) are the TG-DTG curves of PDMS, PMPS and FVMQ, respectively. The analysis shows that the thermal decomposition of PDMS starts at 462.2 ℃ , reaches the maximum decomposition rate at 550.3 ℃ , and when the temperature rises to 655.1 ℃ The thermal decomposition of PDMS ends. PMPS has two sections of thermal decomposition curve, with a slight thermal decomposition in the range of 118.8-391.8 ℃ . The temperature corresponding to the maximum decomposition rate in this interval is 193.2 ℃, and the decomposition rate is only 6.8 %, which may be the additives and fillers in PMPS. Or the volatilization and slight decomposition of part of the matrix at lower temperature. Therefore, this experiment compares the data of the second weight loss section of PMPS with other silicone rubber materials. Immediately after PMPS continued to decompose, the temperature corresponding to its maximum decomposition rate was 518.4 ℃. When the temperature reached 732.9 ℃, the thermal decomposition curve of PMPS tended to be flat and no longer decomposed. The thermal decomposition range of FVMQ is 321.8-585.6 ℃, and its maximum decomposition rate is reached at 479.8 ℃.

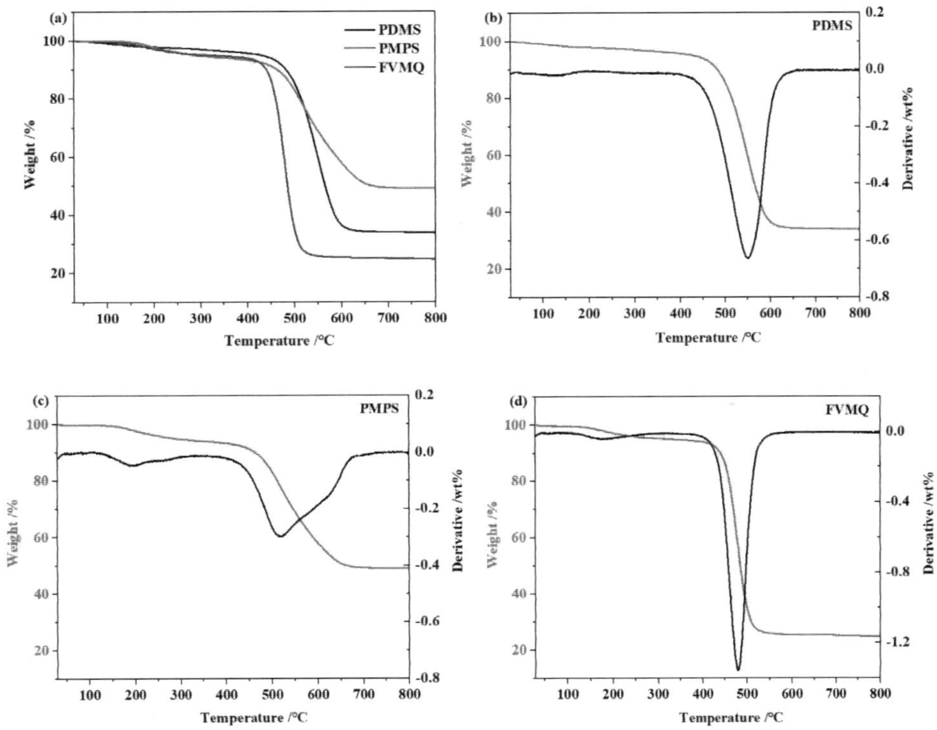

Figure 3. (a): TG images of three kinds of silicone rubber; (b): TG-DTG images of PDMS; (c): TG-DTG images of PMPS; (d): TG-DTG images of FVMQ

Table 3. Thermal decomposition characteristic parameters of three kinds of silicone rubber

	weightless section	Tbegin/°C	Tmax/°C	Tend/°C	v/wt%	w/%
PDMS	1	426.2	550.3	655.1	0.67	33.7
PMPS	1	118.8	193.2	391.8	0.05	93.2

	2	391.8	518.4	732.9	0.30	48.9
FVMQ	1	321.8	479.8	585.6	1.17	24.6

It can be seen from Table 3 that the initial decomposition temperature of the three silicone rubber materials is PDMS(426.2 ℃)>PMPS(391.8 ℃)>FVMQ(321.8 ℃), and the final decomposition temperature is PMPS(732.9 ℃)>PDMS(655.1 ℃)> FVMQ (585.6 ° C). The maximum decomposition rate was PMPS(0.30 wt%)<PDMS(0.67 wt%)<FVMQ(1.17 wt%), and the thermal decomposition residual ratio was PMPS(48.9 %)>PDMS(33.7 %)>FVMQ (24.6 %). In terms of the final decomposition temperature, the maximum decomposition rate and the thermal decomposition residual rate of the material, the thermal stability of PMPS is better.

3.6. Thermal decomposition gas gas chromatography-mass spectrometry (GC-MS) analysis

In this paper, according to the TG results (Table 4), an aluminum foil gas bag was used to collect the thermal decomposition gas of PDMS in the range of 426.2-655.1 °C, and the thermal decomposition gas of PMPS was collected in the range of 118.8-391.8 °C and 391.8-732.9 ° C, respectively. The thermal decomposition gas of FVMQ was collected in the range of 321.8-585.6 ℃.

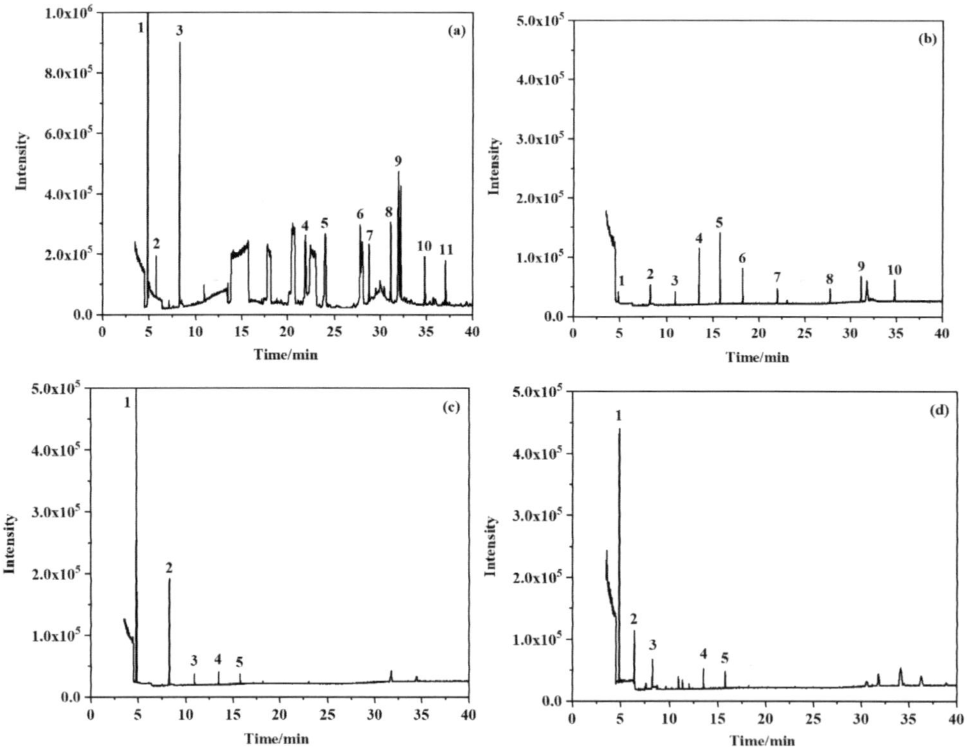

Figure 4. (a): Total ion flow diagram of PDMS at 426.2-655.1 °C; (b): Total ion flow diagram of PMPS at 118.8-391.8 °C; (c): Total ion flow diagram of PMPS at 391.8-732.9 °C; (d): Total ion flow diagram of FVMQ at 321.8-585.6 °C

Figure 4 (a)-(d) are the total ion currents of the thermally decomposed gases of PDMS, PMPS and FVMQ in different temperature ranges. It can be seen from the figure that PDMS has 11 characteristic peaks in the range of 426.2-655.1 ℃, that is, contains 11 characteristic gases; PMPS has 10

characteristic peaks in the range of 118.8-391.8 ℃, and 5 characteristic peaks in the range of 391.8-732.9 ℃; FVMQ There are 5 characteristic peaks in the range of 321.8-585.6 ℃.
Tables 4-6 show the retention time, species and peak area or area ratio of thermally decomposed gases of PDMS, PMPS, and FVMQ in different temperature ranges, respectively.

Table 4. The main characteristic gases of PDMS at 426.2-655.1 °C

serial number	Retention time (min)	belong	molecular formula	Area percentage (%)
1	4.830	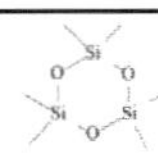	$C_6H_{18}O_3Si_3$	56.21
2	5.765	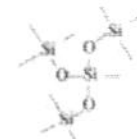	$C_{10}H_{30}O_3Si_4$	0.75
3	8.260	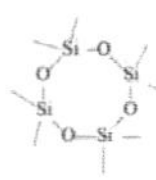	$C_8H_{24}O_4Si_4$	5.31
4	21.995	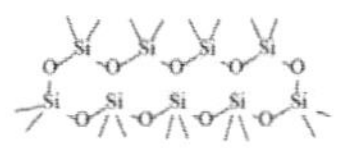	$C_{18}H_{54}O_9Si_9$	5.63
5	24.000		$C_{25}H_{52}O_3Si_2$	7.44
6	27.775		$C_3H_4O_2$	2.75
7	28.775	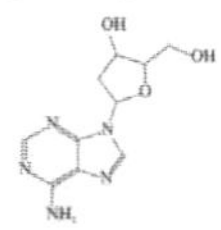	$C_{10}H_{13}N_5O_3$	2.15
8	31.130	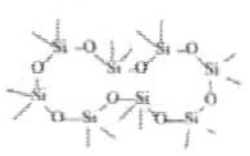	$C_{16}H_{48}O_8Si_8$	2.73
9	31.955	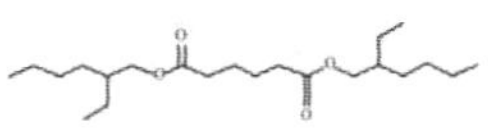	$C_{22}H_{42}O_4$	7.97
10	34.800	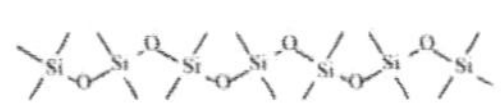	$C_{16}H_{48}O_6Si_7$	2.12
11	37.020	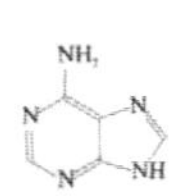	$C_5H_5N_5$	2.01

Table 5. The main characteristic gases of PMPS at 118.8-391.8 °C (Temperature 1) and 391.8-732.9 °C (Temperature 2)

serial number	Retention time (min)	belong	molecular formula	Area percentage (%)
1	4.830	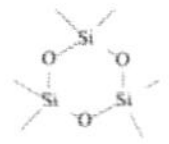	$C_6H_{18}O_3Si_3$	3.65 T1 86.95 T 2

2	8.260	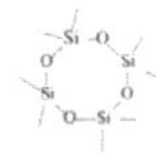	$C_8H_{24}O_4Si_4$	4.93 T 1 9.96 T 2
3	10.915	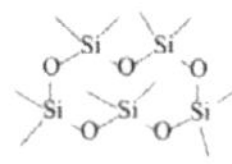	$C_{10}H_{30}O_5Si_5$	3.32 T 1 1.02 T 2
4	13.470	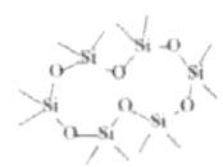	$C_{12}H_{36}O_6Si_6$	16.24 T 1 1.26 T 2
5	15.745	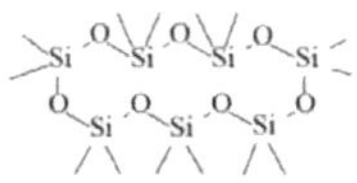	$C_{14}H_{42}O_7Si_7$	18.08 T 1 0.81 T 2
6	18.210	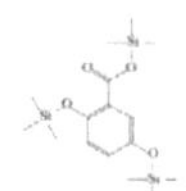	$C_{16}H_{30}O_4Si_3$	12.53 T 1 -
7	21.995	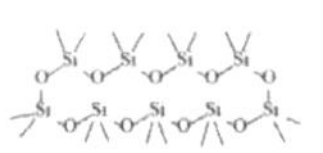	$C_{18}H_{54}O_9Si_9$	7.43 T1 -
8	27.775	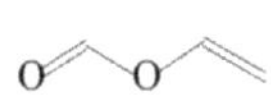	$C_3H_4O_2$	6.60 T1 -
9	31.130	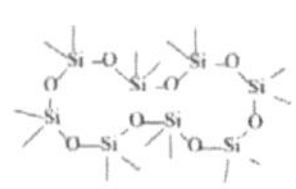	$C_{16}H_{48}O_8Si_8$	12.81 T 1 -
10	34.800	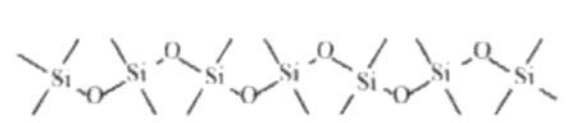	$C_{16}H_{48}O_6Si_7$	14.42 T 1 -

Table 6. The main characteristic gases of FVMQ at 321.8-585.6 °C

serial number	Retention time (min)	belong	molecular formula	Area percentage (%)
1	4.830	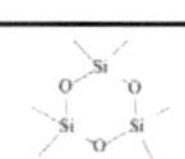	$C_6H_{18}O_3Si_3$	67.13
2	6.435	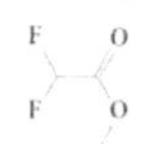	$C_3H_4F_2O_2$	18.88
3	8.260	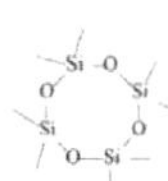	$C_8H_{24}O_4Si_4$	5.85
4	13.470	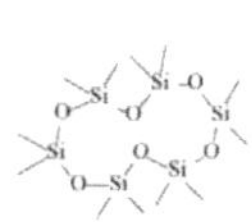	$C_{12}H_{36}O_6Si_6$	4.78

5	15.745	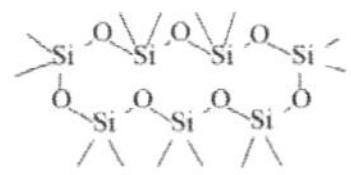	$C_{14}H_{42}O_{7}Si_{7}$	3.36

It can be seen from Table 4 that the characteristic products of thermal decomposition of PDMS are mainly cyclosiloxanes vaporized at high temperature, such as hexamethylcyclotrisiloxane (C6H18O3Si3), octamethylcyclotetrasiloxane (C8H24O4Si4) and ten Octamethylcyclononasiloxane (C18H54O9Si9) and so on. In addition, the characteristic gases of PDMS also include a small amount of long carbon chain substances (C25H52O3Si2, C22H42O4), nitrogen-containing heterocyclic substances (C10H13N5O3, C5H5N5) and carbonyl-containing olefin substances (C3H4O2).
It can be seen from Table 5 that the characteristic gas types and contents of PMPS in different temperature ranges are different. At 118.8-391.8 ℃, the characteristic products of PMPS include various gasified cyclosiloxanes, carbonyl-containing olefins (C3H4O2) and benzene rings (C16H30O4Si3), which may represent the additives and fillers in PMPS. and the volatilization and slight decomposition of the silicone rubber matrix in this temperature range; at 391.8-732.9 ℃, the gas characteristic peaks of PMPS No. 6-10 disappear, and the gas characteristic products are all cyclosiloxane substances, namely C6H18O3Si3, C8H24O4Si4, C10H30O3Si4, C12H36O6Si6 and C14H42O7Si7, and the content of C6H18O3Si3 is much larger than that at 118.8-391.8 ℃, indicating that the PMPS matrix is largely decomposed in this temperature range.
It can be seen from Table 6 that the characteristic gases of FVMQ include C6H18O3Si3, C8H24O4Si4, C12H36O6Si6 and C14H42O7Si7 cyclosiloxanes, as well as fluorine-containing substances (C3H4F2O2).

4. Conclusion

(1) The contact angle test shows that PMPS has the best hydrophobicity, and its contact angle reaches 105.0°, which is suitable for components with higher hydrophobicity requirements inside the high-voltage switchgear; the thermal conductivity test shows that the three silicone rubbers are at room temperature. The thermal conductivity at room temperature is relatively small; the insulation performance test shows that the volume resistivity of PDMS, PMPS and FVMQ at room temperature reaches 1.92×1011, 2.82×1011 and 3.12×1011 Ω/cm, respectively.
(2) DSC shows that the three silicone rubber materials can be used at a high temperature of 300 ℃, showing better thermal stability; PMPS has the best thermal decomposition performance, so it is more suitable for high temperature resistance inside high-voltage switchgear components with higher performance requirements.
(3)GC-MS showed that the thermal decomposition characteristic gases of the three silicone rubber materials all contained cyclosiloxane substances C6H18O3Si3 and C8H24O4Si4. In addition, the thermal decomposition characteristic gases of PDMS include a small amount of long carbon-chain substances, nitrogen-containing heterocyclic substances and carbonyl-containing olefins; the thermal decomposition characteristic gases of PMPS include carbonyl-containing olefins and benzene ring substances; FVMQ The thermal decomposition characteristic of the gas includes fluorine-containing substances.

References

[1] Wang, J., Duan, M. C., Chen, J. J., et al. (2018) Heating Fault Analysis of High Voltage Switch Cabinet Overcurrent Components[J]. HUNAN ELECTRIC POWER,38(02): 18-20.
[2] Guan, W., Zhang, X. X., Cheng, Z., et al. (2017) Experimental Research and Diagnosis on Contaminated Insulation Condition of the Post Insulator in High-voltage Switchgear[J]. High Voltage Apparatus, 53(02): 178-185.
[3] Zhang, Z., Su, D., Chen, Z. Z., et al. (2020) Research on Key Techniques of Wall Bushing in

40.5 kV Ai- insulated Switchgear[J]. High Voltage Apparatus, 56(11): 57-64.

[4] Sun, Q. J., Huang, Y. H., Wu, N., et al, (2018) Research Progress on Performance of Phenyl Silicone Rubber[J]. SILICONE MATERIAL, 32(01): 71-76.

[5] Yang, H. B., Yang, S. G., Zhang, L. P., et al. (2016) Application Status and Research Progress of Silicone Rubber in Electrical Industry[J]. SILICONE MATERIAL, 30(01): 28-31.

[6] Li, D., Du, P., Li, B., et al, (2015) Application of Silicone Rubber Insulation Material in Substation Busbar[J]. Science and Technology Innovation Herald, 12(07): 103.

[7] Long, X. J. (2015) The preparation and properties of halogen-free flame retardant silicone rubber[D]. South China University of Technology.

[8] Luo, P., Ren, Z. Z., Xu, Y., et al. (2013) Aging Condition Analysis of High Voltage XLPE Cables out of Service[J]. TRANSACTIONS OF CHINA ELECTROTECHNICAL SOCIETY, 28(10): 41-46.

[9] Zhang, X. X. (2020) Preparation and Properties of Thermal Conductive Silicone Rubber[D]. Shandong University.

[10] Li, H., Wang, D. H., Qian, X. Q. (2013) Curing kinetics of epoxy resin by non-isothermal DSC method[J]. Journal of Solid Rocket Technology, 36(02): 266-269+284.

Prediction of Ultra-Short-Term power system based on LSTM-Random forest combination model

Ziqiang Fang[1, a], Ziteng Yang[2,b], Haiyang Peng[3,c], Guoqing Chen[4,*]

[1]School of Mathematics and Statistics, Qinghai Normal University , QingHai, China
[2]School of Chemical Engineering, Guangdong University of Petrochemical
[3]Technology, MaoMing, China
[4]School of Mathematics and Statistics, Minnan Normal University, Zhangzhou, China
Department of mathematics, ChengDu JinCheng College, ChengDu, China

*chenguoqing@scujcc.edu.cn; [a]15013579391@163.com;[b] yangziteng1620@163.com; [c]pxxxoc@163.com;

Abstract. The ultra-short-term forecast of the power load system plays an important role in the operation of the power market and economic dispatch. An excellent ultra-short-term power load prediction model can not only greatly reduce the loss of power operation, but also make a reasonable power generation plan accordingly to reduce the waste of power. In this paper, a model applicable to forecasting the ultra-short-term power system is proposed by combining LSTM and Random Forest using the reciprocal error method, based on the power load data of the 10th Teddy Cup Data Mining Competition. The illustrating example shows that the mean absolute percentage error (MAPE) of the combined prediction model of LSTM-Random Forest is 1.40%, the error of the combined model is significantly lower than the uncombined model, implying that the prediction accuracy of the combined prediction model is significantly higher than that of the LSTM model and Random Forest model.

1. Introduction

With the transformation of my country's energy industry and the improvement of the level of informatization, the demand for electricity has increased year by year, and there have been power shortages in many areas. It is necessary to realize reasonable power dispatching, avoid energy waste caused by power resource management, and accurately control the power load. Prediction becomes even more important.

Ultra-short-term load forecasting methods with high accuracy are mainly divided into two categories, one is classical forecasting methods, such as regression analysis, exponential smoothing method [1], moving average method [2], etc. and the other is machine learning method, Such as Random Forest, neural network, support vector machine, GBDT, XGBoost, etc. The former is mostly based on pure mathematical theory, and has great limitations in practical applications, and the prediction accuracy of ultra-short-term power load data with nonlinearity and high frequency is low. The latter machine learning algorithm has the ability to adapt to a large number of unstructured and inaccurate laws, and has strong self-learning ability and complex nonlinear function fitting ability [3]. With the development of computer technology, the shortcoming of the long training time of machine learning has been gradually overcome, and the use of machine learning methods for power load forecasting has gradually emerged, the most typical of which is the neural network method. In the literature [4], based on BP neural network, a load forecasting model that is classified according to different seasons and different weekly periods is constructed. This method effectively improves the forecasting accuracy. However, such classification requires a large amount of data, and there are as many as 168 types of models, which is too complicated. Reference [5] uses the LSTM neural network method for short-term load forecasting, and uses historical data as the feature vector for rolling forecasting. Finally, the prediction accuracy and convergence speed of the LSTM neural network are stronger than the traditional bp neural network. Reference [6] used the reciprocal error method to synthesize the two models of LSTM and XGBoost, and obtained results with higher accuracy than a single model, which further improved the prediction accuracy. Reference [7] proposed a short-term power prediction model based on the fusion of CNN-LSTM and XGBoost, and then used the MAPE-RW algorithm for weight search, which greatly improved the model training speed and prediction accuracy.

In ultra-short-term power load forecasting, few scholars have used traditional mathematical models. In recent years, scholars have mainly used machine learning models to improve forecasting accuracy, and combining multiple models can achieve better forecasting results. Effect. Therefore, this paper will apply the combined model of Random Forest and LSTM to forecast the ultra-short-term power load.

2. Model preparation

2.1. Feature engineering

The data of the power system load forecast analysis contest of the 10th "Teddy Cup" data mining competition is selected for modeling.

In this paper, the power load data of a certain region from January 1, 2018 to August 31, 2021 at 15-minute intervals will be used to judge the prediction accuracy of the power load at 15-minute intervals in the next 10 days by comparing its MAPE, MAE and RMSE indicators.

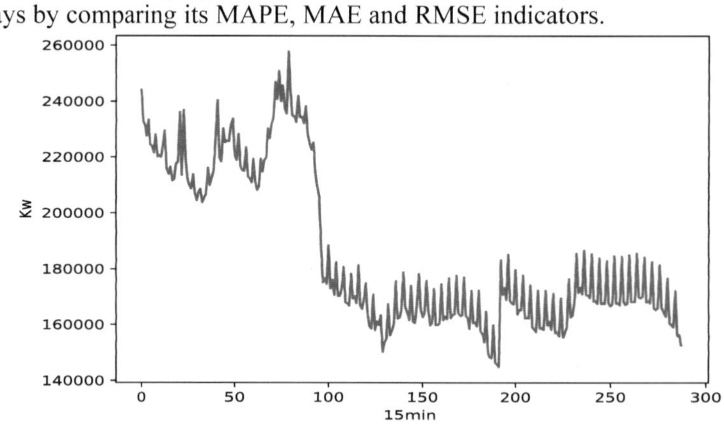

Figure 1. Graph of charge change over 3 days

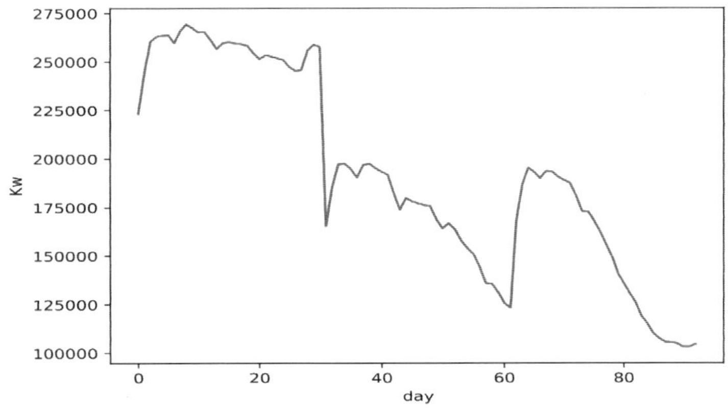

Figure 2. Graph of charge change over 3 months

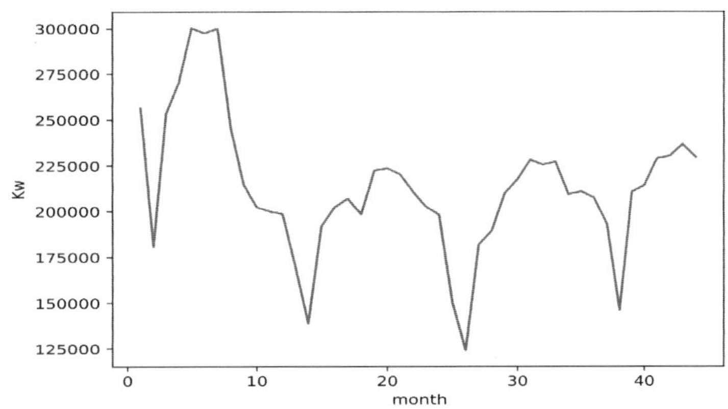

Figure 3. Graph of charge change over 3 years

Firstly, the characteristic construction is carried out according to the given time data. From Figures 1, 2, and 3, it can be noticed that there is a certain law in the daily, monthly and annual power load changes. Therefore, the construction year, season, month, week, hour, minute, etc. seven characteristics. Secondly, I tried to introduce features such as epidemic and weather, and found that introducing such features into the model could not significantly reduce the prediction accuracy of the model, so the features were only constructed in the time dimension.

2.2. Model selection

Based on the research methods of previous scholars, traditional mathematical models are more suitable for predicting medium and long-term power load forecasting, but there are certain shortcomings in predicting ultra-short-term power load. This paper will use machine learning algorithms for prediction. This paper comprehensively uses Random Forest, LSTM, XGBoost, SVM, GBDT methods for prediction, select MAPE (Mean Absolute Percentage Error) as the main indicator to evaluate the prediction accuracy of each model, select MAE (Mean Absolute Error), RMSE (Root Mean Square Error) as secondary evaluation indicators, the calculation formulas of MAPE, RMSE, and MAE are as follows:

$$X_{MAPE} = \frac{1}{n}\sum_{i=1}^{n}\frac{|y_t - \hat{y}_t|}{y_t} \tag{1}$$

$$X_{RMSE}=\sqrt{\frac{1}{n}\sum_{i=1}^{n}(y_i-\hat{y}_i)^2} \tag{2}$$

$$X_{MAE}=\frac{1}{n}\sum_{i=1}^{n}\left|y_t-\hat{y}_t\right| \tag{3}$$

Table 1. Comparison of MAPE, RMSE, ME of each model

Model	MAPE	RMSE/kW	MAE/kW
Random Forest	0.018	7275.138	2913.243
XGBoost	0.027	8360.742	4310.517
LSTM	0.018	7208.120	3058.796
SVR	0.179	44353.197	32644.388
GBDT	0.070	17493.178	11015.615

The MAPE, RMSE, and ME of each model are obtained by calculation. As shown in From 1, it is noted that the prediction accuracy of SVR, GBDT, and XGBoost is not ideal. The prediction accuracy of Random Forest and LSTM is quite close from the index level, so this paper chooses Random Forest combined with LSTM to predict the power load system.

3. Ultra-short term power load forecasting

3.1. Power load forecasting based on LSTM

RNN (Recurrent Neural Network) is a type of recurrent neural network (Recursive Neural Network) RNN is a neural network with memory, including a recurrent network, which makes information persistent. RNN can be understood as multiple copies of the same neural network, each neural network module will pass information to the next neural network module. However, RNN is easy to shovel the problem of gradient explosion and disappearance [8], and LSTM (Long Short-Term Memory) as a variant of RNN [9], can effectively solve this problem. Through the improvement of RNN, the prediction accuracy of LSTM for power load system is significantly higher than other models. The calculation process is roughly as follows:

Step1: First need to decide what information needs to be forgotten in the memory unit, which will be decided through the forget gate:

$$f_t=\sigma(W_f\cdot[h_{t-1},x_t]+b_f) \tag{4}$$

where f_t is the forgetting function, Control the amount of information that needs to be forgotten in the memory unit c_t at the previous moment, σ is sigmoid function, W_f is the forgetting function, $[h_{t-1},x_t]$ represents a new vector formed by merging the input vector x_t and the previous moment h_{t-1} of the hidden layer state, b_f represents the bias term of the forget gate. The value range of f_t is 0-1, When f_t is 0, it means complete forgetting, and when f_t is 1, it means complete retention.

Step2: The input gate will decide what information to keep in the memory unit. This gate contains two small neural network layers:

$$i_t=\sigma(W_i\cdot[h_{t-1},x_t]+b_i) \tag{5}$$

$$\tilde{C}_t=\mathrm{Tanh}(W_C\cdot[h_{t-1},x_t]+b_C) \tag{6}$$

Among them, $\tilde{i}_t$ and f_t have the same structure, which is the amount of information that needs to be saved to determine $\tilde{C}_t$. Equation (6) The unit state $\tilde{C}_t$ is mapped to [-1,1] by Tanh, and $i_t * \tilde{C}_t$ represents the new information to be updated.

Step3: Update unit state, new unit state = old unit state × forget gate value + new information to be updated, namely:

$$C_t = f_t * C_{t-1} + i_t * \tilde{C}_t \tag{7}$$

By combining $\tilde{C}_t$ with C_{t-1} with long-term memory, the new unit state C_t formed can save extremely long-term information, making the model have long-term memory.

Step4: Determine the output layer o_t as:

$$o_t = \sigma(W_o \cdot [h_{t-1}, x_t] + b_o) \tag{8}$$

Step5: Finally, the unit state is processed through tanh (mapped to between -1 and 1), and then multiplied by the outputs, which is the final data result of LSTM, which is shown in formula (9):

$$h_t = o_t * \tanh(C_t) \tag{9}$$

Before modeling, a time series dataset is constructed for the input sequence of the model. For a unit time t (15min), the power load data from time t-5 to t-1 is used as the input X of the model, and the power load data at the current time is used as the output Y.

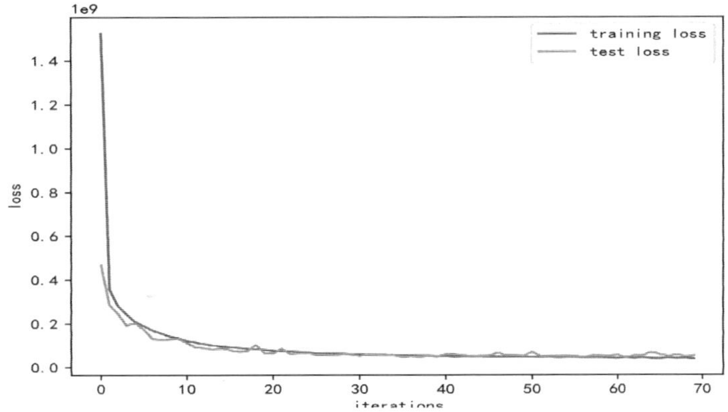

Figure 4. LSTM loss function graph

The loss function results of LSTM are shown in Fig. 4, and LSTM converges when the iteration goes to about 60 times.

3.2. Power load forecasting based on random forest

The CRAT (Classification and Regression Tree) model was proposed by Breiman et al. in 1984 [10], and is often used in classification and regression, while Random Forest is the basic model for ensemble learning [11], applying the idea of bagging to combine multiple decision-making Combine trees. The prediction result of Random Forest is to select the tree with the highest degree of repetition as the final result.

The construction of Random Forest starts with a decision tree, and CRAT consists of two parts, one part is a classification tree and the other part is a decision tree. First, a regression tree needs to be constructed. The regression tree survival algorithm is as follows:

Step1: Input the training set D. In D, apply recursion to divide each area into two sub-areas, and determine the output value of each sub-area to form a binary decision tree.

Step2: Secondly, select the optimal segmentation variable j and segmentation point s through traversal, so that formula (10) takes the minimum value of (j, s).

$$\min[\min_{c_1} \sum_{x_i \in R_1(j,s)} (y_i - c_1)^2 + \min_{c_2} \sum_{x_i \in R_2(j,s)} (y_i - c_2)^2] \tag{10}$$

Step3: Divide the area for (j, s) and determine the corresponding output value:

$$R_1(j,s) = \{x \mid x^{(j)} \le s\}, R_2(j,s) = \{x \mid x^{(j)} > s\} \tag{11}$$

$$\hat{c}_m = \frac{1}{N_m} \sum_{x_i \in R_m(j,s)} y_i, x \in R_m, m = 1,2 \tag{12}$$

Step4: Repeat Step2 and Step3 for the two sub-regions until the conditions are met.

Step5: Divide D into M regions $R_1, R_2, \cdots, R_M$ to generate a decision tree:

$$f(x) = \sum^{M} \hat{c}_m I(x \in R_m) \tag{13}$$

Secondly, the classification tree is constructed. The basic idea is to use the Gini coefficient or out-of-bag error to select the optimal feature and the optimal binary segmentation point of the feature. After the classification and decision tree construction is completed, the CART decision tree is generated. To prevent overfitting problems, CART needs to be pruned. After pruning, the combination of multiple decision trees constitutes a Random Forest.

3.3. Combination model prediction method

For example, after the LSTM and Random Forest in Chapters 4 and 5 have obtained the prediction results of the model, the models are combined, and the combination method is the reciprocal error method, as shown in equations (14) and (15):

$$f_t = \omega_1 f_{1t} + \omega_2 f_{2t}, t = 1,2,\ldots,n \tag{14}$$

$$\omega_i = \frac{\varepsilon_i}{\varepsilon_1 + \varepsilon_2} \tag{15}$$

Among them, formula (14) f_{1t} represents the predicted value of LSTM at time t, ω_1 is the weight coefficient of LSTM, and similarly f_{2t} and ω_2 represent the predicted value of Random Forest at time t and the weight coefficient of Random Forest, respectively. In formula (15), ε_1 and ε_2 represent the errors of LSTM and Random Forest, respectively. When the error of which method is smaller, it will be given a higher weight to improve the prediction accuracy.

4. Analysis of calculation results

Taking the question B of the tenth "Teddy Cup" as the experimental data, the final combined model has a significant improvement in the prediction accuracy of the two independent models, LSTM and Random Forest. The MAPE is 0.014, the RMSE is 6255.030, and the MAE is 2199.531. Each evaluation index of the combined model is significantly higher than that of LSTM and Random Forest. Figure 5 is a comparison chart of the predicted value and the actual value of the combined model, LSTM, and Random Forest.

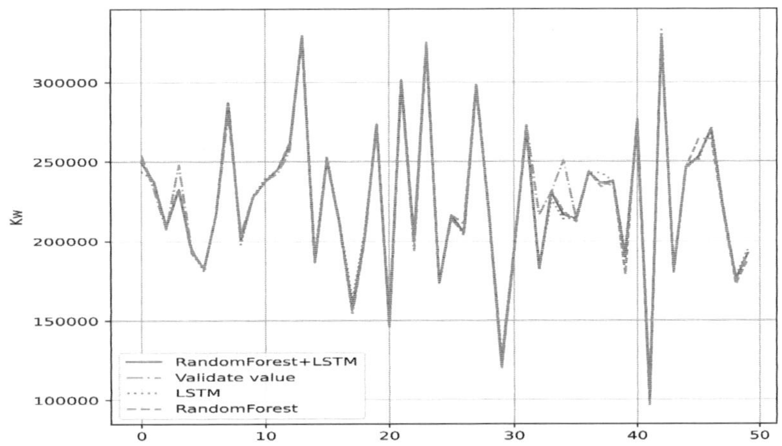

Figure 5. Comparison of prediction results of LSTM, Random Forest and combined models

It can be seen from Figure 5 that the prediction accuracy of the combined model is closer to the real value than LSTM and Random Forest.

5. Conclusion

With the improvement of market demand, the accuracy requirements of ultra-short-term forecasting of power system will increase day by day. In this paper, the reciprocal error method is used to combine LSTM and Random Forest with weight, corrected the problem that a single model has large errors in some time series data, avoids the problem that a single model has large errors in the prediction of individual data, reduces the prediction error of a single model, and effectively improve the accuracy of ultra-short-term power load forecasting.

In the follow-up research, the model can be generalized, and the power load problem can be studied in combination with specific scenarios of industry, agriculture and commerce, by mining the relevant features of its specific application scenarios and combining with more effective information, the prediction accuracy can be improved.

References

[1] Ji, P., Xiong, D., Wang, P., et al. (2012) A study on exponential smoothing model for load forecasting[C]//2012 Asia-Pacific Power and Energy Engineering Conference. IEEE, 1-4.

[2] Zhu, X., Shen, M. (2012) Based on the ARIMA model with grey theory for short term load forecasting model[C]//2012 International Conference on Systems and Informatics (ICSAI2012). IEEE, 564-567.

[3] Zhou, C., Xing, W. Y., Li, Y. L. (2012) A review of load forecasting methods for power systems [J]. Journal of Power Sources, (06): 32-39.

[4] Zhou, D. M., Guan, X. H., Sun, J., Huang, Y. (2002) Research on short-term load forecasting of power system based on neural network [J]. Power Grid Technology, (02): 10-13+18.

[5] Zhang, Y. H., Qiu, C. M., He, X., Ling, Z. N., Shi, X. (2017) A Short-Term Electricity Load Prediction Method Based on LSTM Neural Network [J]. Electric Power Information and Communication Technology, 15(09):19-25.

[6] Chen, Z. Y., Liu, J. B., Li, C., Ji, X. H., Li, D. P., Huang, Y. H., Di, F. C., Gao, X. Y., Xu, L. Z. (2020) Ultra-short-term power load forecasting based on the combined model of LSTM and XGBoost [J]. Power Grid Technology, 2020, 44(02): 614-620.

[7] Zhuang Jiayi, Yang Guohua, Zheng Haofeng, Zhang Honghao. CNN-LSTM-XGBoost short-term power load forecasting method based on multi-model fusion [J]. China Electric Power, 2021,54(05):46-55.

[8] Bengio, Y., Simard, P., Frasconi, P. (1994) Learning long-term dependencies with gradient

descent is difficult[J]. IEEE transactions on neural networks, 5(2): 157-166.
[9] Hochreiter, S., Schmidhuber, J. (1997) Long short-term memory[J]. Neural computation, 9(8): 1735-1780.
[10] Breiman, L., Friedman, J., Olshen, R., et al. (1984) Cart[J]. Classification and Regression Trees, 1984.
[11] Breiman, L. (2001) Random Forests[J]. Machine learning, 2001, 45(1): 5-32.

Preparation and electrochemical properties of vanadium-based anode materials for zinc ion battery

Jiaxin Ren*, Chun Yang , Beibei Zhou

Northeastern University, Liaoning, China

*20191261@stu.neu.edu.cn

Abstract. Although water zinc ion battery (ZIBs) has certain advantages in production and application period, but the speed performance and service time is not strong, always hinders the reusability and rapid charging technology. In order to solve this problem, researchers proposed to improve the reaction kinetics on the basis of stable electrode structure. So this article research using solvent and adding thiourea to design a hybrid structure, the structure is made from manganese doped V6O13 nanobelts and navier-stokes modified porous carbon structure, consisting of one dimensional porous nanometer structure can form diffusion channels, and can provide relatively abundant energy storage sites, and zinc ions doped into V6O13 electronic structure can be further adjustment. In order to effectively control the huge volume expansion and optimize the electrical conductivity, n-S modified carbon structure with electrical conductivity can be used to protect the nanoribbons and effectively improve the electron transport rate in the electrochemical reaction. This electrode structure can not only accelerate the diffusion rate and reaction kinetics of zinc ions, but also show cyclic stability and high rate performance. This study mainly analyzed the application effect of reduced graphene oxide coated with sea urchin-like VO2 as anode material of water zinc ion battery, so as to provide effective basis for material preparation and electrochemical performance research.

1. Introduction

With the steady development of modern electronic products and new energy powered transportation vehicles, social residents have higher and higher demands for electronic products, and put forward higher requirements for electrochemical energy system development and design. From the perspective of sustainable economic development, it is found that non-renewable resources are in serious shortage in recent years, and the damage of natural environment is more and more serious. Therefore, researchers begin to design and study green renewable energy to replace traditional non-renewable energy, such as common solar energy, geothermal energy, wind energy, etc. [1-3] Through the development and storage of green new energy, it can not only reduce the excessive loss of energy resources, but also lay a foundation for the follow-up development. Therefore, scholars of various countries are determined to study various energy storage devices, and high-performance electrochemical energy system technology has been rapidly developed. At present, zinc ion battery (ZIBs) while set a positive role in the industrialized production, have application in all kinds of electronic products, but the lithium resources shortage and increased cell design cost directly, and the resulting fire case more and more, so research scholars began to large quantities of sodium ion battery (SIBs) and potassium ion battery (PIBs). From the perspective of practical application, the output of zinc metal in various countries is very rich, the practical cost is low, and the theoretical pen capacity is very high, which can reach 820mA h G-1. At the same time, the water electrolyte not only has a high safety factor, but also has a strong ionic conductivity, which can help zinc ion diffusion at high speed. In addition, zinc ion battery has a low REDOX potential, which can reach 0.762V relative to standard hydrogen electrode. Zinc anode also has strong chemical stability in the air. Designers do not need to operate in the glove box in actual installation, and the overall process is very simple. Therefore, researchers began to study the application value and direction of water rechargeable ZIBs in the field of energy storage in a large range. Vanadium-based compounds, as common cathode materials of zinc ion batteries, have been paid more attention by researchers because of their rich oxidation states. In the past decades, most researchers have proved that vanadium compounds can be used as anode materials for zinc ion batteries. In essence, various structures of vanadium oxide are composed of multiple V-O coordination polyhedra connected by common vertices, common edges or common planes. Common polyhedra include regular octahedron, square cone and tetrahedron, etc., which will change with the valence state of V. At present, a large number of vanadium based compound materials have been reported as positive materials of water ZIBs, such as VO2, V2O5, V2O3, vanadate, etc. For example, Cai et al. proposed ni-mediated electronic structure engineering of VO2 nanoribbons, which is mainly used to improve zinc ion storage performance of water ZIBs. Since the introduction of Ni in the VO2 lattice can improve the fixed electron configuration, improve the surface reactivity, and effectively improve the reaction kinetics of zinc ions, this study can meet the actual energy storage requirements. The specific crystal structure of the design is shown in Figure 1 below. In this paper, based on the understanding of the application status of water zinc ion battery, according to the design of various vanadium based anode materials, the actual electrochemical performance of deep exploration.[4-6]

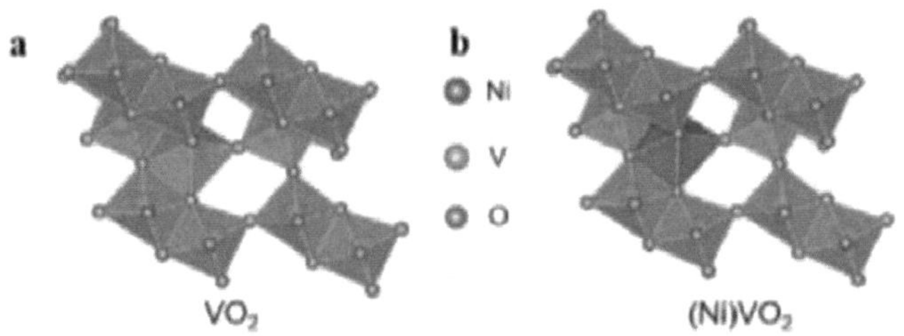

Figure 1. Crystal structure diagram

2. Method

2.1. Preparation materials

First, urchin-like VO2. Disperse 1mmol V2O5 in 30 ml of deionized water, then drop one ml of H2O2 one by one. After the solution turned red, 5 mL of glycerol was slowly added and stirred continuously for 5 min. 5 mg of sodium dodecane base sulfate was dispersed in the solution and ultrasound was performed for 5 min. The obtained mixture was placed in a 50mL high pressure reaction kettle lined with Teflon, heated at 180°C for 8 h, centrifuged several times with deionized water and ethanol, and dried for 8 h under vacuum condition of 60 °C. Vo2-sdbs was finally obtained, and the samples without SDBS were labeled as VO2.

Second, VO2-SDBS@r GO. 50 mg of VO2-SDBS were dispersed in 5 mL of GO solution. After ultrasonic treatment for 30 minutes, the suspension was balanced by stirring at room temperature for 12 hours. Finally, it was placed in an oven at 80°C to dry overnight.

Third, assemble the battery. The button battery was assembled in the air to test its electrochemical performance. The active substance, acetylene black and polyvinylidene fluoride were mixed in a ratio of seven to two to one, and the appropriate amount of N-methyl-2-pyrrolidone was used as a solvent to make the corresponding slurry. The slurry was evenly coated on a stainless steel container and dried at

60°C for 8 hours. High purity zinc band and glass fiber were applied to the anode and diaphragm respectively. The electrolyte was 3M trifluoromethane zinc sulfate solution.

2.2. Analysis of material characteristics

According to the analysis of VO2-SDBS@r GO synthesis schematic diagram as shown in Figure 2 below, methyltriol can be used as a reducing agent in this process to provide large quantities of hydrophilic group -O H for material preparation, while SDBS can be regarded as a structural guiding agent to induce the synthesis of sea-urchin-like precursors by solvothermal method. At the same time, rGO wrapped in the outer layer of the precursor can be regarded as both a protective agent of the structure and a conductive agent.[7-8]

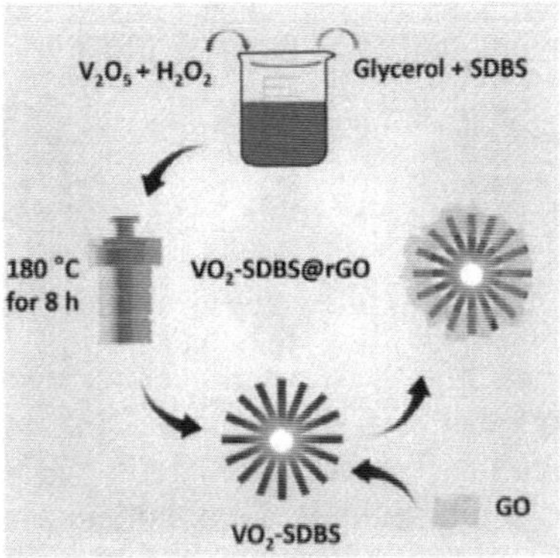

Figure 2. Synthesis diagram of VO2-SDBS@r GO

In order to scientifically prepare relevant samples, researchers proposed to carry out relevant characterization tests in practical exploration. After obtaining data information such as XRD patterns and characteristic peaks of samples, it was determined that all elements existed and could be evenly distributed through comparative observation. Then it was proved that VO2-SDBS@r GO samples had been successfully produced. At the same time, the Raman spectra of VO2-SDBS and VO2-SDBS@r GO in the wavelength range of 100 to 2000cm-1 and the XRD patterns of VO2-SDBS@r GO should be analyzed. The specific results are shown in Figure 3 and 4 below. The peak position of the former is 138.4cm-1, 191.9cm-1, 279.6cm-1, 403.8cm-1, 523.7cm-1, 690.4cm-1, 992.6cm-1, which is affected by the V-O bond in VO2. In addition to the peak related to VO2, there are two peaks in 1348.3cm-1 and 1591.2cm-1, which correspond to the D-band and G-band of RGO. The stronger wind in THE D-band proves that rGO has more defects.[9-10]

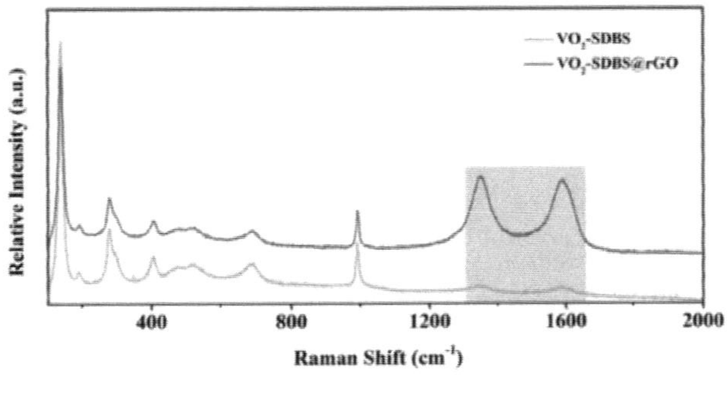

Figure 3. Raman spectra

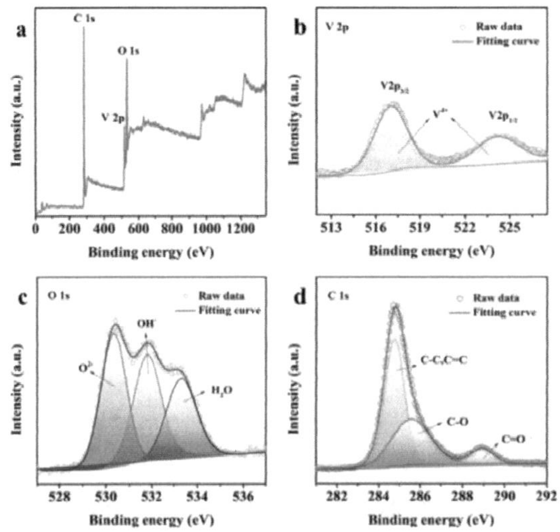

Figure 4. XRD patterns of VO2-SDBS@r GO

In order to better compare the wettability between various electrode materials and electrolytes, this paper also tested the contact Angle between three samples and 3mol L-1 Zn (CF3SO3) 2 solution. The final results showed that VO2, Vo2-SDBS and VO2-SDBS@r GO were all hydrophilic materials with higher wettability. The reason is that S D B S contains hydrophilic groups. However, rGO is a hydrophobic material, so VO2-SDBS@r GO has lower wettability.

3. Result analysis

3.1. Electrochemical performance

To evaluate the electrochemical performance of the sample VO2-SDBS@r GO, it can be regarded as the cathode material of the new ion battery and tested and analyzed after assembly into the button battery. The specific results are shown in Figure 5 below. Figure (a) represents the curve of VO2-SDBS@r GO cyclic voltammetry at a scanning speed of 0.2MVS-1, Figure (b) and (c) represent the charge and discharge curve at a current density of 1Ag-1, Figure (d) and (e) represent the CURVE of GCD and rate performance at a different current density, Figure (f) represents the cycle life performance of VO2-SDBS@r GO when the current density is 8Ag-1.

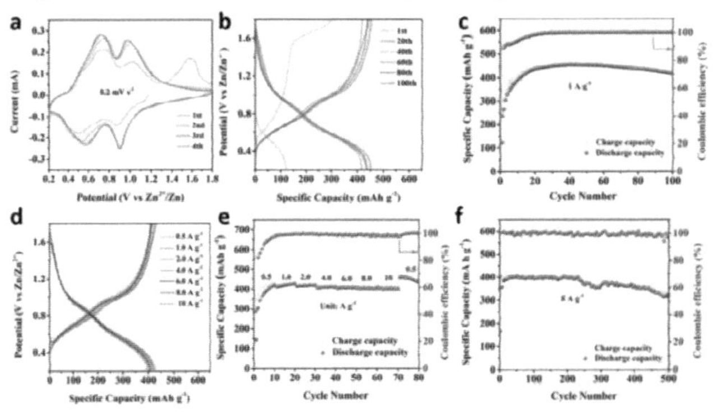

Figure 5. Electrochemical performance test

Figure A shows two obvious REDOX peaks, and the SHAPE of VC curve in the first ring is different from that in the following three rings. There are two reduction peaks in the cathode scanning curve of the first ring at 0.8V and 0.52V. In the anodic scanning curve, there are three oxidation peaks of 0.71V, 1.02V and 1.59V, which are directly related to the activity of the material. Meanwhile, in Figure E, when the current density reaches 0.5Ag-1, 1.0Ag-1, 2.0Ag-1, 4.0Ag-1, 6.0 ag-1, 8.0Ag-1 and 10Ag-1, The reversible capacities of 437.8mA h g-1, 426.8mA h g-1, 420.7mA h g-1, 413.3mA h g-1, 410.0mA h g-1, 408.9mA h g-1, 402.8mA h g-1 can be obtained. The retention of the actual capacity can reach 92%, which proves that it has a strong rate performance. In addition, the test analysis of the high rate constant cycle performance of the sample VO2-SDBS@r GO electrode shows that after 500 cycles at the current density of 8Ag-1, the capacity retention can reach 79.8%, and the highest discharge specific capacity is 406.6mA h g-1.

3.2. Charge storage mechanism analysis

According to the analysis of VO2-SDBS@r GO non-in-situ XRD patterns as shown in FIG. 6 below, it was found that new phases appeared during charge and discharge, which were in the corresponding crystal planes of the characteristic peaks of 12.29° and 30.10°. At the same time, the characteristic peak of VO2 moves to a smaller Angle during discharge and returns to the original Angle when charging. This process proves that there is a reversible reaction in the electrode material during charge and discharge. In other words, the distance between the electrode materials can be increased by adding zinc ions in the discharge process, and the distance between the electrode materials can be restored by releasing zinc ions in the charging process. At the same time, under different charge and discharge states, there are some differences in the morphology of the sample. For example, when the discharge reaches 0.2V, the surface of the material will appear nanoparticles, which is directly related to the new phase during the discharge. In the case of 1.8V charge, the nano ions are not obvious and the actual lamination is more compact. This result also proves that the material has reversibility during charge and discharge. As the number of cycles continues to rise, the original VO2 will change into a stable lamellar structure with a rough surface, and the actual thickness will be controlled between 50 and 100nm.[11-13]

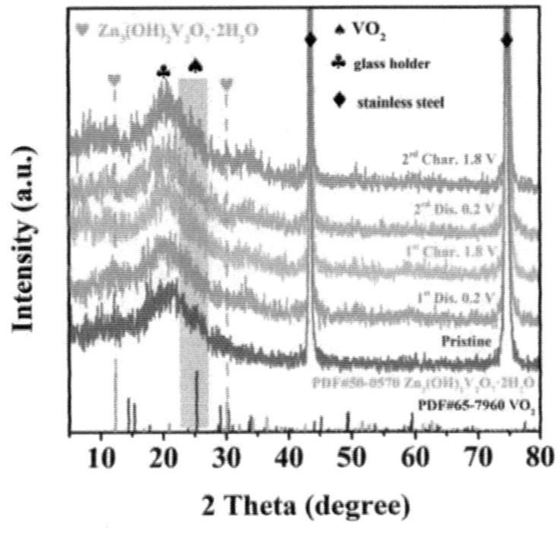

Figure 6. Unin-situ XRD patterns of VO2-SDBS@r GO

4. Conclusion

To sum up, in the steady development of modern science and technology, simple solvothermal reaction is used to produce vanadium oxide materials, and the form and content of traditional battery production is changed. N-s modified carbon and REDOX graphene composite structure are regarded as cathode materials to produce more effective water-borne zinc ion batteries. It can not only adapt to the application value of zinc ion in electrochemical reaction more quickly according to the unique layered structure, but also understand the micro morphology and structure form of positive electrode according to the characterization test analysis. Through electrochemical analysis of material preparation, it can be found that the prepared water material can still provide 417.7mA h g-1 reversible capacity after 100 cycles at the current density of 1.0Ag-1, and the actual capacity retention can reach 92%. At the same time, in practice, two button batteries can be used to successfully drive the digital clock, the actual time can reach 72 hours. All research scholars, therefore, is to attach importance to drainage preparation of zinc ion battery vanadium battery anode materials and the corresponding electrochemical performance, pay attention to the situation, combined with practical technology development and material reserves deep to explore the application of more valuable materials and production technology, only in this way can we truly achieve the sustainable development of resources construction goal.

Reference

[1] Mao, X. X. (2021) Preparation and electrochemical properties of manganese and vanadium anode materials for zinc ion battery [D]. Guangxi university, DOI: 10.27034 /, dc nki. Ggxiu. 2021.001070.

[2] Wang, X. (2021) Preparation and electrochemical properties of vanadium anode materials for zinc ion battery [D]. Jilin university. DOI: 10.27162 /, dc nki. Gjlin. 2021.001861.

[3] Zhang, J. (2021) Preparation and electrochemical properties of manganese/vanadium anode materials for zinc ion battery [D]. Harbin industrial university. DOI: 10.27061 /, dc nki. Ghgdu. 2021.002463.

[4] Wu, F. D. (2021) Preparation and electrochemical properties of manganese anode materials for zinc ion battery [D]. Harbin industrial university. DOI: 10.27061 /, dc nki. Ghgdu. 2021.003785.

[5] Guan, C. (2021) Hydrothermal synthesis of nanomaterials based on sodium vanadate [J]. Shenyang university of technology. DOI: 10.27322 /, dc nki. Gsgyu. 2021.000898.

[6] Wang, X. (2021) Structural design and performance optimization of anode materials for vanadium based water zinc ion battery [D]. Shandong university. DOI: 10.27272 /, dc nki. Gshdu. 2021.000325.

[7] Tang, W. (2021) Study on the positive electrode of zinc ion battery with mesoporous spinel type vanadium material [D]. Hubei university of technology. DOI: 10.27131 /, dc nki. Ghugc. 2021.000207.

[8] Zhou, S. H., Wu, X. W., Xiang, Y. H., Zhu, L., Liu, Z. X., Zhao, C. X. (2021) Progress in Chemistry, 33(04):649-669.

[9] Luo, H. (2021) Preparation and electrochemical properties of vanadium anode materials for zinc ion battery [J]. Harbin industrial university. DOI: 10.27061 /, dc nki. Ghgdu. 2021.000392.

[10] Zhou, S. H. (2020) Preparation, modification and electrochemical properties of Manganese anode materials for zinc ion battery [D]. Xiangtan university. DOI: 10.27426 /, dc nki. Gxtdu. 2020.002014.

[11] Wang, S. W., Gao, H. G., Zheng, H. Y., Wang, F., Luo, W., Wu, S. D., Zhang, Y. (2020) Drainage of zinc ion battery vanadium oxide anode materials research progress [J]. Journal of chemistry, 2020 turned (10) : 891-896 + 939. DOI: 10.14159 / j.carol carroll nki. 0441-3776.2020.10.004.

[12] Yin, B. S. (2020) Zinc ion battery preparation and electrochemical properties of vanadium battery anode materials research [D]. Harbin industrial university. The DOI: 10.27061 /, dc nki. Ghgdu. 2020.004972.

[13] Huang, J. T., Zhou, J., Liang, S. Q. (2021) Acta Physico-Chimica Sinica, 37(03):33-55. (in Chinese)

Journal of Physics: Conference Series **2387** (2022) 012035 doi:10.1088/1742-6596/2387/1/012035

Research on control and management mechanisms in Time-sensitive networks

Ying Chang*, Hongxue Yang, Qinghua Zhu

School of Telecommunication Engineering, Beijing Polytechnic, Beijing, 100176, China

*changying@bpi.edu.cn

Abstract. In recent years, with the development of the new generation of network information and communication technology, the internet of things, the internet of vehicles and the industrial internet technology have been booming. At the same time, many time-sensitive businesses have sprung up. The demand of these businesses for data transmission is developing towards the direction of high rate, low delay and low jitter. The current internet technology is facing great challenges. However, due to the inherent problems of forwarding mode and control management mechanism, traditional ethernet is difficult to meet the real-time, deterministic and reliability demands of time-sensitive flows. In the face of the urgent demand for the deterministic data transmission of time-sensitive businesses, time-sensitive network technology emerges at the historic moment.

1. Introduction

The time-sensitive network task force is committed to enhancing the real-time transmission capability of the network flow through the formulation of the real-time ethernet protocol standards [1]As a set of standards published by the time-sensitive network, IEEE 802 1Qbv proposes the time-aware flow mechanism and the gate control scheduling mechanism [2]. The combination between the two can separate the time-sensitive flow from the best effort flow, ensuring the certainty of delay and shaking of the time-sensitive data transmission [3]. The time-aware flow mechanism can distribute all time-sensitive flow to the same buffer queue of the highest priority. The gate control scheduling mechanism opens and closes the gate of corresponding queue on schedule to control the predictable data transmission. But IEEE 802.1Qbv does not provide the algorithm for gate control list configurations [4]. Nor does it designate a specific algorithm for calculation of when time-sensitive flow should be transmitted by the network interface controller (NIC) of the source host. Therefore, design of the time-sensitive network scheduling algorithm remains a general research interest.

Besides, the time-sensitive network, while ensuring the steady transmission of time-sensitive flows, usually causes a great waste of broadband. (See Figure 1) On the one hand, in order to prevent the exclusive transmission window of the time-sensitive flow from being influenced by the best effort flow which has not yet been transmitted previously, the time-aware shaping mechanism offers a protective belt to ensure the link to be available upon the start of transmission [5]. But excluding the data frames that are already transmitted, no other data frames will be transmitted again, which can result in full waste of the remaining frequency band. On the other hand, when the time for the time-sensitive flow to arrive at the port of the interchanger is sparse, and when the transmission time slot is inadequate to schedule other flow, link bandwidth will also be wasted. Both conditions are resulted from the discontinuous transmission of the time-sensitive flow. If transmission can be finished more compactly through the time adjustment of the time-sensitive flow scheduling, opening of the interchanger port will be less often, which can largely reduce the bandwidth waste [6,7].

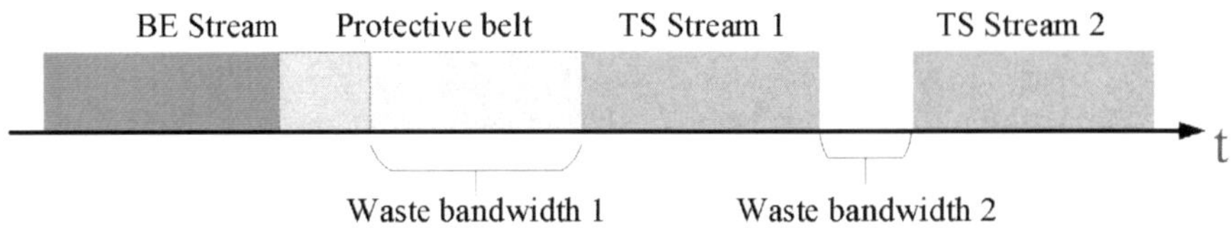

Figure 1. Bandwidth waste caused by time-sensitive flow transmission.

Scheduling of the time-sensitive flow is similar to the question of binary knapsack in operational research. Generally, integer linear programming or satisfiability modulo theories are used to solve questions of the kind. Sha et al. [8] analyzed key parameters that can affect the real-time certain behaviors under the impact of IEEE 802.1Qbv. Meanwhile, based on its common parameter configurations, they also deduced constraints to be satisfied during the flow scheduling process to ensure the certainty of the end-to-end delay and delay variation. Falk et al. [9] came up with the integer linear planning formula to solve the question of joint route and scheduling under the no-queue arbitrary isochronous communication scenario. Additionally, they assessed the scalability of the formula using a large number of cases. Dürr et al. [10] mapped the problem of time-sensitive network scheduling to the nowait jobshop scheduling problem (NWJSP), and employed the tabu search algorithm to work out the optimal scheduling plan for the time-sensitive flow. From the above literature review, one can observe that all these scholars use the typical integer linear planning solver or the modern heuristic algorithm, or just consider the scheduling scenes of the small-scale network and time-sensitive flow, or verify the feasibility of the algorithm only in the problem of large-scale time-sensitive flow scheduling. But the algorithm itself is limited in improving the problem-solving performance and scalability of large-scale flow scheduling plans. Therefore, this paper applies an improved simulated annealing genetic algorithm to the problem of periodic time-sensitive flow scheduling for configurations of the time-sensitive data transmission time and the gate control time list of interchanger port. After that, the scheduling compaction algorithm is used for compaction adjustment of the time-sensitive flow transmission time.

2. Problem modeling

The objective of the time-sensitive network scheduling algorithm is to identify the time for various source hosts to send the time-sensitive flow to the network as well as the time for different queue gates on the interchanger outlet port that constitute the gate control list. In this paper, the integer linear planning is employed for modeling of the scheduling problem. The set of all interchanger outlet ports in the network is written as $P=\{P_1,P_2,\ldots,P_m\}$. The time-sensitive flow set that is to be scheduled is written as $F=\{F_1,F_2,\ldots,F_n\}$. The time for the time-sensitive flow to be set to the network via the source host is $T=\{T_1,T_2,\ldots,T_n\}$. Each data flow is corresponding to a series of transmission operations, which include transmission operations of different data flows on the outlet port of each interchanger along their route path. The j operation of the i data flow is denoted by $O_{i,j}$. So the transmission operation set of the i data flow passing n_i interchangers can be represented by $O_i=\{O_{i,1},O_{i,2},\ldots,O_{i,n}\}$. Among them, every operation consists of a binary group (P,t) that records the outlet port serial number and the starting time of transmission.

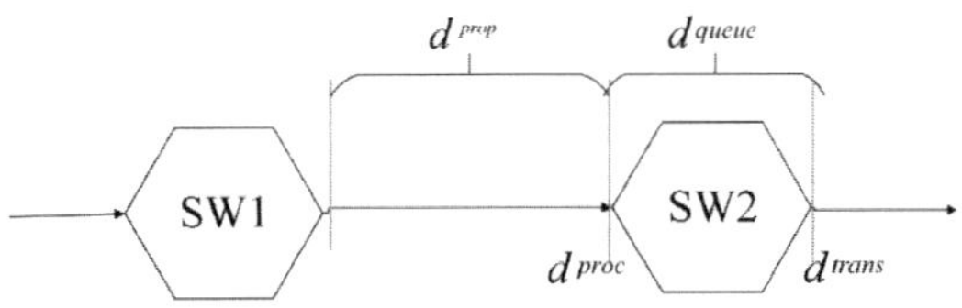

Figure 2. Composition of network delay

As shown in Fig. 2, network delay can generally be divided into four types, including propagation delay, d^{prop}, of data packet along the chain, processing delay, d^{prop}, necessary for the interchanger to receive the data packet and identify the forwarding port, queueing delay, d^{queue}, of the data packet in the outlet port buffering queue, and transmission delay, d^{trans}, necessary for the data packet to transmit to the link. In order to minimize the port-to-port delay, d^{trans}, of the time-sensitive flow upon the generationof the initial scheduling plan, we require the data packet to be immediately forwarded without queueing on the interchanger outlet port. Under this condition, the queueing delay, d^{queue}, is set to be 0. To simplify the research issue, we assume all links and nodes of the whole network to have the same propagation delay, d^{prop}, and processing delay, d^{proc}. The transmission delay, d^{trans}, is linked to the link transmission rate. The transmission delay set and the transmission operation set are corresponding to each other. So the j transmission delay of the i can be written as $d_{i,j}^{trans}$. Since the propagation, processing and transmission of every data packet in the network is organized strictly in accordance with the set sequence, we can deduce the cumulative network delay, $C_{i,j}$, consumed by the j transmission of F_i, which can be written as below:

$$C_{i,j} = j * \left(d^{prop} + d^{proc}\right) + \sum_{k=1,2,\ldots,j} d_{i,k}^{trans} \tag{1}$$

The starting time, $t_{i,j}$, of the operation, $O_{i,j}$, can be written as below:

$$t_{i,j} = T_i + C_{i,j} - d_{i,k}^{trans} \tag{2}$$

Furthermore, the time for F_i to finish scheduling can be written as S_i below:

$$\begin{aligned} S_i &= T_i + (n_i + 1)\left(d^{prop} + d^{proc}\right) + \sum_{k=1,2,\ldots,n_i} d_{i,k}^{trans} \\ &= T_i + D_{i,n_i} + \left(d^{prop} + d^{proc}\right) \end{aligned} \tag{3}$$

In terms of a scheduling plan, x, that is already known, the total time used by it to finish the time-sensitive flow scheduling or S_x can be defined as the time for the last flow reaching the target host to finish scheduling:

$$S_x = \max\left\{S_i \middle| i \in \{1, 2, \ldots, n\}\right\} \tag{4}$$

After definition and deduction of relevant variables, we can deduce the optimization goal of time-sensitive flow scheduling, that is, to minimize total time, S_x, required for completion of scheduling. In the following that, the constraint of this question is considered. To finish the transmission scheduling of the time-sensitive flow as soon as possible, and to increase its transmission continuity and compactness, we allocate the sending time as early as possible for every flow upon the configuration of time slot. At the same time, the time for a series of transmission operation of the flow along its transmission path is also identified. Since the outlet port of the interchanger can only transmit one flow only each time, the transmission time slot allocated for the flow requires the satisfaction of the following core constraint, that is, outlet ports of different flows cannot be overlapped in time during the transmission process. In other words, the starting time for the next scheduling flow to transmit on the outlet port should not be later than the transmission completion time of the former scheduling flow on this port. Therefore, the constraint that the flow sending time should meet can be deduced. With two transmission operations, $O_{i,k}$ and $O_{j,l}$, on the same port as an example, the constraint can be specified as below:

$$T_j - T_i \geq S_{i,k} - \left(S_{j,l-1} + d_{prop} + d_{proc}\right) \tag{5}$$

To prevent all transmission operations on the same port from influencing each other, the transmission conflict of all ports should be detected. To the end, the transmission conflict set, G, is introduced:

$$G = \left\{\left(O_{i,k}, O_{j,l}\right) \middle| P_{O_{i,k}} = P_{O_{j,l}} \wedge t_{O_{i,k}} < t_{O_{j,l}}\right\} \tag{6}$$

Therefore, the integer linear planning of time-sensitive flow scheduling can be written as below:

$$\begin{aligned} &\min\ S_x \\ &\text{s.t.}\ \ T_j - T_i \geq S_{i,k} - \left(S_{j,l-1} + d_{prop} + d_{proc}\right),\ \ \forall \left\{O_{i,k}, O_{j,l}\right\} \in G \end{aligned} \tag{7}$$

3. Algorithm design

The final scheduling plan is generated by two steps, including the generation of the initial plan and the adjustment of the scheduling plan. The realization process involves three parts, including scheduling sorting algorithm, time-slot allocation algorithm, and scheduling compression algorithm. The algorithm can be performed as in Figure 3. First, the flow reserved information, link capacity, routing list and so on are input. Second, the scheduling sorting algorithm is used to work out the flow scheduling sequence to ensure the overall scheduling to be completed in the shortest time. The time-slot allocation algorithm is used to compute the time required for the completion of scheduling under each sorting plan. In other words, sending time that is as early as possible is allocated for the time-sensitive flow in the given scheduling sequence under the constraint. Additionally, time for the start and completion of different transmission operations is solved. After that, the scheduling compression algorithm is employed for adjustment. Specifically, the flow transmission operations where transmission gaps might exist and can be postponed are properly put off to generate the final scheduling plan.

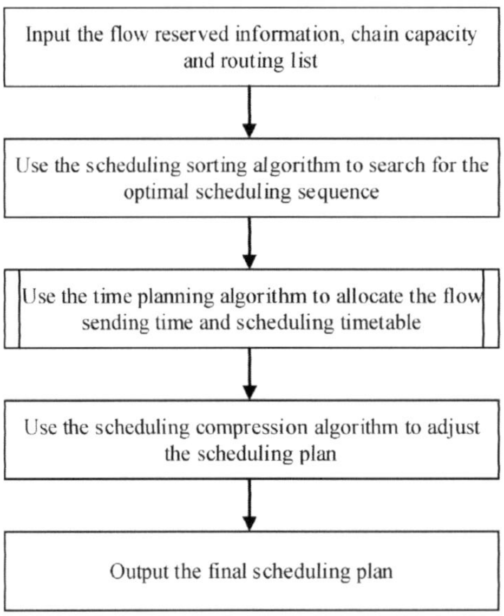

Figure 3. Scheduling algorithm flow chart.

3.1. Scheduling sorting algorithm

(A) Algorithm design thinking

The scheduling sorting algorithm constitutes a main part of the scheduling algorithm. In the current research work, the modern heuristic algorithm demonstrates favorable performance in solving NP-hard problems. But restricted by the searching mechanism, the typical heuristic algorithm still has great room for improvement of its problem-solving performance and scalability [11]. Therefore, this paper designs the scheduling sorting algorithm based on the improved simulated annealing genetic algorithm.

(B) Algorithm design process

The value of the fitness function is a standard to judge every individual in a population, namely the quality of the scheduling plan under each scheduling sequence. The higher the value of the fitness function is, the higher the quality of the solution is, and also the stronger the individual can get adapted to the natural selection process. The optimization objective of the algorithm is to minimize the total scheduling time, S_x, of the plan x. S_x can be given by the time-slot allocation algorithm, so the fitness function can be written in the reciprocal form:

$$f(x)=\frac{1}{S_x-\delta} \tag{8}$$

Where, δ is a constant, which is used to control the difference of fitness of individuals. When this constant is large, individuals are significantly differentiated from each other, which is beneficial for the selection of excellent individuals. When the constant value is small, the individual difference of the population can be reduced to avoid falling into the local optimum. A relatively direct sorting and coding is adopted to cope with the sequence of the flow scheduling. In other words, every flow is encoded, and the scheduling sequence is presented as a series of gene coding. For example, the gene sequence "3526471" means that the transmission time-slot is first allocated to No. 3 time-sensitive flow and then No. 5 flow is scheduled. The purpose of operation selection is to choose individuals with superior performance by certain method from the parent population and then pass them down to the next generation. According to the natural selection process, individuals with a high fitness have a higher chance to be chosen and to survive, while individuals with a low fitness might be eliminated. In this paper, the roulette wheel selection operator is employed, whose basic idea is to ensure that the probability of individuals to survive is directly proportional to its fitness function value. Specifically, the fitness function of every individual in the population is computed, which can be written as $f(S_i)(i=1,2,\ldots,m)$, of which m is the number of individuals in a population. The probability for every individual to be inherited to the next generation is computed, namely $p(x_i)$:

$$p(x_i)=\frac{f(x_i)}{\sum_{k=1,2,\ldots,m} f(x_k)} \tag{9}$$

Thereby, the cumulative probability, $P(x_i)$, can be obtained:

$$P(x_i)=\sum_{j=1,2,\ldots,i} p(x_j) \tag{10}$$

Within [0, 1], a random number r_i is generatedly by the uniform distribution. If $r<q[1]$, then the individual 1 is chosen; otherwise, the individual which can substantiate $q[h-1]<r\leq q[h]$ will be chosen.

Chromosomal chiasma: Interlace operation exchanges in some way part of genes from the two pairs of gene sequences which are chosen from the population by the crossover probability. Thereby, two brand-new individuals are generated. As a main operator of the genetic algorithm, crossover operator can endow the algorithm with the global searching ability. In this paper, the two-point cross-over method is chosen. (See Figure 4.) Two locus are randomly chosen from the gene sequence by the uniform distribution probability. Part of genes from the filial generation 1 are directly inherited from the gene segments between two gene locus from the chromosome of parent 2. The remaining part inherits the unrepeated genes from the chromosome of parent 1 by the original sequence. But an opposite operation is adopted for filial generation 2.

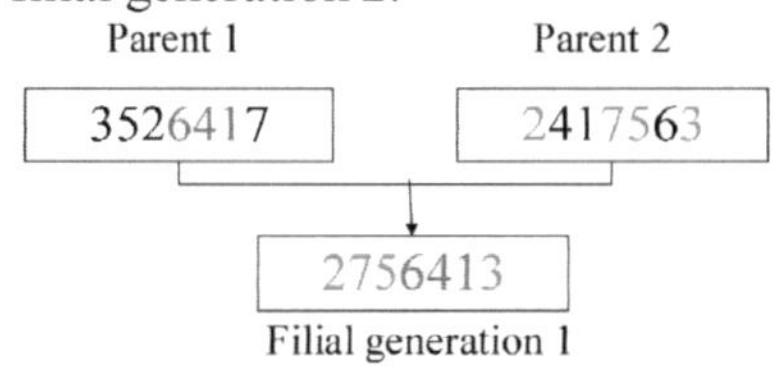

Figure 4. Two-point crossover.

Gene mutation: According to the mutation probability, the individuals which are on the gene mutation list are chosen. The value of one or multiple genes from the gene sequence is changed into other gene value so as to obtain new individuals. The mutation operator is a method that can assist the crossover operator to generate new individuals. It equips the genetic algorithm with certain global searching capability and also maintain the variety of gene sequences for the population. Considering the coding characteristics of sorting, this paper adopts inverse mutation, which is shown in Fig. 5. It randomly chooses two gene locus randomly from the chromosome of individuals. Then, the gene value of the two gene locus is exchanged to generate new individuals.

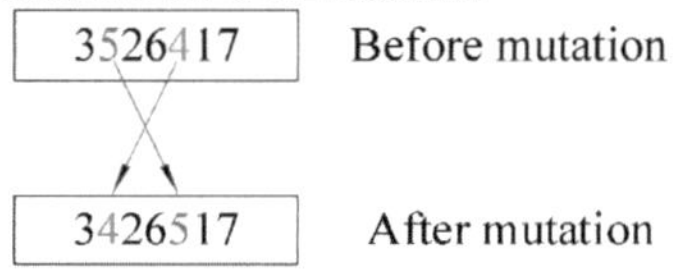

Figure 5. Inverse mutation

Simulated annealing: According to simulated annealing, it accepts a new solution which is relatively inferior to the current solution by certain probability. At the core of simulated annealing is that Metropolis accepts the use of the criterion. In this paper, the criterion is applied to choose new individuals generated by interlace operation and mutation operation. Whenever a new individual is generated by interlace and mutation operation, and if the fitness function value of the new individual is larger than that of the original one, then the new individual will be accepted. On the contrary, if the fitness function value of the new individual is smaller than that of the original individual, the new individual will be accepted by the probability, q. q can be given by the following equation:

$$q = \exp\left[-\frac{\left(f_{old} - f_{new}\right) * \theta}{K}\right] \tag{11}$$

Where, f_{old} and f_{new} represent the fitness function value between the original individual and the new individual. θ is a constant, which is used for scaling of the fitness function; K represents the temperature parameter of the current iteration. At the end of each iteration, $K' = K * \alpha$ is used for cooldown operation; α is the cooldown coefficient. The significance of this probability equation lies in that a solution with a poorer performance is highly likely to be accepted when the initial temperature of iteration is relatively high. Consequently, the solution obtained is deviated from the local optimum. In the late period of iteration, when the temperature drops, the space to accept the poorer solution will keep on narrowing down, making it possible to search for the optimal solution within a smaller local area and ensure stable convergence of the algorithm.

(C) Algorithm operation flow

The algorithm operation flow is shown below:

First, the flow size, flow routing list, link capacity, and other information are input to calculate initialize the flow transmission and operation port list as well as the transmission and operation delay list. Meanwhile, the number of individuals in the population, maximum iteration times, cooldown coefficient, scaling coefficient, initial temperature and other parameters are initialized. Through random arrangement, the initial population is generated. The time-slot configuration algorithm is conducted on the individual sequence. The flow sending time is allocated in accordance with the time-sensitive flow scheduling sequence. Additionally, the overall scheduling time and the fitness function value of individuals are computed. Next, natural selection, chromosome crossover, and gene mutation are conducted on the population. Metropolis is used for optional acceptance of new individuals generated by crossover and mutation. Following that, the simulated annealing is performed for cooldown. At last, the optimal individual is found from the whole population. The iteration continues until it reaches the convergence standard and the optimal scheduling plan is output. The pseudo-code involved in the realization process of the algorithm is presented in Algorithm 1.

Algorithm 1 Scheduling Sorting Algorithm Based on Simulated Annealing Genetic Algorithm

Input: F, C, O^P
Output: x_{best}, T, G^t
function (Scheduling sorting algorithm: F, C, O^P)
  Initialize parameters, including g, m, R_c, R_m, K, α
  Calculate the transmission operation delay list, O^d
  Generate the initial population, X
  while (fails to satisfy the convergence condition, do)
    for (every individual in the population, x_i do)
      Allocate the transmission time and transmission operation timetable for the individual, xi, by the time-slot allocation algorithm (xi, O^P, O^d)
      Calculate the fitness function value, $f(x_i)$, of the individual, x_i
    end for
    Choose the optimal individual, x_{best}, from the population and keep it
    Conduct a natural selection on the population
    Generate new individuals to be chosen through crossover operation of the population by the crossover probability, R_C
    Generate new individuals to be chosen through mutation operation of the population by the mutation probability, R_m
    Accept new individuals selectively by the Metropolis standards
    Cool down $K \leftarrow K \times \alpha$
  end while
end function

3.2. Time-slot allocation algorithm

In terms of the specific time-sensitive flow scheduling sequence, the time-slot allocation algorithm based on the greedy algorithm is used to set the sending time as early as possible for every flow. The algorithm is performed as below. According to the input flow scheduling sequence, the algorithm can cope with each flow in turn. The greedy algorithm first assumes that the flow is sent into the network when the scheduling period starts. Since the flow transmission process does not line up, the time for the flow to start transmission from the outlet port of each interchanger can be identified. It is detected whether each transmission operation can be overlapped with the transmission duration of the former flow. If there is any overlapping, the difference between the completion time of the former flow transmission and the starting time of the current flow transmission is calculated, which is also the time for the current operation to put off. After the conflict test is conducted on all transmission operations, the maximum time for all operations to put off is adopted to put off the sending of the current flow. This can help allocate the transmission time-slot for every flow by the greedy algorithm and ensure it to meet the no-queue condition. A striking characteristic of this algorithm is that, among the set of complete ordering flows given by the scheduling sorting algorithm, the distribution of the flow initial sending time is subject to the influence of flows scheduled previously. This attribute can be further applied to incremental scheduling of the time-sensitive flow. When new flows to be scheduled are added, the time-slot allocation algorithm can be directly used to allocate the transmission time-slot therefor without influencing the original flow scheduling process.

Algorithm 2 Time-slot Allocation Algorithm Based on Greedy Algorithm

Input: x, O^P, O^d
Output: T, O^t, S_x, G^t
function (Scheduling sorting algorithm, O^P and O^d)
    for (every flow, F_i do, in the flow scheduling sequence, x)
        Assume the flow transmission time to be 0
        for (every transmission operation, $O_{i,j}$ do, of the flow F_i)
        Calculate the starting time of the transmission operation
         if (the transmission starting time and the transmission of the former flow on the port are in conflict), then
        Calculate the delay required to avoid the conflict caused by the operation
        end if
    end for
    the flow transmission delay is the maximum delay of all operations
      Update the flow transmission operation timetable, O^t
      Update the gate control time-table, G^t
   end for
end function

3.3. Scheduling compression algorithm

The purpose of the scheduling compression algorithm is to put off the sending time and transmission starting time of some flows. This can reduce the interval for the interchanger outlet port to transmit time-sensitive flows, so the time-sensitive flow queue gate can continuously transmit more flows upon each opening. In order to realize scheduling compression, it is necessary to relax the conditions on flow transmission no-queueing. But the adjustment of the plan should not change the flow scheduling sequence and total time required for completion of scheduling in the original plan. Two time-sensitive flows are transmitted subsequently on the outlet port of the interchanger. Since flow 2 arrives late, transmission of flow 2 starts not immediately after the end of flow 1 transmission but a little while later. The period of time on the port without any work is known as the slack time. At the moment, we will postpone the transmission operation of flow 1 that arrives at the port first, ensuring that the port can continue transmitting flow 2 after the completion of flow 1 transmission. This can reduce the waste of the bandwidth and there will be more flow scheduled before flow 1 transmission.

Algorithm 3 Scheduling Compression Algorithm

Input: T, O^t, G^t
Otput: T, G^t
function (Scheduling compression algorithm, T, O^t, G^t)
    while (there is operation that can postpone do)
        for (every operation, G_t^i, on every port from the gating timetable, G_t)
            if the operation satisfies the delay condition, then
                Put down the slack time of the operation
            else

```
            this operation is recorded as the operation that cannot be postponed
          end if
      end for
      for each flow Fi do
          Calculate the real slack time of all operations
            Postpone the first operation that can be delayed of this flow. If the first flow is postponed, the transmission time, Ti, should be delayed
          Update the flow transmission operation timetable, Q^t
          Update the gating timetable G^t
      end for
    end while
end function
```

4. Algorithm performance test and analysis

In this section, performance test and experimental analysis are conducted on the scheduling algorithm of this paper. First, the random scale-free network is generated to test and assess how the scheduling sorting algorithm based on the simulated annealing genetic algorithm under different complexities of network topography and different scales of time-sensitive flows outperforms the traditional simulated annealing algorithm and genetic algorithm. Additionally, the improvement achieved by the simulated annealing genetic algorithm in convergence performance and evolutionary performance is verified. Solid evidence is found to prove the favorable scalability of the simulated annealing genetic algorithm in addressing the question of time-sensitive flow scheduling and improving the schedulability of time-sensitive flows. Second, the role of the scheduling compression algorithm in reducing the number of cases in which the time-sensitive flow queue gate opens and the bandwidth waste resulted from the transmission interval.

Test and analysis of simulated annealing genetic performance

In this part, the set of random network topography and time-sensitive flows is generated. The time-sensitive network scheduling algorithm based on the simulated annealing genetic algorithm is analyzed and assessed. In the real network, the degree distribution of network topography in the Internet usually obeys the power-law distribution. A large number of hub nodes with lots of links exist. The newly-added nodes tend to be connected with nodes whose number of links is higher. Considering these characteristics, the Barabási Albert (BA) model is used to simulate the time-sensitive network topography. The NetworkX complex network software package provided by Python is used to generate the random network topography. It is assumed that the number of network nodes is n and the number of sides added each time is m. According to the BA model, each time a new node link with m sides is connected to the existing network. The probability for the new node to be connected with the existing node, i, is $P_i = k_i \big/ \sum_j k_j$, where k_i denotes the degree of the node i. Here, we set m to be 1. During each iterative growth of the network, a new node is connected to an existing node according to the degree of the current node so that there is only one path existing between each two arbitrary nodes. As shown in Figure 6, NetworkX is used to generate a random network topography with 100 nodes by the aforesaid rule. In the random network topography thus generated, the network marginal node whose degree is 1 is chosen as the communication host. The port-to-port route is given by the classical Dijkstra algorithm.

Figure 6. A random scale-free network with 100 nodes

In choosing network parameters, this paper assumes that all routes in the network are the same in length and their capacity is all 1Gbps. The data packet size is 375B. The propagation delay of a single data packet on the network link is d^{prop} =375B/1Gbps = 3μs; the processing delay on the interchanger is $d^{proc} = 5us$; the transmission delay on the outlet export is $d^{trans} = 3us$. Concerning the reserved size of each flow, a number is randomly chosen from 1 to 3 at an even probability as the number of data packets included in this flow. So, the corresponding transmission delay is 3μs, 6μs and 9μs, respectively. To start with, the problem-solving performance of the algorithm while scheduling different scales of time-sensitive flows is tested under the network topography with 300 nodes. The optimal solution that can be obtained by the scheduling algorithm which is based on the simulated annealing genetic algorithm, traditional genetic algorithm, and simulated annealing algorithm are compared. Fig. 8 displays how the total scheduling time of the plan provided by the three kinds of scheduling algorithms changes with the growing number of flows to be scheduled.

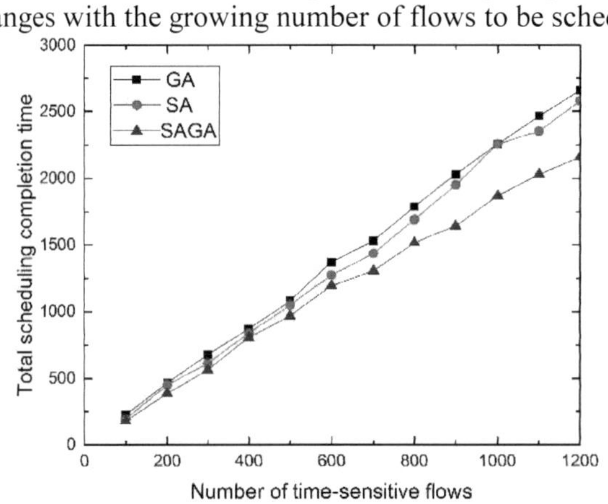

Figure 7. Curve graph of solution obtained by three algorithms that is changing with the scale of flows to be scheduled

From Figure 7, it can be seen that the simulated annealing genetic algorithm outperforms the traditional genetic algorithm and the simulated annealing algorithm in obtaining the solution. The scheduling plan that is obtained by the algorithm can finish the scheduling of the time-sensitive flow in a shorter time. Additionally, as the scale of the time-sensitive flow to be scheduled increases, the excellent problem-solving performance of the simulated annealing genetic algorithm appears more significant. When the number of flows to be scheduled reaches more than 1,200, the total scheduling completion time of the scheduling plan obtained by the simulated annealing genetic algorithm can reduce the delay by 18.9% and 17.2%, compared with the optimal solution obtained by the genetic algorithm and the simulated annealing algorithm, respectively.

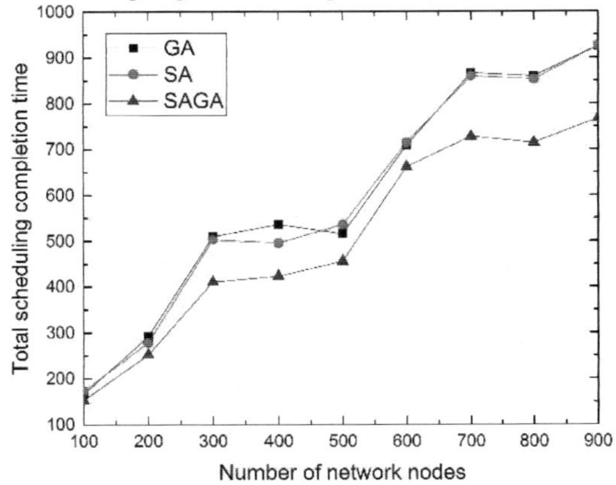

Figure 8. Curve graph of solution obtained by three algorithms that is changing with the scale of network topography

In order to further test the performance of the algorithm in solving the time-sensitive flow scheduling in the large-scale network topography, this paper uses different number of nodes to construct the random network topography of different scales. Since this algorithm does not consider the issue of routing, a major factor influencing the algorithm's problem-solving performance is the number of time-sensitive flows to be scheduled. So as the network topography becomes increasing complex, the number of randomly-generated flows to be scheduled increases as well. The test results are shown in Figure 8. Under different complexity scenarios of network topography, the simulated annealing genetic algorithm is significantly better than the traditional genetic algorithm and the simulated annealing algorithm in problem-solving performance. The total scheduling completion time of the scheduled plan proposed in this paper can reduce that obtained by the genetic algorithm and simulated annealing algorithm by 14.4% and 13.6%, respectively.

Test and analysis of scheduling compression algorithm performance

How the scheduling compression algorithm can reduce the cases of gate opening and the time wasted between transmissions is tested under the network topography with 300 nodes. During the test process, the scheduling sorting algorithm and the time-slot allocation algorithm are first used to generate the sensitive-flow scheduling plan at the initial stage. Later, this plan is put in the scheduling compression algorithm to postpone the slack flow transmission operation. At last, the bandwidth waste caused by the discontinuous transmissions of time-sensitive flows and the number of cases of gate opening in the test process of the initial plan and the scheduling process of the final plan are tested, respectively.

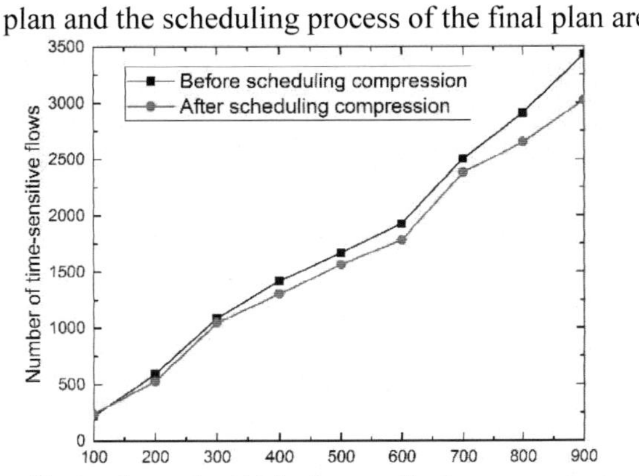

Figure 9. Number of cases in which the gate opens before and after scheduling compression.

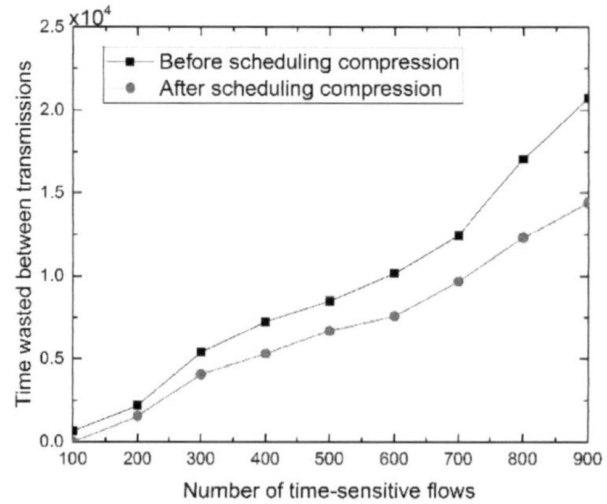

Figure 10. Comparison of time wasted between transmissions before and after scheduling compression

Figure 9 shows how the number of cases in which the time-sensitive flow queue gate opens on the interchanger port of the whole network changes. Comparison sheds light on that the plan after scheduling compression can effectively reduce the number of cases of gate opening. Thereby, each time the gate of the time-sensitive flow queue is opened by the interchanger outlet port, multiple time-sensitive flows can still be transmitted continuously. Thereby, the best effort flow can get more transmission space. Figure 10 compares the time wasted before and after the scheduling compression. The computing of the time wasted between transmissions is based on the interval of transmitting time-sensitive flows on the port, which is the sum of the idle time of the interchanger in scheduling the time-sensitive flow. After adjustment of the initial plan by the scheduling compression plan, the time wasted between transmissions is significantly reduced by 19.6% to 43.4%. This can not only enhance the continuity of the time-sensitive flow scheduling, but also reduce a large number of bandwidth resources available.

5. Conclusions

To sum up, this paper uses the integer linear planning approach for modeling of time-sensitive flow scheduling. The optimization goals and constraints of scheduling algorithm design are identified. On that basis, this paper applies the improved simulated annealing genetic algorithm to solving the time-sensitive flow scheduling plan, and uses the greedy algorithm to address the allocation of time-sensitive transmission time-slot allocation. Through integration of the scheduling sorting algorithm and the time-slot allocation algorithm, the initial scheduling plan is generated. Specifically, the use of the simulated annealing genetic algorithm can not only improve the algorithm's search performance of the optimal solution, but also, to some extent, enhances the scalability of the algorithm. The idea of the greedy algorithm can allocate the earliest sending time for each time-sensitive flow so that the whole scheduling process can be completed earlier and all together, which can increase the schedulability of data flows. At the same time, this paper attempts to cope with the discontinuity of time-sensitive transmission flows in the initial scheduling plan. The scheduling compression algorithm is designed for adjustment and optimization of the scheduling plan. Part of the flow transmissions which can be postponed on the outlet port of the interchanger is appropriately postponed to improve the compactness of the time-sensitive flow scheduling and reduce the bandwidth waste.

Different scales of time-sensitive flow scheduling are considered under the randomly-generated scale-free network topography of different scales. The simulated annealing genetic algorithm under different scenarios displays favorable problem-solving solution. Meanwhile, when the network scale and scale of flows to be scheduled is large, the simulated annealing genetic algorithm outperforms the classical heuristic algorithm in having a more stable convergence efficiency and better solution.

Thanks to the adjustment and optimization by the scheduling compression algorithm, the total number of cases of gate opening happening to the time-sensitive flow queue and the time wasted between transmissions can be effectively reduced, which can verify the validity of this algorithm in improving the bandwidth utilization rate.

Acknowledgments
The authors would like to thank the associate professor Zhaoming Lu who is working at Beijing University of Posts and Telecommunications (P.R China) in particularly for the support in thesis examination and guidance. The authors are also grateful to the teachers of OAI WORKSHOP. (http://www.opensource5g.org/) for their invaluable support in deploying the system and in providing experimental data and technical documentation. This work was supported by the general project of science and technology program of Beijing Education Committee (NO.KM202010858005), and 2022X040-KXZ.

References
[1] Dias, A. L., Sestito, G. S., Turcato, A. C and Brandão, D. 2018 Panorama, challenges and opportunities in PROFINET protocol research *2018 13th IEEE International Conference on Industry Applications (INDUSCON)*, pp. 186-193.
[2] Zhang, C., et al. 2020 Packet-size aware scheduling algorithms in guard band for time sensitive networking. *CCF Transactions on Networking, 3.*
[3] IEEE Standard for Local and metropolitan area networks--Bridges and Bridged Networks--Amendment 29: Cyclic Queuing and Forwarding *IEEE 802.1Qch-2017 (Amendment to IEEE Std 802.1Q-2014 as amended by IEEE Std 802.1Qca-2015, IEEE Std 802.1Qcd(TM)-2015, IEEE Std 802.1Q-2014/Cor 1-2015, IEEE Std 802.1Qbv-2015, IEEE Std 802.1Qbu-2016, IEEE Std 802.1Qbz-2016, and IEEE Std 802.1Qci-2017)*, vol., no., pp.1-30, 28 June 2017.
[4] Zhao, L., He, F., Li, E and Lu, J. 2010 Comparison of Time Sensitive Networking (TSN) and TTEthernet *2018 IEEE/AIAA 37th Digital Avionics Systems Conference (DASC)*, pp. 1-7.
[5] Rostan, M., Stubbs, J. E and Dzilno. D. 2010 EtherCAT enabled advanced control architecture *2010 IEEE/SEMI Advanced Semiconductor Manufacturing Conference (ASMC)*, pp. 39-44.
[6] ISO/IEC/IEEE International Standard - Telecommunications and information exchange between information technology systems - Requirements for local and metropolitan area networks - Part 1CM: Time-sensitive networking for fronthaul *ISO/IEC/IEEE 8802-1CM:2019(E)*, vol., no., pp.1-66, 23 Aug. 2019.
[7] Arestova, A., Baron, W., Hielscher, K. S. J and German, R. 2022 ITANS: Incremental Task and Network Scheduling for Time-Sensitive Networks *IEEE Open Journal of Intelligent Transportation Systems*, vol. 3, pp. 369-387.
[8] Sha, L., Sathaye, S. S and Strosnider, J. K. 2002 Scheduling real-time communication on dual-link networks *Real-time Systems Symposium IEEE.*
[9] Falk, J, Dürr, F and Rothermel, K. 2018 Exploring Practical Limitations of Joint Routing and Scheduling for TSN with ILP *2018 IEEE 24th International Conference on Embedded and Real-Time Computing Systems and Applications (RTCSA)*, pp. 136-146.
[10] Dürr, F and Nayak, N. G. 2016 No-wait Packet Scheduling for IEEE Time-sensitive Networks (TSN) *The 24th International Conference ACM,* pp. 203–212.
[11] Chen, Y., Liu, Y., Zhao, J and Zhu. Q. 2020 Mobile Edge Cache Strategy Based on Neural Collaborative Filtering *IEEE Access*, vol. 8, pp. 18475-18482.

A Two-level clustering chain energy heterogeneous routing protocol for WSN

Mei Hu[1, 2], Yongxi Wang[1,*]

[1]Department of Electronic and Information Engineering, Lanzhou Institute of Technology, Lanzhou
[2] hum@lzit.edu.cn.

*wangyx@lzit.edu.cn

Abstract. In view of the disadvantages of high energy consumption, short life cycle, low **throughput** and no application correlation in heterogeneous WSN, a two-level clustering chain routing protocol is proposed under the heterogeneous WSN of intelligent water-saving irrigation system of ecological agriculture in Hexi Gobi. The irrigation subnets are divides into clusters according to the actual layout in the wireless network and the nodes in the cluster form the chain on the basis of the minimum distance. The two-level cluster headers in reserve are selected by rate of the remaining energy of terminal nodes in cluster vs. average energy in local region. Eventually the level clusters heads and the two-level cluster headers are chosen according to the largest ratio of the remaining energy to the distance between the terminal nodes and the base station. Results in simulation show that the proposed algorithm is superior to PEGASIS and SEP in energy consumption, life cycle and throughput.

1. Introduction

WSN (Wireless Sensor Network) is different from the traditional wireless network, which is limited in bandwidth, memory and other resources. Especially its limited energy resources directly affect the life cycle and information quality in the sensor network[1]. The clustering routing algorithms in WSN are good methods of saving energy consumption and prolonging life cycle of the network all the time[2]. In recent years the clustering routing protocols with high efficiency, scalability and portability are used for heterogeneous scene, which have acquired certain results. However there are problems such as low energy nodes being cluster heads, energy holes and without application relevance in WSN.

PEGASIS[3] (Power-Efficient Gathering in Sensor Information Systems) is a single-chain clustering routing protocol. Each node simply delivers data to the close nodes and eventually data are sent to the cluster header in a multi-hop manner, which greatly reduces the communication cost in the course of clustering. Compared with LEACH, it further increases the survival time of network. Because there is only a cluster head elected in a round, the death of nodes will cause paralysis of the entire network and too long single chain will increase the delay of data transmission. It is not suitable to extensively apply in practice.

SEP[4] (Stable Election Protocol) proposes the clustering routing protocol for energy heterogeneous networks. Depending on the initial energy, the nodes are divided into ordinary nodes with lower energy and senior nodes with higher energies. The common nodes become the cluster heads once with $(1+\alpha\cdot m)/p_{opt}$ in the interval of round, while the advanced nodes are the cluster heads with greater probability 1+α in the same interval. The heterogeneity of node initial energy is taken into account in SEP protocol, so this effectively balances the energy between nodes. Although the selected cluster heads according to initial energy to some extent extend the network's life cycle, the advanced nodes with higher probability are to be cluster heads, which restricts energy consumption in the network. Meanwhile the senior nodes far away from BS will be in death at first, which shortens the life of the network.

In conclusion, the single-chain clustering in PEGASIS and the energy heterogeneous clustering of SEP diminish the energy consumption and prolong life cycle of network in various aspects. But application relevance of the network is out of consideration. After the characteristics of WSN and intelligent water-saving irrigation system of ecological agriculture in Hexi Gobi are considered, a two-level clustering chain energy heterogeneous routing protocol based on PEGASIS and SEP is constructed to effectively manage energy of the nodes, extend the network's life and save the economic cost. It provides the prerequisite for ecological agriculture with the large-scale application of WSN in Hexi Gobi [5].

2. The structure of system

Because the irrigation sub-networks of intelligent water-saving irrigation system of ecological agriculture in Hexi Gobi consisting of pump station, control valve, irrigation trunk pipe, branch pipe and dropper pipe are laid in straight line[6~7]. So the irrigation subnet L×L is divided into two equal rectangular areas, which are low energy area (DNQ) and high energy area (GNQ) [8~9]. The same number of ordinary terminal nodes with lower energy and the senior terminal nodes with higher energy is respectively located at the DNQ region nearest to BS and the GNQ region far away from BS, as shown in figure 1.

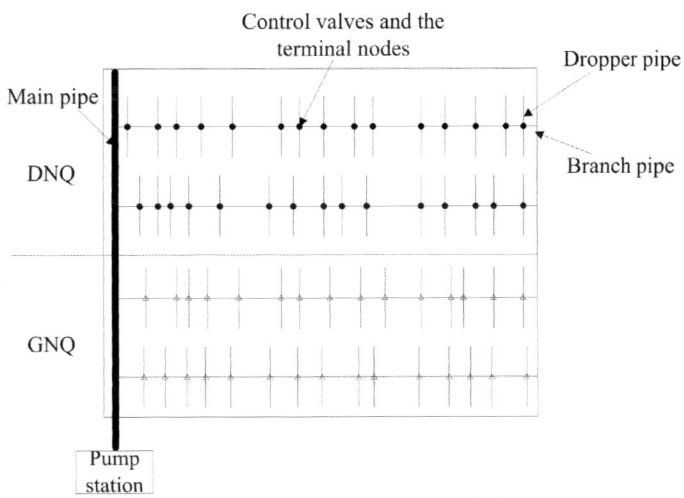

Figure 1. Heterogeneous systems model

3. The two-level clustering chain energy heterogeneous routing protocol

The protocol goes in rounds including clustering stage, chain stage, selecting cluster heads stage and data communication stage.

3.1. Clustering stage

In order to improve the efficiency of clustering algorithm each irrigation subnet is divided into m fixed equal segment according to the actual layout of intelligent water-saving irrigation system. Each segment is a cluster, as shown in figure 2.

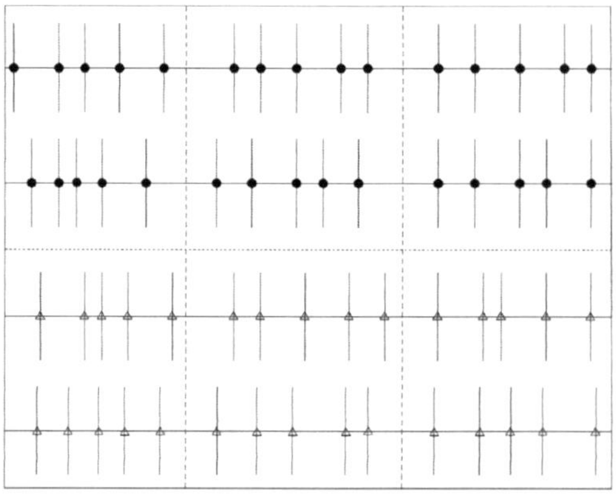

Figure 2. Clustering Chain topology

3.2. Chain stage

3.2.1. Chain stage in clusters

After divided into clusters, the network begins the initial chains in the clusters. The nodes in each region send a "RTS" frame to BS, including identifier, location, energy and region and so on. After BS receives the frame, the distances between all nodes and BS are calculated and then "ACK" frame is returned to the nodes, it includes the distance and location. The nodes farthest away from BS are to be first chain nodes.

The request frame "RTS" is broadcasted by the first node h (x_h, y_h) in chain to obtain the distance from each neighbor node j (x_j, y_j) to the first node, as shown in formula 1. After the close nodes j (x_j, y_j) return the "ACK" frame, if distances between the first node and neighbor nodes are minimum, they join the chain. And then the nodes in the chain will no longer join the chain. Follow the above protocol to loop until all nodes are traversed in the cluster.

$$d_{(H,j)} = \sqrt{(x_h - x_j)^2 + (y_h - y_j)^2} (j = n_1, n_2, n_3 \cdots) \tag{1}$$

3.2.2. Chain stage inter-cluster

At this stage, the two-level cluster heads can only be chains in the DNQ and GNQ. When the "RTS" frame are broadcasted by the level cluster head selected from the two-level cluster heads and the "ACK" frame are returned by the two-level cluster heads, the two-level cluster heads which are closest from the level cluster heads can form a chain. Follow this method until all the two-level cluster heads are traversed.

3.3. Selecting cluster heads stage

3.3.1. Selecting the two-level cluster heads stage

In the first round of network operation, the nearest node away from BS in each region is selected as the two-level cluster heads. In every round of communication, energy consumption of the cluster head and far away from BS are comprehensively considered. The nodes whose ratio of the remaining energy of terminal nodes in the cluster to the average energy of local region is bigger are taken as the backup two-level cluster heads. While the terminal nodes whose ratio of the remainder of energy to the distance between terminal nodes and BS is bigger are regarded as the final two-level cluster heads. The coordinates of terminal nodes in the DNQ and GNQ regions and the base station are (x_{D_CH}, y_{D_CH}), (x_{G_CH}, y_{G_CH}), (x_{BS}, y_{BS}). The judgment threshold is as shown in formulas 2~3.

$$\begin{cases} W_{res}(j) \geq W_{AG}(r) \\ p_{G\_CH}(i) = \dfrac{W_{res}(j)}{\sqrt{(x_{G\_CH} - x_{BS})^2 + (y_{G\_CH} - y_{BS})^2}} \end{cases} \tag{2}$$

$$\begin{cases} W_{res}(i) \geq W_{AD}(r) \\ p_{D\_CH}(i) = \dfrac{W_{res}(i)}{\sqrt{(x_{D\_CH} - x_{BS})^2 + (y_{D\_CH} - y_{BS})^2}} \end{cases} \tag{3}$$

The average energy of all nodes in the present round of DNQ and GNQ regions is $W_{AD}(r)$ 、 $W_{AG}(r)$, as shown in formulas 4~5.

$$W_{AG}(r) = \frac{1}{N_2} \sum_{i=1}^{N_2} W_{(Gi)}(r) \tag{4}$$

$$W_{AD}(r) = \frac{1}{N_1} \sum_{i=1}^{N_1} W_{(Di)}(r) \tag{5}$$

3.3.2. Selecting the level cluster heads

Selecting the level cluster heads is similar to the two-level cluster heads in clusters, but it is not necessary to consider the candidate level cluster heads. The coordinates two-level cluster heads in DNQ and GNQ regions and the base station are (x_{Di}, y_{Di}), (x_{Gi}, y_{Gi}), (x_{BS}, y_{BS}). According to the threshold formulas 15 ~ 16 the largest value between P_{Di} and P_{Gi} will be the level cluster heads.

$$p_{Gi} = \frac{W_{res}(i)}{\sqrt{(x_{Gi} - x_{BS})^2 + (y_{Gi} - y_{BS})^2}} \tag{6}$$

$$p_{Di} = \frac{W_{res}(i)}{\sqrt{(x_{Di} - x_{BS})^2 + (y_{Di} - y_{BS})^2}} \tag{7}$$

3.4. Data communication stage

The CSMA-CA mechanism is adopted to prevent data conflicting and retransmission in the data communication stage[10~11]. Before signal transmission, channels are sensed for the same carrier. If not, the channel is free and will directly transmit the data. If the carrier exists, the channel will be randomly withdrawn for a time and then detect channels.

The two-level cluster heads transmit, fuse data which are collected and received, then send data to the level cluster heads. It again merges and transmits the collected and received data to the base station.

4. Simulation Study

4.1. Simulation parameters

When the protocol is performed by simulation software NS2, the initial energy W0 of the normal terminal nodes are 0.5J. The number of sensor nodes M is 100. The number of simulation rounds is 5,000. The frame size is 100bit. The base station coordinates is (50, 100). The received and fused data is 4000 bit. The ratio α of the senior terminal nodes to terminal nodes is 0.5. The energy factor β is 1. Irrigation subnet is divided into 3 fixed equal segments. The energy consumption in the free space and multipath channel are 10pJ/bit/m^2 and 0.001310pJ/bit/m^4. Consumed energy for sending, receiving and fusing per bit data are 50nJ/bit, 5nJ/bit.

4.2. Simulation results study

4.2.1. Energy consumption in network

Since energy consumption is firstly considered in the large-scale application of the sensor network, the two-level clustering chain energy heterogeneous routing protocol is evaluated on the basis of network energy consumption. It shows that the energy consumption of PEGASIS is close to the SEP in figure 3.While the energy consumption in improved routing algorithm is significantly less than both of them, indicating that the protocol is better in energy expenditure.

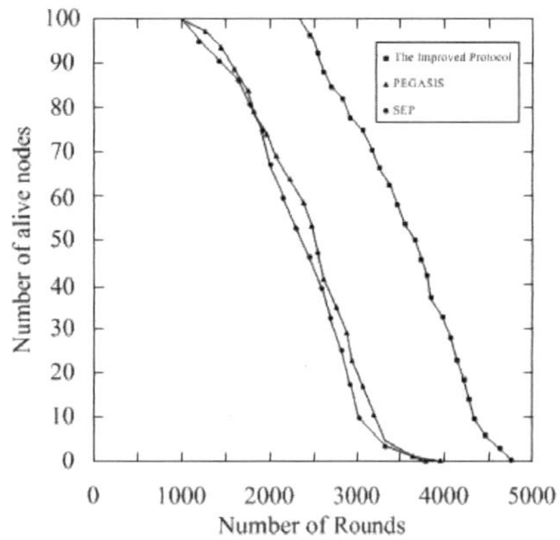

Figure 3. The Energy consumption of network

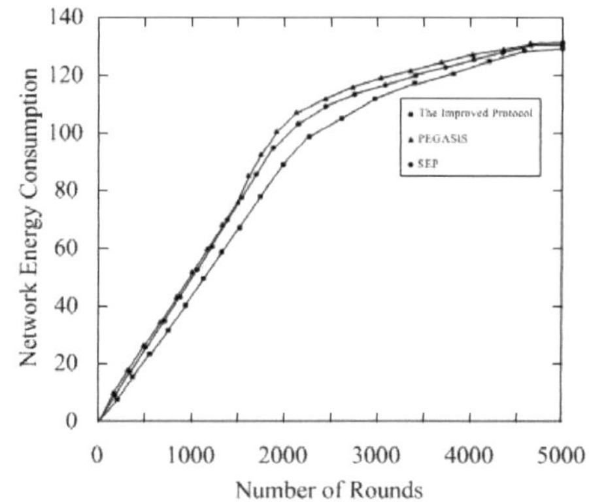

Figure 4. The Number of alive nodes

4.2.2. Life cycle in network

To further evaluate performance of the improved protocol, the two-level clustering chain energy heterogeneous routing protocol compared with PEGASIS and SEP are simulated in light of the number of alive nodes, as shown in figure 4. When the network operates about 1200 rounds, the nodes die significantly in SEP and the nodes of PEGASIS begin to die. When the network runs about 2300 rounds, the nodes in the SEP and PEGASIS die over one half. Yet the nodes in the two-level clustering chain energy heterogeneous routing protocol die initially. Therefore, the algorithm has better characteristic of extending the life cycle.

4.2.3. Throughput in network

The two-level clustering chain energy heterogeneous routing protocol has superior performances in both network energy consumption and life cycle, thus this provides a theoretical basis for its application of intelligent water-saving irrigation system in the ecological agriculture of Hexi Gobi. Meanwhile the routing protocol is analyzed in terms of network throughput to guarantee the quality of data transmission in the system. The improved protocol has a higher network throughput than the PEGASIS and SEP as shown in figure 5, so it is more efficient to transmit data.

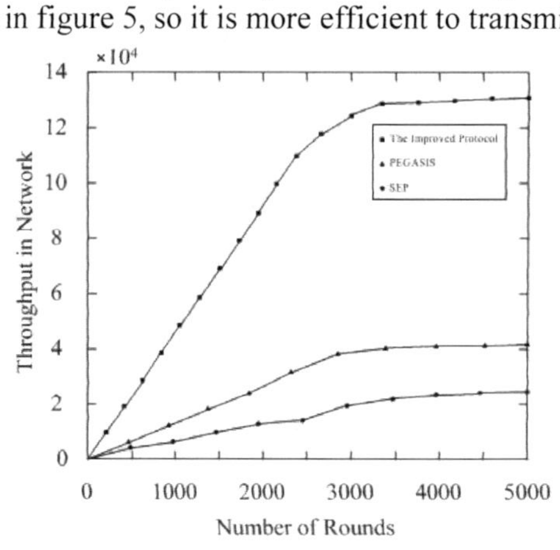

Figure 5. Network throughput

5. Conclusions

Combining the chain structure of PEGASIS and the heterogeneity of SEP, The two-level clustering chain energy heterogeneous routing protocol is proposed in the heterogeneous environment of intelligent water-saving irrigation system of ecological agriculture in Hexi Gobi. The network is divided into clusters according to the actual layout of the wireless network. The nodes far away from

BS being the first nodes join the chain in the chain stage according to the nearest distance between the cluster and the terminal nodes. And then the cluster headers are selected at a given threshold to ensure that the nodes with the maximum energy serve as the cluster headers at all levels. Experiments results show that the improved algorithm compared with the PEGASIS and SEP is better in aspects of network energy consumption, throughput and life cycle. Therefore this protocol can be widely applied in the intelligent water-saving irrigation system of ecological agriculture in Hexi Gobi.

Acknowledgments

This work was supported by the Innovation Foundation Projects in Gansu Province Department of Education No. 2021A-159. The supported project—Multi-jump Clustering Heterogeneous Routing Protocol for WSN in Hexi Gobi Ecological Agriculture.

References

[1] Li, Z. X., Xie, R., Luo, K et al. (2022) Minimizing Outage Probability Driven Wireless Backhaul Scheme in Heterogeneous Networks[J]. China Communications. 19(06):248-262

[2] Song, X., Huang, X., Gao, Y. M. et al. (2022) Study on Cluster-Based Energy Saving Routing Protocols for Wireless Sensor Networks[D]. The Journal of China Universities of Posts and Telecommunications, 01:102-112.

[3] Lindesy, S., Raghavendra, C. (2002) PEGASIS: power-efficient gathering in sensor information system[C]//Proc. of 2002 IEEE aerospace conference. 1-6.

[4] Smaragdakis, G., Matta, I., Bestavros, A. (2004) SEP: A stable election protocol for clustered heterogeneous wireless sensor networks [J]. Proceeding of 2nd International Workshop on Sensor and Actor Network Protocol and Applications (SANPA).

[5] Xie, J. L., Gao, W. J. (2020) Heterogeneous Network Selection Optimization Algorithm Based on a Markov Decision Model[J]. China Communications, 17(02): 40-53.

[6] Zhang, Y., Zhang, H. L., Ji, H.,et al. (2020) Distributed cooperative deployment strategy in multi-UAVs assisted heterogeneous networks[J]. The Journal of China Universities of Posts and Telecommunications, 12: 11-19.

[7] Wang, X. W., Zhang, Y. X., Ma, Y. X.,et al. (2020) The Development Status and Countermeasures of Gobi Agriculture in Gansu Province Gansu[J]. Agricultural Science and Technology, 7:71-75.

[8] Kaur, S., Mir, R. N., Khamparia, A., Rani, P., Gupta, D., Khanna, A. (2021) Heterogeneous load balancing clustering protocol for Wireless Sensor Networks[J]. Cognitive Systems Research, 70:10-17.

[9] Elfouly, F. H., Ramadan, R. A., Khedr, A. Y., Yadav, K., Azar, A. T., Abdelhamed, M. A. (2021) Efficient Node Deployment of Large-Scale Heterogeneous Wireless Sensor Networks[J]. Applied Sciences, 11(22): 10924-10931.

[10] Osamy, W., Salim, A., Khedr, A. M. (2020) An information entropy based-clustering algorithm for heterogeneous wireless sensor networks[J]. Wireless Networks, 26(4):1869-1886.

[11] Abidi, W., Ezzedine, T. (2020). Effective clustering protocol based on network division for heterogeneous wireless sensor[J]. Computing, 102(2): 413-425.

Research on single cell membrane algorithm and engineering application based on membrane computing theory

Qi Song[1], Yourui Huang[1] *, Jiachang Xu[3], Wenhao Lai[1,2], Shanyong Xu[1], Tao Han[1], Xue Rong[1]

[1] School of Electrical and Information Engineering, Anhui University of Science and Technology, Huainan, Anhui Province, 232000, China
[2]Joint National-Local Engineering Research Center for Safe and Precise Coal Mining, Anhui University of Science and Technology, Huainan, Anhui Province, 232000, China
[3]School of Computer Science and Engineering, Anhui University of Science and Technology, Huainan, Anhui Province, 232000, China
First author's e-mail: 2019100052@aust.edu.cn

*Corresponding author's e-mail: hyr628@163.com (Yourui Huang)

Abstract. Membrane computing is a new computing paradigm with great significance in the field of computer science. The Multi-membrane search algorithm (MSA) is proposed based on the membrane computational population optimization theory. It showed excellent performance in the test. This paper further studies the performance characteristics of a single individual (Single Cell Membrane Algorithm, SCA) of MSA. SCA can generate adaptive solution sets for problems of different dimensions. Through transcription and reprocessing rules, new weakly correlated feasible solutions are formed for global search and local exploration. This paper is based on the unimodal Sphere function and the multimodal Rastrigr function, at dim=3, 5, 10, 20, 30, 40, 50, 60, 70, 80, 90, 100, 300, 500, 1000 and Q=1.00, 0.75, 0.50, 0.40, 0.30, 0.20, 0.10, 0.005, 0.025, 0.010, the SCA was optimized for 1000 iterations. Analyze the impact of the key parameter Q of SCA on the search performance of the algorithm in problems of different dimensions. The results show that under the set conditions, SCA has better performance when Q is 0.010 and 0.025 in the unimodal function test. In the multimodal function test, SCA has better performance when dim≤100 and Q≤0.200, and when dim>100 and Q≥0.200. In addition, this paper employs one engineering problem: I-beams to perform engineering tests on SCA and obtain results superior to other algorithms participating in the comparison. The test and comparison results show that SCA can also be used as a derivative algorithm of MSA, and has good performance.

1. Introduction

In production practice, it is necessary to select an optimal solution in the case of multiple feasible solutions. For example, for the complex equivalent function existing in engineering design, within the acceptab range, there are many feasible solutions, how to select parameters to minimize the design cost under the condition that the requirements are met [1]. In the process of resource allocation, there are various allocation schemes, how to use limited resources for reasonable allocation and maximize benefits [2]. Problems like this are still numerous and boil down to optimization problems.

From a quantitative point of view, the optimization problem is to establish an equivalent mathematical model for the object to be optimized, solve the functions in the model, and find the optimal solution where all feasible solutions. Under the condition of constraints, obtaining the optimal solution of the function to be optimized is often the key to solving the problem. The mathematical method used to solve this kind of problem is called optimization algorithm. When the computer has been widely used, the solution of some large-scale optimization problems can be realized on the computer through computing software, which makes the optimization method more widely used than ever before [3][4][5]. Since the objective function of nonlinear engineering design optimization problem often contains many local optimal solutions, the designer is always concerned about finding the global optimal solution. To solve such problems, researchers continue to seek inspiration from the fields of biology [6][7], physics [8], and sociology [9], and develop a variety of algorithms suitable for solving complex problems, that is, meta-heuristic algorithms.

Metaheuristics are proposed relative to optimization algorithms. Optimization algorithms can obtain the optimal solution to a problem, while metaheuristics are algorithms built on intuition or experience. It can give a feasible solution to the problem at an acceptable cost (referring to computing time and space), and the degree of deviation of the feasible solution from the optimal solution is not necessarily predicted in advance [10][11]. The construction of most meta-heuristic algorithms is based on relevant knowledge or experience in specific fields [12]. Since practical engineering problems are often very complex and contain a large number of locally optimal solutions, the search mechanism of meta-heuristic algorithms is inspired by natural biological behavior. Because this biological behavior after tens of millions of years of natural selection often has excellent validity in a class of problems, it has been greatly developed in recent years. For example, the Particle Swarm Optimizer (PSO) [13], the Whale Optimization Algorithm (WOA) [14], the Grey Wolf Optimizer (GWO) [15], etc. have been proposed by simulating biological group behavior. By simulating physical phenomena, the Multi-Verse Optimizer (MVO) [16], the Black Hole Optimizer Algorithm (BHOA) [17], and the Simulated Annealing (SA) [18] are proposed respectively. Inspired by sociological phenomena, the Imperial Competition Algorithm (ICA) [19] and the Cultural Algorithm (CA) [20] are proposed.

In general, meta-heuristic optimization methods achieve optimization through two stages: a global search stage and a local exploration stage. In the global search stage, the algorithm needs to explore as much as possible within the definition domain to determine the possible region where the global optimal solution exists; in the local exploration stage, it needs to further search in the region where the global optimal solution may exist, and then infinitely approach the optimal solution. In the limited number of iterations, whether the balance between global search and local exploration can be maintained will directly affect the solution performance of the algorithm [15].

Membrane computing is a non-traditional distributed parallel computing model inspired by the structure and function of cells and is an important branch of meta-inspired optimization methods. Its goal is to draw inspiration from the biological behavior of cells and apply it to computer science [21]. The optimization algorithm based on membrane computing design has been widely used in control science [22][23], electrical engineering [24][25], biology [26], medicine [27], computer [28][29], and other fields [30][31], and achieved good results. However, according to the NFL (No Free Lunch) theory, the performance of the optimization algorithm is equivalent due to the mutual compensation of all possible functions [32][33]. This means that a membrane computational optimization algorithm can be adapted to solve a class of problems, but cannot be effective for all. The NFL theory keeps membrane computing open and developing in the field of optimization. Previous membrane computing researchers only applied the limited characteristics of cell membranes, and is still a lot of room for development in membrane computing, which can solve emerging new problems.

Based on the survival of the fittest in cell groups, sexual reproduction and the evolutionary laws of single-cell individuals, MSA is designed by abstracting the life activities of individual cells and groups [34]. For MSA individuals, the intergroup communication behavior begins after the division of old cells and ends before the birth of new cells, so the search behavior of individuals has the characteristics of independent and identical distribution in the region, which enables a single MSA individual to perform independent optimization behaviors. In this paper, we further investigate the single entity of the MSA algorithm: the single cell membrane algorithm (SCA) and the parameter Q optimization properties.

2. Algorithm mathematical model

2.1. Cell generation mathematical model

When designing the algorithm, to simulate various organelle states inside the cell, we use a matrix to represent the cells of SCA, and each row of the matrix represents an organelle:

$$SCA = \begin{pmatrix} X_1 \\ X_2 \\ \vdots \\ X_i \\ \vdots \\ X_I \end{pmatrix} \quad (1)$$

Where： X_i represents the ith organelle; I represents the number of all organelles in the SCA, J represents the dimension of the candidate solution to the problem, $I = J+1$.

Since each organelle of SCA contains a feasible solution, the internal structure of the organelle X_i is as follows:

$$X_i = \begin{pmatrix} x_i^1 & x_i^2 & \cdots & x_i^j & \cdots & x_i^J \end{pmatrix} \quad (2)$$

Where: x_i^j represents the jth dimension of the feasible solution contained in the ith organelle.

In order to show the internal structure of the SCA algorithm more intuitively, we represent x_i^j in the interior of SCA:

$$SCA = \begin{pmatrix} X_1 \\ X_2 \\ \vdots \\ X_i \\ \vdots \\ X_I \end{pmatrix} = \begin{pmatrix} x_1^1 & x_1^2 & \cdots & x_1^j & \cdots & x_1^J \\ x_2^1 & x_2^2 & \cdots & x_2^j & \cdots & x_2^J \\ \vdots & \vdots & \ddots & \vdots & \ddots & \vdots \\ x_i^1 & x_i^2 & \cdots & x_i^j & \cdots & x_i^J \\ \vdots & \vdots & \ddots & \vdots & \ddots & \vdots \\ x_I^1 & x_I^2 & \cdots & x_I^j & \cdots & x_I^J \end{pmatrix} \quad (3)$$

At the initial moment, all the variables in the solution set of SCA are randomly generated in the definition domain, which can avoid the influence of the initial value setting on the performance test of the algorithm.

SCA initial generation rules:

$$x_i^j = rand \times (ub_j - lb_j) + lb_j \quad (4)$$

Where: *rand* represents a random number between (0,1); ub_j represents the maximum value of the jth dimension parameter; lb_j represents the minimum value of the jth dimension parameter.

2.2. Enzyme catalysis and cellular energy rules

Enzymes play an important catalytic role in cells, and their activity directly affects the protein synthesis efficiency of cells. The search process of the SCA algorithm imitates the enzyme activity characteristics in the cell life cycle through the Q value and generates reactive enzymes that vary with the number of iterations.

Mathematical model of reactive enzyme activity:

$$\alpha = Q \times \ln \frac{t}{T} \tag{5}$$

Where: a represents the enzyme activity value of SCA; t represents the current iteration number; T represents the maximum iteration number.

Since chemical reactions in organisms are carried out cooperatively, the role of enzymes is achieved through catalytic rules, and in cells, the presence of *HSP* can not only improve the reaction efficiency, but also evolve independently, and the evolution process and its enzymatic reactions can be carried out simultaneously. Based on this, we design the energy mathematical model of SCA:

$$HSP = -\frac{e^{\alpha} - 1}{e^{\alpha} + 1} \tag{6}$$

Where: *HSP* represents the amount of energy within the SCA that affects the search.

During the process of cell secretion of proteins, under the action of *HSP*, organelles obtain energy and can process macromolecules. The reaction enzyme activity a can affect the *HSP* value, which increases with the number of iterations, and the *HSP* gradually decreases from the initial value.

2.3. Transcription and reprocessing rules

When SCA generates a new feasible solution, first, the optimal solution in the cell is transcribed, and then the transcription result is further processed by different organelle membranes to generate different new solutions.

Transcription Rules:

$$x_i^j = x_i^{best} + HSP \times (rand\text{-}0.5) \times (ub_i - lb_i) \quad (j \neq i) \tag{7}$$

Reprocessing rules:

$$x_i^j = rand \times (x_j^{best} - x_i^j) + x_i^j \quad (j = i) \tag{8}$$

Where: *rand* represents a random number between 0-1; x_j^{best} represents the jth dimension variable of the optimal solution in the current iteration times.

2.4. Processing collection rules

SCA collects and combines the parameters processed by all organelles in each iteration into a new solution. The design was inspired by the storage function of the cell vacuole, so X_I was called the tonoplast of the SCA. This setting can help SCA quickly generate solutions when the initial solution set is not effective. Make a large move to get out of the current unfavorable position.

Processing collection rules:

$$SCA = \begin{pmatrix} X_1 \\ X_2 \\ \vdots \\ X_i \\ \vdots \\ X_{I-1} \\ X_I \end{pmatrix} = \begin{pmatrix} \boxed{x_1^1} & x_1^2 & \cdots & x_1^j & \cdots & x_1^J \\ x_2^1 & \boxed{x_2^2} & \cdots & x_2^j & \cdots & x_2^J \\ \vdots & \vdots & \ddots & \vdots & \ddots & \vdots \\ x_i^1 & x_i^2 & \cdots & \boxed{x_i^j} & \cdots & x_i^J \\ \vdots & \vdots & \ddots & \vdots & \ddots & \vdots \\ x_{I-1}^1 & x_{I-1}^2 & \cdots & x_{I-1}^j & \cdots & \boxed{x_{I-1}^J} \\ x_I^1 & x_I^2 & \cdots & x_I^j & \cdots & x_I^J \end{pmatrix} \longrightarrow SCA = \begin{pmatrix} X_1 \\ X_2 \\ \vdots \\ X_i \\ \vdots \\ X_{I-1} \\ X_I \end{pmatrix} = \begin{pmatrix} \boxed{x_1^1} & x_1^2 & \cdots & x_1^j & \cdots & x_1^J \\ x_2^1 & \boxed{x_2^2} & \cdots & x_2^j & \cdots & x_2^J \\ \vdots & \vdots & \ddots & \vdots & \ddots & \vdots \\ x_i^1 & x_i^2 & \cdots & \boxed{x_i^j} & \cdots & x_i^J \\ \vdots & \vdots & \ddots & \vdots & \ddots & \vdots \\ x_{I-1}^1 & x_{I-1}^2 & \cdots & x_{I-1}^j & \cdots & \boxed{x_{I-1}^J} \\ \boxed{x_1^1} & \boxed{x_2^2} & \cdots & \boxed{x_i^j} & \cdots & \boxed{x_{I-1}^J} \end{pmatrix} \tag{9}$$

The collection rule of X_I collects the variables generated in the processing rule of the aforementioned organelle and generates a new solution. Without adding a new solution load, the SCA operator is allowed to obtain as many degrees of freedom as possible in multiple dimensions, and the algorithm can be prevented from falling into a regional optimum as much as possible.

Pseudo code:

Create random *SCA by Eq* (4)

Initialize Q, s, and S
Update a indexed by *Eq* (5)
Update *HSP* indexed by *Eq* (6)
while($s \leq S$)
Evaluate the fitness of all X_i
Update the fitness of X^{best}
for each membrane X_i indexed by I
for each parameter x^j indexed by J
$rand1 = random\ [0,1]$
$rand2 = random\ [0,1]$
$if: i \neq I$
$if: \ i = j$
$x_i^j = rand1 \times (x_j^{best} - x_i^j) + x_i^j$
else
$x_i^j = x_j^{best} + HSP \times (rand2 - 0.5) \times (ub_i - lb_i)$
end if
else
$x_i^j = x_j^j$
end if
end for
end for
Update a indexed by Eq (5)
Update HSP indexed by Eq (6)
Update s indexed by s=s+1
End while

3. Performance Testing

3.1. Test function introduction
This section adopts two types of functions: unimodal function and multimodal function, and conducts multiple sets of tests for the SCA algorithm. The test experiments are mainly used to explore and prove the comprehensive performance of SCA: the computing power of SCA for unimodal and multimodal functions of different dimensions under different Q value settings.
We selected the unimodal Sphere benchmark function and the multimodal Rastrigr benchmark function based on the characteristics of the functions. The Sphere function has only one peak, which can facilitate the horizontal comparison of the convergence speed of the operation results of different dimensions of the SCA algorithm. The Rastrigr function has multiple local optima, which can detect the global search ability of the algorithm, and it is more challenging to the convergence ability of the algorithm in higher dimensions.
Sphere function:

$$F_1(x) = \sum_{i=1}^{n} x_i^2 \tag{10}$$

Where: $x \in [-100,100]$, the optimal value is $f(x)=0$;

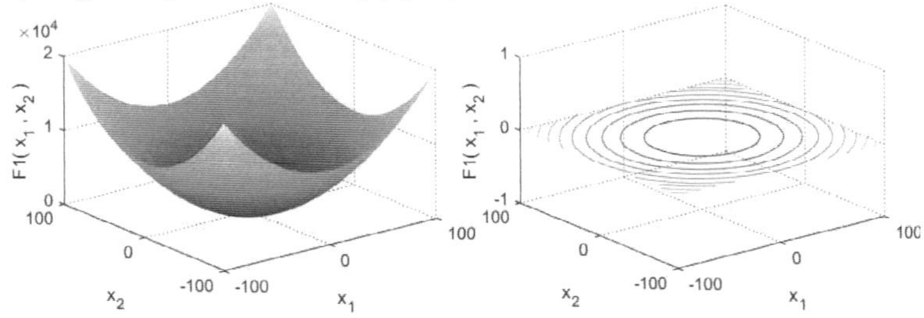

(a) (b)

Figure 1. Sphere function (a) 3D stereogram (b) plane projection diagram.

From Figure 1, the two-dimensional stereo image of the Sphere function is like the bottom shape of a valley. The function is continuous and has only one minimum value, and there is no local optimal solution.
Rastrigr function:

$$F_9(x) = \Sigma_{i=1}^{n} [x_i^2 - 10\cos(2\pi x_i) + 10] \tag{11}$$

Where: $x \in [-5,5]$, the optimal value is $f(x)=0$;

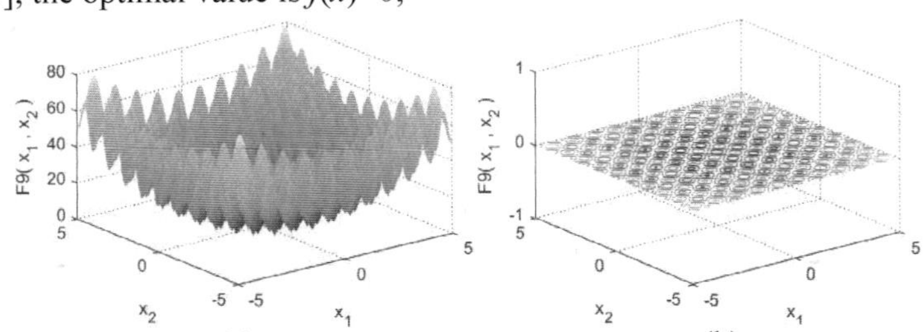

(a) (b)

Figure 2. Rastrigr function (a) 3D stereogram (b) Plane projection.

From Figure 2, the Rastrigr function and the Sphere function have commonalities in the overall trend, which are similar to the bottom shape of the valley, but the Rastrigr function has a wide range of regional minimums in the definition domain. This kind of regional minimum value is often a test of the optimization algorithm's ability to maintain a balance between global searchability and regional exploration ability. If the iteration step size is too small, even if the number of iterations is extremely high, the optimization algorithm will fall into the local optimal solution because it cannot step out of the local minimum region, thus affecting the search performance.

3.2. Test parameter setting

The experimental platform is built on the hardware environment of Inter i9-10900F and 32G memory and Windows 10 operating system, and uses Matlab R2019a software to set and realize various parameters of the experiment.
In the simulation, 1000 iterative tests were performed under the same parameters of the same algorithm under test, and each iterative test was repeated 10 times, and the best time was selected for display analysis.
The simulation process parameters are set as follows:
Q=1.00、0.75、0.50、0.40、0.30、0.20、0.10、0.005、0.025、0.010;

dim=3、5、10、20、30、40、50、60、70、80、90、100、300、500、1000
Since the initial feasible solutions of the algorithm are randomly generated, the starting point of each simulation is different, which is due to the characteristics of the algorithm itself and has nothing to do with the parameter settings.

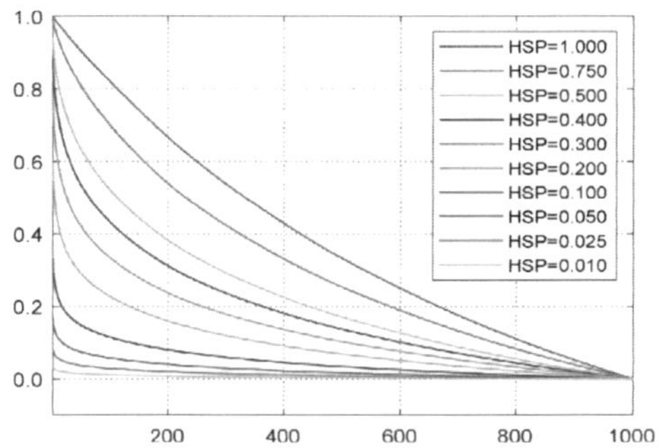

Figure 3. Change curve of HSP under different Q values (1000 iterations).

Figure 3 shows the value of *HSP* under different *Q* values, and we can find its trend characteristics: Among the ten values selected in the experiment, the starting point of *HSP* is proportional to the value of *Q*, and the decreasing speed of *HSP* is not constant, but changes with the iteration step size. However, for high values of *Q*, such as when *Q*=1 and *Q*=0.750, it can be intuitively found from the fig that the curvature of the descending curve changes slightly, which means that the step size changes approximately uniformly.
In medium-sized *Q* values, such as *Q*=0.500, *Q*=0.400, *Q*=0.300, and *Q*=0.200, it can be seen from the fig that the initial *HSP* value is large, but in the first 50 iterations, the change range of *HSP* is obviously higher than the number of subsequent iterations. This design can make the algorithm obtain a higher convergence speed in the early stage, slow down the convergence speed in the later stage, reduce the convergence step size, and relatively improve the solution accuracy of the algorithm.
In the value of small *Q*, such as *Q*=0.100, *Q*=0.050, *Q*=0.025 and *Q*=0.010, we can find that the step size of each iteration is small.

3.3. Test Results and analysis

Table 1. The 1000 iterations result of SCA in Sphere function test

Dim	1.000	0.750	0.500	0.400	0.300	0.200	0.100	0.050	0.025	0.010
3	2.900E-04	1.255E-03	1.908E-04	1.360E-04	5.204E-05	6.169E-06	5.749E-06	1.258E-06	5.632E-07	8.561E-08
5	1.166E-03	2.821E-03	1.171E-03	4.277E-04	1.690E-04	1.146E-04	3.715E-05	1.231E-05	3.753E-06	4.279E-07
10	2.880E-02	2.697E-02	9.239E-03	1.168E-02	2.620E-03	8.562E-04	4.693E-04	1.259E-04	2.828E-05	5.700E-06
20	3.484E-01	2.329E-01	1.192E-01	7.769E-02	2.910E-02	1.809E-02	4.981E-03	9.028E-04	3.251E-04	4.167E-05
30	1.854E+00	9.833E-01	5.799E-01	2.502E-01	1.539E-01	3.712E-02	1.134E-02	5.150E-03	1.065E-03	7.455E-05
40	3.106E+00	2.470E+00	7.828E-01	6.900E-01	4.094E-01	1.353E-01	3.293E-02	9.850E-03	2.558E-03	3.831E-04
50	9.000E+00	4.651E+00	1.504E+00	1.761E+00	7.476E-01	3.308E-01	8.806E-02	1.489E-02	4.032E-03	9.373E-04
60	1.399E	7.319E	4.215E	2.177E	1.538E+0	5.189E-01	1.828E-	3.551E	1.195E-	1.174E-

	+01	+00	+00	+00	0		01	-02	02	03
70	2.315E+01	1.469E+01	7.172E+00	2.660E+00	2.270E+00	1.053E+00	2.579E-01	4.899E-02	1.498E-02	2.463E-03
80	3.820E+01	2.023E+01	1.152E+01	5.501E+00	3.322E+00	1.582E+00	4.184E-01	8.456E-02	2.999E-02	3.797E-03
90	5.048E+01	3.258E+01	1.746E+01	9.913E+00	4.858E+00	2.380E+00	5.164E-01	1.196E-01	3.756E-02	5.436E-03
100	8.688E+01	4.810E+01	1.854E+01	1.391E+01	8.015E+00	3.052E+00	8.338E-01	2.271E-01	5.857E-02	8.255E-03
300	3.619E+03	1.991E+03	7.580E+02	5.668E+02	3.443E+02	1.544E+02	3.403E+01	8.477E+00	2.315E+00	3.204E+01
500	2.112E+04	1.333E+04	5.507E+03	3.429E+03	2.021E+03	8.767E+02	2.107E+02	5.416E+01	1.454E+01	3.755E+04
1000	2.229E+05	1.478E+05	7.107E+04	4.642E+04	2.526E+04	1.170E+04	3.042E+03	7.354E+02	2.067E+02	3.412E+05

We use shading to mark the optimal value. It can be seen from Table 1 that the test results obtained by SCA in 1000 iterations of the unimodal sphere function are at *dim*=3, *dim*=5, *dim*=10, *dim*=20, *dim*=30, *dim*=40, *dim*=50, *dim*=60, *dim*=70, *dim*=80, and *dim*=90, the result of *Q*=0.010 is the best where the 10 parameters compared, and the result of *Q*=0.025 is the second-best; while in *dim*=100 and *dim*=300, When *dim*=500 and *dim*=1000, the result of *Q*=0.025 performs the best in the test.

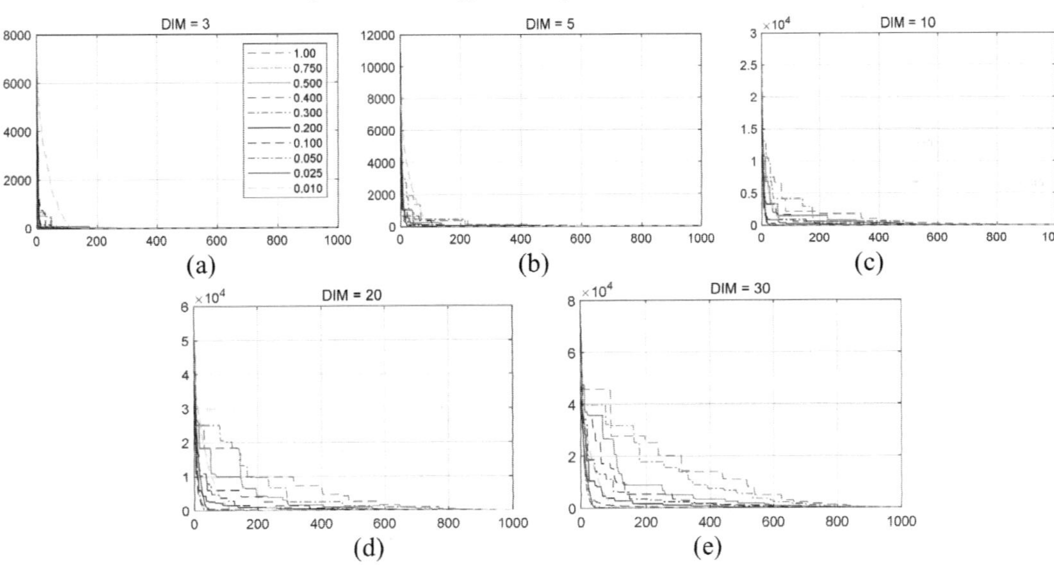

Figure 4. The 1000-time optimization process of SCA in the lower dimension test of the Sphere function. (a)Dim=3(b) Dim=5 (c) Dim=10 (d) Dim=20 (e) Dim=30

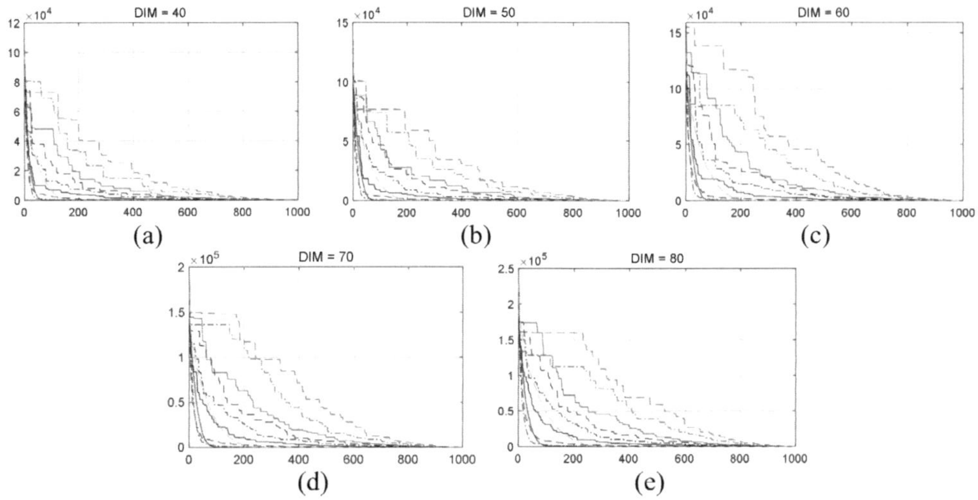

Figure 5. The 1000-time optimization process of SCA in the intermediate dimension test of the Sphere function. (a)Dim=40 (b) Dim=50 (c) Dim=60 (d) Dim=70 (e) Dim=80

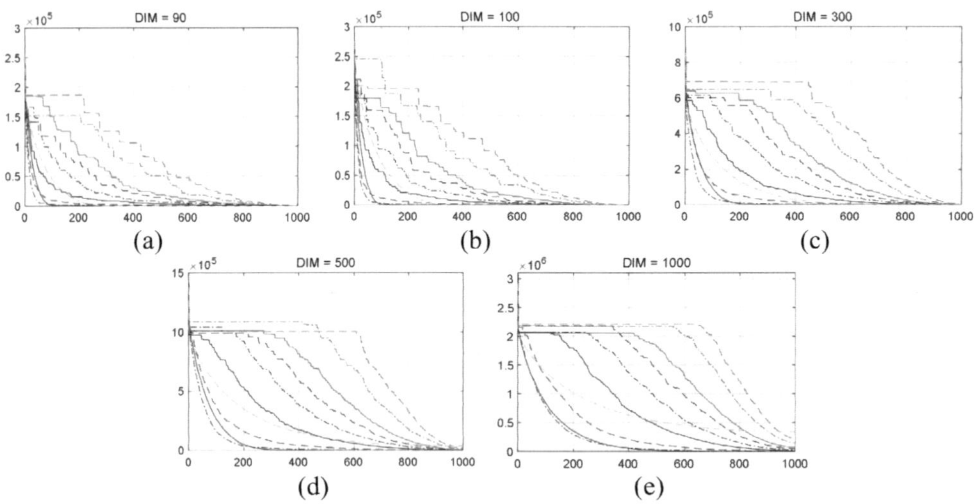

Figure 6. The 1000-time optimization process of SCA in the higher-dimensional test of the Sphere function. (a)Dim=90 (b) Dim=100 (c) Dim=300 (d) Dim=500 (e) Dim=100

From Figure 4, Figure 5, Figure 6 and the results, we can see that when Q=0.010, its value is constantly converging, but the convergence speed is not the fastest in the early stage. Since it is the lowest Q value in the test, its *HSP* is also lower than other Q values in the later stage. When the number of iterations is sufficient, the calculation accuracy can be effectively improved. Since there is no local optimal solution, the balance between global search and local exploration has an influence on the results, but it is weaker than the influence of iteration times and Q value on the solution accuracy.
At Q=0.010, the convergence curve of SCA is not always the fastest, compared to other smaller Q parameters. When *dim*=3, *dim*=5, *dim*=10, *dim*=20, *dim*=30, the convergence result is gradually improved after 100 iterations. When *dim*=50, *dim*=60, *dim*=70, *dim*=80, *dim*=90, the convergence results improved after the 150th iterations. When *dim*=100, it takes more than 300 iterations to gradually improve the iterative convergence effect. This is because SCA adopts a greedy mechanism, and the sphere function is an unconstrained unimodal function, and the slow optimization result of

Q=0.010 requires continuous accumulation of iterations to achieve a competitive effect. When *dim*≥100, the performance of *Q*=0.025 is stronger than that of *Q*=0.010, because the number of iterations is still insufficient relative to the dimension of solving the problem, and the fitness value of the test function is the multi-dimensional accumulation of the solution set. SCA needs to increase the *Q* value parameter to obtain a higher *HSP* value, improve the efficiency of transcription rules, and avoid the weakening of reprocessing rules in calculation. At the same time, it can be seen from the fig that although *Q*=0.025 in 1000 iterations, *dim*=300, *dim*=500, *dim*=1000 obtains better results than *Q*=0.010. However, if the number of iterations continues to increase, it can be speculated that *Q*=0.010 will gradually surpass the result of *Q*=0.025.

Table 2. The 1000 iterations result of SCA in Rastrigr function test

Dim	1.000	0.750	0.500	0.400	0.300	0.200	0.100	0.050	0.025	0.010
3	7.85E-05	1.32E-04	1.33E-04	9.42E-06	4.17E-05	1.66E-05	1.99E+00	3.31E-07	1.99E+00	1.99E+00
5	1.01E+00	3.21E-03	1.99E+00	1.99E+00	9.96E-01	2.99E+00	1.99E+00	2.09E+01	7.96E+00	4.97E+00
10	1.40E+01	8.97E+00	1.39E+01	1.69E+01	1.29E+01	2.29E+01	2.09E+01	1.69E+01	2.19E+01	3.28E+01
20	2.81E+01	4.69E+01	4.39E+01	5.77E+01	3.68E+01	6.47E+01	6.17E+01	5.67E+01	7.16E+01	7.56E+01
30	9.78E+01	5.63E+01	8.77E+01	7.17E+01	7.67E+01	4.38E+01	1.04E+02	1.36E+02	1.28E+02	1.58E+02
40	1.48E+02	1.43E+02	9.79E+01	1.29E+02	9.57E+01	9.86E+01	1.49E+02	1.97E+02	1.84E+02	1.81E+02
50	1.64E+02	2.18E+02	1.56E+02	1.84E+02	2.01E+02	1.71E+02	2.28E+02	1.94E+02	2.48E+02	2.16E+02
60	2.81E+02	2.14E+02	2.39E+02	2.55E+02	1.94E+02	2.44E+02	2.57E+02	2.65E+02	2.86E+02	2.45E+02
70	2.22E+02	2.88E+02	2.63E+02	2.87E+02	2.40E+02	3.21E+02	3.33E+02	3.17E+02	3.58E+02	3.22E+02
80	4.11E+02	3.61E+02	3.86E+02	3.44E+02	3.12E+02	3.36E+02	3.25E+02	3.75E+02	3.97E+02	3.68E+02
90	4.72E+02	3.96E+02	4.31E+02	3.25E+02	4.41E+02	4.81E+02	4.19E+02	4.37E+02	4.59E+02	4.51E+02
100	5.38E+02	4.39E+02	4.53E+02	4.67E+02	4.54E+02	4.00E+02	4.39E+02	4.57E+02	5.37E+02	4.77E+02
300	2.46E+03	2.27E+03	2.02E+03	1.82E+03	1.90E+03	1.78E+03	1.68E+03	1.54E+03	1.40E+03	1.55E+03
500	4.83E+03	4.37E+03	4.11E+03	3.81E+03	3.39E+03	3.27E+03	2.93E+03	2.78E+03	2.74E+03	2.51E+03
1000	1.08E+04	1.04E+04	9.50E+03	9.21E+03	8.47E+03	8.23E+03	6.99E+03	6.15E+03	5.82E+03	5.32E+03

From Table 2, we can find that the results of SCA in the multimodal Rastrigr function 1000 iterations test, with *Q*=0.200, *dim*=100 as the boundary. When *dim*≤100, the test performance of SCA when *Q*≥0.200 is generally better than that when *Q*<0.200; when *dim*>100, the test performance of SCA when *Q*<0.200 is generally better than that when *Q*≥0.200.

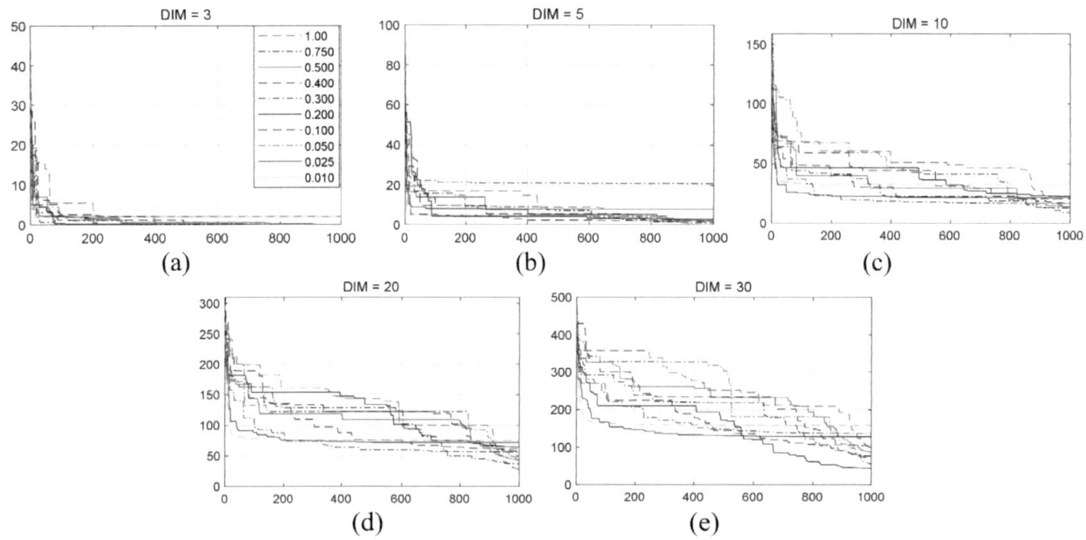

Figure 7. The 1000-time optimization process of SCA in the lower dimension test of the Rastrigr function. (a)Dim=3(b) Dim=5 (c) Dim=10 (d) Dim=20 (e) Dim=30

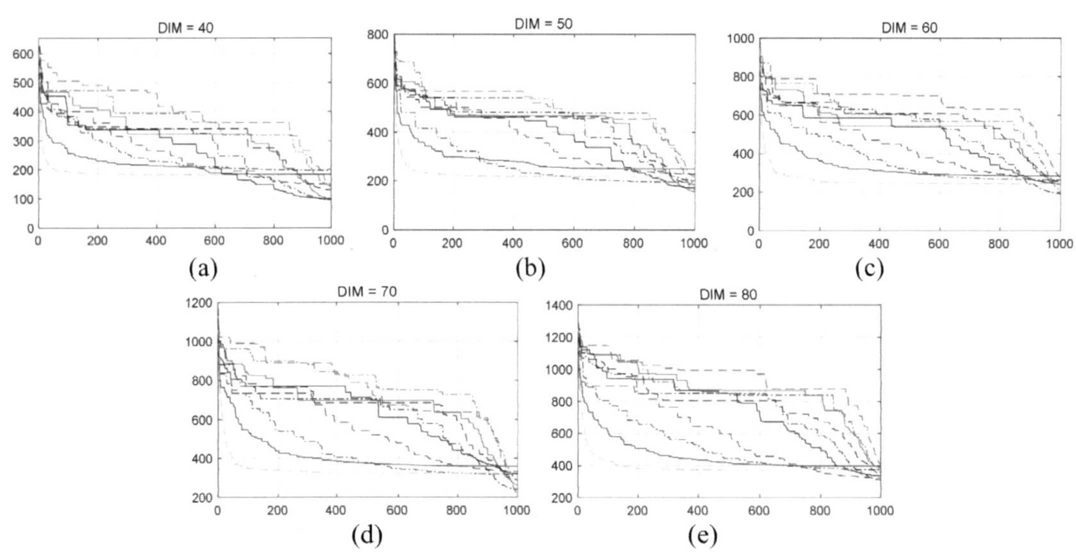

Figure 8. The 1000-time optimization process of SCA in the intermediate dimension test of the Rastrigr function. (a)Dim=40 (b) Dim=50 (c) Dim=60 (d) Dim=70 (e) Dim=80

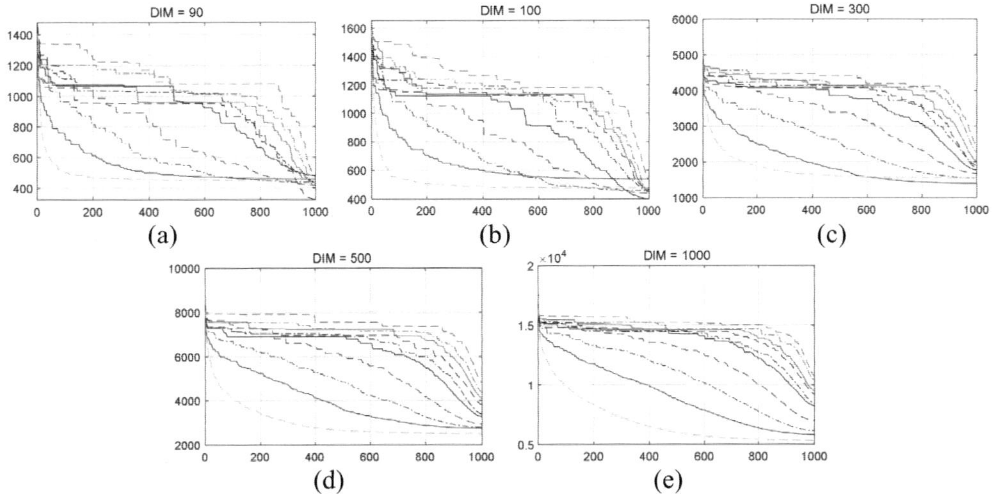

Figure 9. The 1000-time optimization process of SCA in the higher-dimensional test of the Rastrigr function. (a)Dim=90 (b) Dim=100 (c) Dim=300 (d) Dim=500 (e) Dim=100

From Figure7, Figure8, Figure9, the convergence curve of SCA in each dimension when Q=0.200(black line) is used as a reference. When $dim \leq 100$, it can be found that when Q<0.200, the convergence curve of SCA can drop rapidly in the early stage, but because the value of *HSP* is too small, it is easy to fall into the local optimal solution in the later stage of the algorithm. When $Q \geq 0.200$, the convergence curve of SCA is slower in the early stage than that when Q<0.200, but it will gradually surpass in the later stage, and there is a step-like decline in the whole convergence process.

When dim>100, it can be found that with the increase of dimension, the optimization results obtained with Q<0.200 are generally better than those obtained with $Q \geq 0.200$. This is because the Q value is too large, resulting in a large change in the solution set in a single iteration. In the solution of high-dimensional problems, the change of the one-dimensional solution has little effect on the fitness value obtained from the overall solution set, and it is difficult to obtain obvious convergence.

In the multimodal Rastrigr function, due to the existence of the local optimal solution, the faster convergence speed of the fitness curve in the early stage does not mean that the global optimal solution is found, or it may fall into the local optimal solution. With a sufficient number of iterations, SCA's reprocessing rules and processing collection rules will occur multiple times during the iterative process. This provides more opportunities for SCA to jump out of the local optimal solution, and improves its ability to solve in higher dimensions.

4. Engineering problem application

In this section, we use the I-beam optimization problem and tubular column optimization problem in the engineering domain to evaluate the performance of SCA for practical applications. Each problem was run independently multiple times using the optimization algorithm, the optimal results were selected, and the results were compared with different corresponding algorithms in the recently published literature.

I-beams are widely used in construction projects due to their superior mechanical properties and good economic benefits. For the optimal design of the I-beam, the goal is to minimize the vertical deflection of the I-beam while satisfying both the cross-sectional area and stress constraints under a given load. The width of the flange $b=x_1$; the height of the section $h=x_2$; the thickness of the web $tw=x_3$; the thickness of the flange $tf=x_4$; the above four are variables. When the length of the beam is L=5200cm and the elastic coefficient E=523.104KN/cm2, the maximum vertical deflection of the beam is $f(x)=PL3/48EI$.

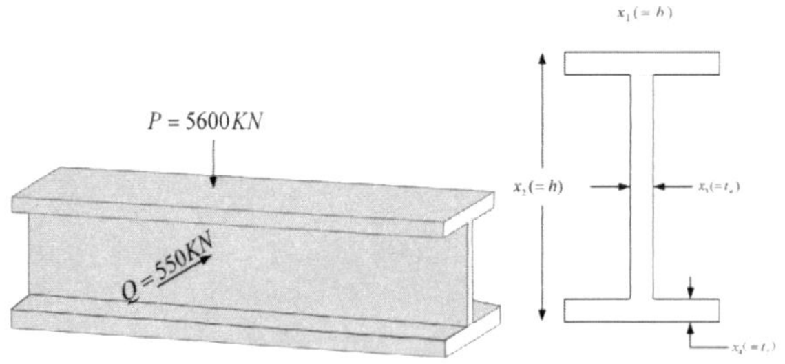

Figure 10. I-beam equivalent example diagram.

The objective function and constraints of the I-beam design problem are expressed as follows:
Related variables:

$$\vec{x} = [x_1, x_2, x_3, x_4] \tag{12}$$

Engineering Equivalent Function:

$$f(x) = \frac{5000}{x_3(x_2-2x_4)^3/12 + (x_1x_4^3/6) + 2bx_4(x_2-x_4/2)^2} \tag{13}$$

Constraints:

$$g_1(X) = 2x_1x_3 + x_3(x_2-2x_4) \le 300 \tag{14}$$

$$g_2(X) = \frac{18x_2\times10^4}{x_3(x_2-2x_4)^3 + 2x_1x_3(4x_4^2+3x_2(x_2-2x_4))} + \frac{15x_1\times10^3}{(x_2-2x_4)x_3^2+2x_3x_1^3} \le 56 \tag{15}$$

variable definition domain:

$$10 \le x_1 \le 50 \tag{16}$$

$$10 \le x_2 \le 80 \tag{17}$$

$$0.9 \le x_3 \le 5 \tag{18}$$

$$0.9 \le x_4 \le 5 \tag{19}$$

Table 3. I-beam optimization results

Algorithm	Variable				Cost
	x_1	x_2	x_3	x_4	
SCA	80	49.9836	0.9	2.322557	0.013074
SNS [35]	80	50	0.9	2.3217	0.013074
GWO [15]	80	50	0.9	2.3217	0.0131
IIARSM [35]	79.99	48.42	0.9	2.4	0.0131
AEFA-C [35]	79.9671	49.99	0.9	2.3164	0.0131

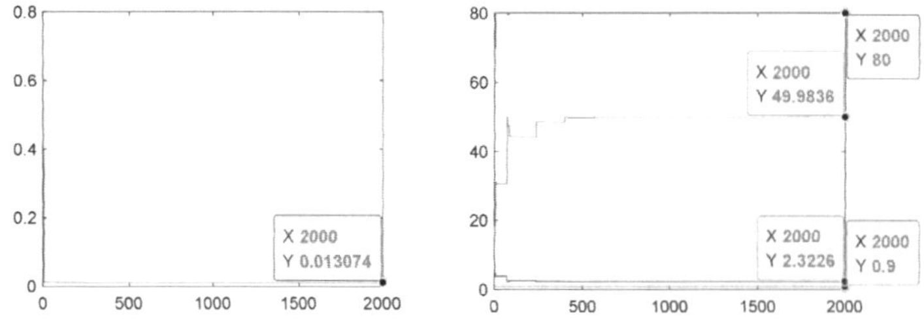

(a) (b)

Figure 11. I-beam optimal cost search history(a) Fitness history (b) Variable search history

Table 3 lists the SCA and various other algorithms to solve this problem, where IARSM: Improved Adaptive Response Surface Method. EMGO-FCR: Global Optimization Algorithm Using Fuzzy Clustering Based on Ensemble Metamodel. AEFA-C: Artificial Electric Field Algorithm. SCA is better than other methods. In the testing process, the parameter used by SCA is Q=0.1, and the number of iterations is 2000.

5. Conclusion and prospects

Membrane computing is inspired by cells and is widely used in optimization computing. In this paper, an in-depth study of the single-individual SCA of the MSA algorithm is carried out. Through the performance test of the unimodal Sphere function and the multimodal Rastrigr function, the influence of the selection of the key parameter Q value on the search performance of the algorithm in the multi-dimensional case is discussed. The following conclusions are drawn:

1. SCA can generate a new solution that is weakly correlated with the optimal solution under the current number of iterations through transcription rules. The Q value setting can adjust the variation range of weak correlation. For different problems, setting a reasonable Q value can effectively improve the optimization ability of the algorithm.
2. The reprocessing rules of SCA not only have a certain correlation with the current optimal solution, but also have a certain degree of freedom, which can effectively improve the algorithm convergence speed and resist the local optimal solution. The processing and collection rules have sufficient degrees of freedom, which can effectively improve the randomness of the algorithm solution set and prevent the algorithm from falling into a local optimal solution. SCA has the design of multiple organelles, has high adaptability to multi-dimensional problems, has excellent balance of global search and local exploration in high-dimensional optimization problems.
3. This paper uses classic engineering to verify the practical application ability of SCA. SCA achieved a result of 0.013074 on the I-beam design, and SCA found a lower cost design compared to the other algorithms involved in the comparison. Users can use SCA to solve similar problems.

For future work, we will explore the performance improvement methods of SCA and MSA based on the work of this paper.

Acknowledgments

Funding: “This research was funded by the National Natural Science Foundation of China (No.61772033) 、 the Open Research Grant of Joint National-Local Engineering Research Center for Safe and Precise Coal Mining (NO. EC2021006) and Anhui University of Science and Technology Introduction High-level Talents Scientific Research Fund (NO. 2021yjrc02)”

References

[1] Lü, X. Q. et al. (2020) Energy management of hybrid electric vehicles: A review of energy optimization of fuel cell hybrid power system based on genetic algorithm[J]. Energy Conversion and Management, 205(C): 112474-112474.

[2] Wang, J., Guo, S. (2020) Overview of Regional Integrated Energy System Operation Optimization Based on Comprehensive Demand Side Respons[J]. IOP Conference Series Earth and Environmental Science, 2020, 512(1): 012013.

[3] Gao, Y., Yang, Q. W., Wang, X. F., Li, J. H., Song, Y.J. (2022) A Review of New Swarm Intelligence Optimization Algorithms [J]. Journal of Zhengzhou University (Engineering Edition), 43(03): 21-30.

[4] Wang, P., Zhou, Y., Luo, Q., Han, C., Niu, Y., Lei, M. (2020) Complex-valued encoding metaheuristic optimization algorithm: A comprehensive survey Neurocomputing.

[5] Osaba, E., Carballedo, R., Diaz, F., Onieva, E., Masegosa, A. D., Perallos, A. (2018) Good practice proposal for the implementation, presentation, and comparison of metaheuristics for solving routing problems. Neurocomputing, 271, pp. 2-8.

[6] Gang, Y. (2022) Chaotic Butterfly Optimization Algorithm based on Particle Swarm Optimization[J]. International Core Journal of Engineering, 8(3).
[7] Hashim, F. A., Hussien, A. G. (2022) Snake Optimizer: A novel meta-heuristic optimization algorithm[J]. Knowledge-Based Systems, 242.
[8] Sang, T. T. et al. (2021) Forecasting of excavation problems for high-rise building in Vietnam using planet optimization algorithm[J]. Scientific Reports, 11(1): 23809-23809.
[9] Borji, A. (2007) A New Global Optimization Algorithm Inspired by Parliamentary Political Competitions[C]. Artificial Intelligence Mexican International Conference on Advances in Artificial Intelligence. Springer-Verlag.
[10] Dokeroglu, T., Sevinc, E., Kucukyilmaz, T., Cosar, A. (2019) A survey on new generation metaheuristic algorithms. Computers & Industrial Engineering, 137, Article 106040.
[11] Zhao, Z. Q., Huang, D. S. (2007) A mended hybrid learning algorithm for radial basis function neural networks to improve generalization capability, Applied Mathematical Modelling, Volume 31, Issue 7, Pages 1271-1281, ISSN 0307-904X.
[12] Neumann, F., Witt, C. (2010). Combinatorial Optimization and Computational Complexity. In: Bioinspired Computation in Combinatorial Optimization. Natural Computing Series. Springer, Berlin, Heidelberg.
[13] Cao, Z. Y., Yan, Y. C., Tang, K. (2022) Path optimization of open collaborative innovation of energy industry in urban agglomeration based on particle swarm optimization algorithm. Energy Reports, Volume 8, Pages 5533-5540, ISSN 2352-4847.
[14] Mirjalili, S, Lewis, A. (2016) The whale optimization algorithm. Advances in Engineering Software, 95, pp. 51-67.
[15] Mirjalili, S., Mirjalili, S. M., Lewis, A. (2014) Grey wolf optimizer. Advances in Engineering Software, 69, pp. 46-61.
[16] Mirjalili, S., Mirjalili, S. M. (2015) Hatamlou A. Multi-Verse Optimizer: a nature-inspired algorithm for global optimization[J]. Neural Computing and Applications, 27(2):495-513.
[17] Bouchekara, H. R. E. H. (2014). Optimal power flow using black-hole-based optimization approach. Applied Soft Computing, 24, 879–888.
[18] Mamano, N., & Hayes, W. B. (2017) Sana: simulated annealing far outperforms many other search algorithms for biological network alignment. Bioinformatics, 33 (14), pp. 2156-2164.
[19] Atashpaz-Gargari, E., Lucas, C. (2008) Imperialist competitive algorithm: An algorithm for optimization inspired by imperialistic competition[C]// IEEE Congress on Evolutionary Computation. IEEE.
[20] Robert, G., Chung, C. (1996) A self-adaptive approach to representation shifts in cultural algorithms[C]. Proceedings of IEEE International Conference on Evolutionary Computation. Nagoya: [s.n.], 94-99.
[21] Gheorghe, P. (2010) A quick introduction to membrane computing, The Journal of Logic and Algebraic Programming, Volume 79, Issue 6, Pages 291-294, ISSN 1567-8326.
[22] Bie, D., Miguel, A. (2019) Gutiérrez-Naranjo, Zhao J, Zhu Y. A membrane computing framework for self-reconfigurable robots[J]. Natural Computing,18(3).
[23] Mahalingam, K., Rama, R., Sureshkumar, W. (2017) Robot Motion Planning Inside a Grid Using Membrane Computing[J]. International Journal of Imaging and Robotics™,17(1), 33-51.
[24] Huang, Y., Li, J., Han, T., Xu, S. (2021) Path Planning Algorithm for Underground Coal Mine Robot Based on Membrane Computing [J]. Industrial and Mining Automation, 47(11): 22-29.
[25] Xu, J., Huang, Y., Liu, Y., Zhang, X. (2020) Attitude Optimization Control of Unmanned Helicopter in Coal Mine Using Membrane Computing[J]. Mathematical Problems in Engineering.
[26] Complexity Research; Researchers from Southwest Jiaotong University Detail Findings in Complexity Research (A Review of Membrane Computing Models for Complex Ecosystems and a Case Study on a Complex Giant Panda System) [J]. Journal of Technology,2020.
[27] Yahya, R. I., Shamsuddin, S. M., Yahya, S. I., Alsalibi, B., Al-Khafaji, G. K. (2018). Membrane

Computing for Real Medical Image Segmentation[J]. ARO-The Scientific Journal of Koya University,2018,6(2), 27-38.

[28] Fan, S., Paul, P., Wu, T., Rong, H., Zhang, G. X. (2020) On Applications of Spiking Neural P Systems[J]. Applied Sciences, 10(20).

[29] Jiang, S., Wang, Y., Xu, J., Xu, F. (2019) The Computational Power of Cell-like P Systems with Symport/Antiport Rules and Promoters[J]. Fundamenta Informaticae, 164(2-3).

[30] Pérez-Hurtado, I., Pérez-Jiménez, M. J., Zhang, G., Orellana-Martín, D. (2018) Simulation of Rapidly-Exploring Random Trees in Membrane Computing with P-Lingua and Automatic Programming[J]. International Journal of Computers Communications & Control, 13(6).

[31] Lian, Y., Ping, G. (2016) Design a membrane system for matrix multiplication[J]. Optik,127(20).

[32] Koppen, M., Wolpert, D. H., Macready, W. G. (2001) Remarks on a recent paper on the ''no free lunch" theorems. IEEE Trans Evol Comput, 5(3):295–296.

[33] Christensen, S., Oppacher, F. (2001) What can we learn from No Free Lunch? A first attempt to characterize the concept of a searchable function Proc. Genetic and Evolutionary Computation Conference GECCO-2001, Morgan Kaufmann, San Francisco, CA (2001), pp. 1219-1226.

[34] Song, Q., Huang, Y., Lai, W., Han, T., Xu, S., Rong, X. (2021) Multi-membrane search algorithm. PLoS ONE 16(12): e0260512.

[35] Bayzidi, H., Talatahari, S., Saraee, M., Lamarche, C. P. (2021) "Social Network Search for Solving Engineering Optimization Problems." Computational Intelligence and Neuroscience, edited by Radu-Emil Precup, vol. 2021, Hindawi Limited, Sept. pp. 1-32.

Design and implementation of ECC combined with OPT encryption algorithm

Xiaotian Yang[a], Ran Ma[b], Fei Gao*

Tibet University School, Tibet Lhasa
[a]778920383@qq.com; [b]824041393@qq.com;

*E-mail:zdkx@utibet.edu.cn

Abstract. 【Objective】 Combining ECC and OPT, a novel elliptic curve algorithm was proposed, which can effectively resist the reverse calculation of quantum computer and ensure the security of private key. 【Method】 On the basis of elliptic curve encryption algorithm, OPT algorithm was introduced to protect the private key. 【Result】 The signature and verification of messages can be successfully completed by the repeated algorithm on Python platform. Finally, the algorithm is proved to be safe under Shor algorithm and quantum computer attack through algorithm theory. 【Conclusion】 ECC combined with OPT encryption algorithm can effectively resist reversible computation of functional quantum devices, prevent eavesdroppers from forging signatures, and ensure the security of modern public key cryptosystems.

1. Introduction

The development of communication and encryption technology has always been the focus of the development of human society. Nowadays, encryption technology is everywhere, such as: daily life, the network world, encryption plays an indispensable role in our world.But there is encryption and decryption, in the protection of privacy at the same time, there are also people covet the privacy of others.The competition between encryption technology and eavesdropping technology can be best described as "The priest climbs a foot, the devil climbs ten.".RSA, a traditional encryption algorithm, is widely used in online payment, communication, email and other fields due to its simplicity and security.But what makes RSA encryption "ever" powerful and secure is that RSA is based on large number factorization or the difficulty of solving discrete logarithms[1].For example, it is easy to multiply two prime numbers, but difficult to find the factors of a large number.Therefore, RSA and other encryption algorithms based on public keys are generally considered to be secure.However, with quantum computing, these complex mathematical problems proved to be easily solved in polynomial time.Functional quantum devices can perform reversible calculations and forge signatures very quickly, and when practical quantum computers become available to eavesdroppers, it will break the security of almost all modern public-key cryptosystems[2][3].

Quantum computers were proposed in 1980 by Feynman[4]In 2016, IBM launched the 5 qubit superconducting quantum platform[5],The international research and development of quantum computer has entered a breakthrough stage.At present, quantum computers are developing rapidly and attracting worldwide attention. Many countries take the research and development of quantum computers as their national development strategy.Although the development of quantum computer will bring new breakthroughs in science, medicine, finance and other aspects in the future, the arrival of

quantum computer also brings unprecedented challenges to traditional cryptography,In October 2019, Google officially published in Nature: Quantum computers in scalable ways[7],The Shor's algorithm is implemented.Shor's algorithm[6]It was proposed in 1994 by mathematician Peter Shor,It's a quantum algorithm for the decomposition of large numbers,The widely used public key encryption algorithm can be broken with quantum computer.And a study by Craig Gidney of Google and Martin Ekera of KTH Royal Institute of Technology in Stockholm, Sweden, showed that it was possible to calculate 2048-bit RSA integers in 8 hours using 20 million noisy qubits[8],This study shows that Shor algorithm has become a reality, indicating that the development of quantum computers will have a huge impact on the whole Internet privacy and data security.
OPT[9][10] encryption algorithm has the characteristics of random key generation, one-time encryption and high efficiency of encryption and decryption operation, which can effectively resist the attack of quantum computer. However, it also has a serious defect, that is, the key is required to be as long as the plaintext, and it is difficult to share the key.In combination with ECC algorithm and OPT algorithm both reference random number to ensure algorithm security, we use OPT algorithm and 256-bit random number encryption to protect private key.The main contributions of this paper are as follows:

- The ECC algorithm is improved by introducing OPT algorithm and combining the advantages of the two algorithms to achieve the coexistence of security and efficiency.It can effectively resist the new encryption algorithm of quantum computer and ensure that the private key is not threatened by quantum computer.
- We use the existing scheme to implement the algorithm in Python platform, build a simple key signature scheme, and verify the feasibility of the scheme through experiments.
- The security of the algorithm is verified by theoretical analysis.

2. ECC combined with OPT elliptic curve encryption algorithm

2.1. The relevant knowledge

Finite field: finite field is the basis of elliptic encryption curve, finite field is the set of finite number of elements, the set is defined as Fp.Meet the following properties;Contains two operations (point addition operation and dot product operation) four propertiesClosure, that is, if a and b are both members of set Fp, then a+ b is also members of set Fp;The identity of addition, that is, the existence of 0 is a+0=a;Multiplicative identity, that is, if there is 1, you get a dot 1 is equal to a;The additive inverse, a is a member of the set Fp, minus a is a member of the set Fp, a plus minus a is equal to 0;So if a is a member of set Fp then a minus 1 is also a member of set Fp.Fp means that there are only P elements and p is prime.The element set is {0,1,2..., p-1}, element addition such that a+b= c (modp).
Elliptic curve:likey**2=x**3+a*x+b .Figure 1 shows two common elliptic curves.

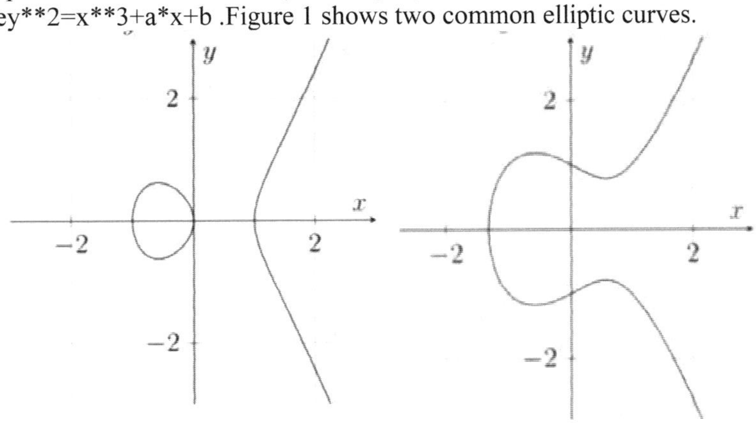

Figure 1. Elliptic curve

2.2. The relevant knowledge

Elliptic curve is widely used because of its point addition operation, point addition: elliptic curve takes any two points P, Q.P= （x1， x2） Q=(x2， y2) So P plus Q is the line that goes through P and Q,Find its third intersection with the elliptic curve, R, and then make that point symmetric with respect to the x axis, R,See Figure 2 below.And the addition of points satisfies the addition closure, addition invertibility, addition associative law and addition distributive law.

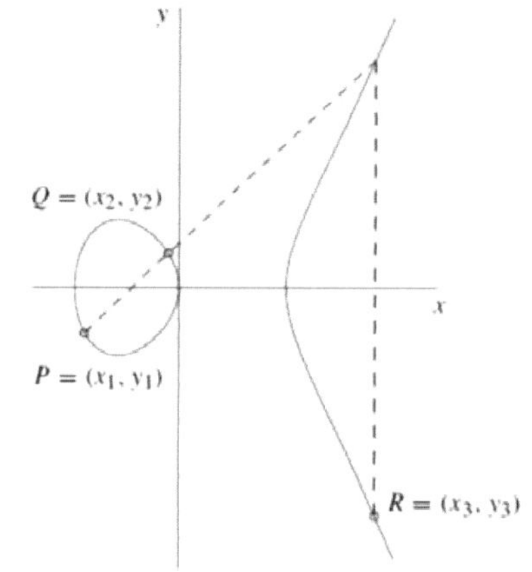

Figure 2. Addition of points on an elliptic curve

Discrete logarithm: Discrete logarithm is the core of elliptic encryption algorithm.The inverse process of scalar multiplication in elliptic curves can be regarded as a discrete logarithm problem.The scalar multiplication of elliptic curves is similar to the associative law of addition of elliptic curves.The popular interpretation of elliptic encryption algorithm is P=eG, the asymmetric algorithm, e*G→P step is easy to get, on the contrary, given P and G it is difficult to deduce e.

Digital signature: Similar to the signature on a document in real life to prove the identity of the signer, the sender achieves password conversion of the digital unit through a specific signature algorithm. This conversion allows the receiver of the data unit to know the source and integrity of the data, and prevents others from forging it.Digital signature verification: the decryption process corresponding to the encryption transformation during signature verification.Digital signatures can provide message authentication, message integrity, and non-repudiation.

2.3. Solution Architecture

ECC algorithm combined with OPT algorithm flow chart is shown in Figure 4： First, a new private key is generated through OPT algorithm (we call it "pseudo private key"), then the message and pseudo private key are signed, and the public key holder decrypts the obtained public key and received signature.Verify the decrypted hash value and the hash value formed by the message to determine whether the authentication is successful.

The "pseudo-key" elliptic algorithm can be thought of as a triplet of (G, E, D).As shown in Figure 3, G is the key generation algorithm, G(pk, sk),F(sk)→sk',G consists of pk and sk, where pk is the public key, sk is the private key, and sk'is the pseudo private key,First, the public key pk and private key sk are generated through the key generation algorithm G, and then the pseudo private key sk 'corresponding to the private key is generated through algorithm F.User A broadcasts the public key, and user B receives the public key pk from user A.User B uses E(E stands for encryption algorithm, for example, E(pk, m) consists of pk and m, and m refers to the plaintext input) to generate A ciphertext c and send it to user A.User A uses algorithm D, which is the decryption algorithm D(sk ', c) to obtain the corresponding private key and ciphertext as input, and output plaintext D(sk ',c)→

M.Obtain the message m that user B wants to send.The private key is solved by algorithm F to obtain F-1(sk ')→sk(F-1 is the inverse of function F).

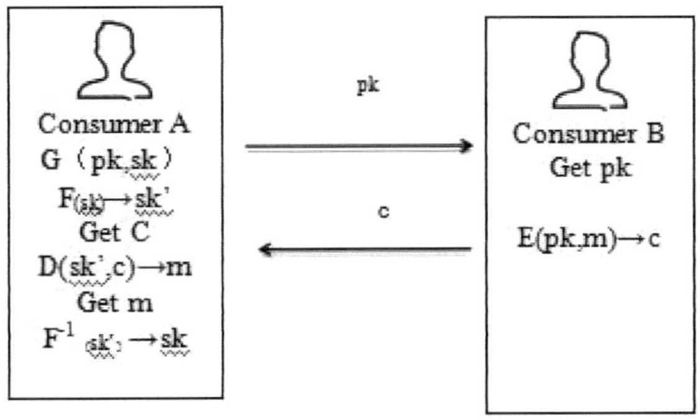

Figure 3. Algorithm schematic diagram

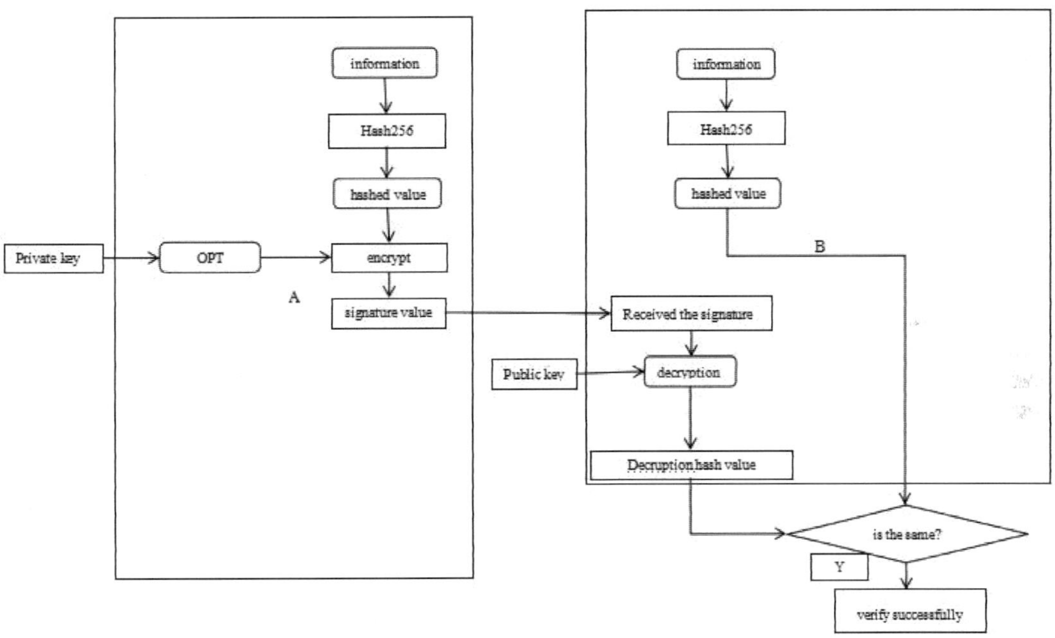

Figure 4. Algorithm flow chart

2.4. Algorithm design principle

The design principle of the algorithm is divided into three parts: pseudo key generation and verification, message signature and signature validity verification.

Pseudo-key generation and verification:

- Pseudo-key generation: $e' = e \oplus k$, pseudo-key e' is generated by xor with a 256-bit random number k. e is the private key provided by the signer, k is the generated random number, e ' is the pseudo-private key, and $\oplus$ is defined as xor operation.
- Generate the corresponding public key: e'G=P, where P is the public key (P is a coordinate point) and G(x,y) is the starting point. Generate the corresponding public key through the pseudo key e' and the starting point.

- (Randomly select a new coordinate point: $kG=R$, the coordinate of R is (x2,y2), and R is a random number k corresponding to the generated coordinate point. Let r =x2 for convenient observation, then the seat of R is marked as (r,y2).
- Discrete logarithm generation: let $uG+vP=kG$ and $uG+vP=kG$ satisfy the discrete logarithm problem, u and v are two prime numbers of large number decomposition.
- $uG+vP=kG \rightarrow vP=(k-u)G \rightarrow P=[(k-u)/v]G \rightarrow e'G=[(k-u)/v]G \rightarrow e'=(k-u)/v$, $e=e' \oplus k \oplus k$.
- Any point (u, v)(u, v) given by the signer) satisfying the above equation proves to be the holder of the private key e.

Message signatures:

- $u=z/s$ $v=k/s$,Let z represent the data processed by the hash function.
- $e'=e \oplus k$.
- The $uG+vP=R=kG \rightarrow uG+ve'G=kG \rightarrow u+ve'=k \rightarrow z/s+re'/s=k \rightarrow (z+re')/s=u \rightarrow s=(z+re')/k$.r and s are used as message signatures.

Signature validation:

- To calculate $u=z/s, v=r/s$.
- $uG + vP=R$ is calculated by determining whether the x coordinate of R is equal to r, which means the signature is valid.

2.5. Algorithm implementation

The general outline of the algorithm is shown in Figure 5:

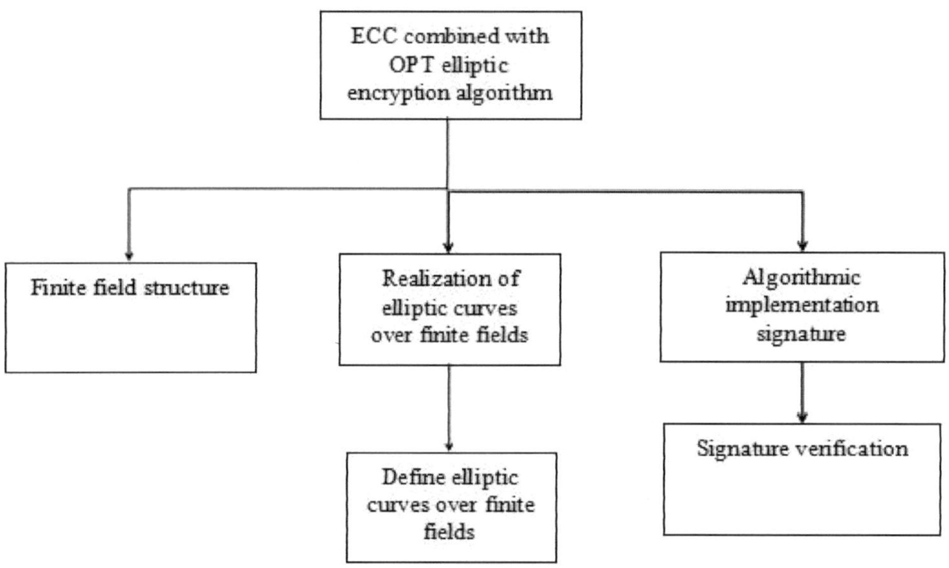

Figure 5. Overall outline of algorithm design

Part of the pseudo-code based on Python code is as follows:

- Point addition on an elliptic curve pseudo-code: four cases of point addition are discussed in the implementation process：1) The two points are perpendicular to the X-axis.2) The lines of the two points are not perpendicular to the X-axis, and they are two different points.3) The two points are the same.4) Tangent line perpendicular to X-axis:

```
Function< __add__>[self, other]:
Begin
  if<self.a != other.a or self.b != other.b>Then{
     raise ValueError/}
```

```
if< self.x is None>Then{
        output other/}
   if <other.x is None>Then{
        output self/}
       if <self.x == other.x and self.y != other.y>Then{
       output Function<__class__>[None, None, self.a, self.b]/}
   if <self.x != other.x>Then{
       s←(other.y-self.y)/(other.x-self.x)
       ←s**2 -self.x-other.x
       y←s*(self.x-x)-self.y
      output Function<__class__>[x, y, self.a, self.b]/}
   if <self == other>Then{
      s←(3 * self.x ** 2 + self.a) / (2 * self.y)
      x←s ** 2 - 2 * self.x
      y←s * (self.x - x) - self.y
      output Function<__class__>[x, y, self.a, self.b]
   if <self == other and self.y == 0*self.x>Then{
      output Function<__class__>[x, y, self.a, self.b]/}
End
```

- Elliptic curve dot product algorithm pseudo code：

```
Function<__rmul__>[self, coefficient]:
Begin
   coef← coefficient
   current ←self
   output Function<__class__>[x, y, self.a, self.b])
   while< coef> do{
      if <coef & 1>Then{
         result += current
         current += current
          coef >>= 1/}
   output result/}
End
```

- The elliptic curve used to define an elliptic curve over a finite field is y**2=x**3+7,The prime numbers required by a P finite field, where N is the order of the group:

```
    A←0
    B←7
    P←2**256 - 2**32 - 977
    N←0xfffffffffffffffffffffffffffffffebaaedce6af48a03bbfd25e8cd0364141
Function<S256Field>[FieldElement]:
Begin
Function<__init__>[self, num, prime=None]:
Begin
Function<FieldElement>.Function<__init__>[num<-num, prim←P]
End
End
Function<S256Point>[Point]:
Begin
Function<__init__>[self, x, y, a←None, b←None]:
     a, b←S256Field(A), S256Field(B)
         if <type(x) == int>Then{
        Function<S256Field>.Function<__init__>[x←S256Field(x), y←S256Field(y), a←a, b←b) ]/}
         else:
        {Function<S256Field>.Function<__init__>[x←x, y←y, ←a, b←b]/}
End
```

- Verify the signature pseudocode:

```
Function <verify>[ z, sig]:
Begin
    1/s←sig.s**(N - 2)%N
      u←z/s
      v←r/s
      total ← u*G+v*P
      output total.x==sig.r
End
```

- Implement message signing pseudocode:

```
  Function <sign>[self, z]:
Begin
    k←Function<deterministic_k>[z]
      r←(k * G).x
      1/k←(k**,N - 2)% N
  S←(z+re)/k
      if <s > N / 2>Then{
      S←N - s
      output Signature(r, s)}
 End
```

- Signature validity verification pseudocode:

```
Begin
  P←S256Point(Px,Py )
  Z←z
  R←r
  S←s
  u←( Z *S**N-2%N )% N
  v← (R* S**N%N) % N
  output((u*G + v*R).x.num == R)
End
```

3. Algorithm analysis

3.1. Feasibility verification

The validity of the pseudo-key algorithm is verified by the code.The process is as follows:

1. Experimental message signing: "my secret" as the key input.The decimal value of key E generated by the hash algorithm is:62971298242950415662486979275162298594154135681004836692467839909933090737920.
2. Use my Message as the signature content.The signature hash z generates a hexadecimal fo: 0x231c6f3d980a6b0fb7152f85cee7eb52bf92433d9919b9c5218cb08e79cce78.
3. The generated random number k corresponds to the decimal number: 53141233081433290542791408994650223676848194257993344528424053017128121420544.
4. The hexadecimal value of the pseudo-key e ' generated by the xor key and the random number k is: 0xfe44a2dd99975ae11dc4ff8edc3115c822be60969d6f4c8625330ef8e54e4400.
5. Hexadecimal representation of r (the x coordinate of R) : 0x6238767416e7318ce2fd44b42008bbb9c166e17596fe1a00d9caf4fc41bc202e.
6. The hexadecimal representation of s is,s=(z+re)/k(r and s are signed messages): 0xb59bac6df30fed1aa0a3487b1ec3d19e7e9a3c870cc3375e724b7251a264a9ec.
7. The hexadecimal coordinate of the corresponding public key P is : (0caaecdde9c869a116766f6d2a9a63d2ca5c34cefcf5f31f85fa3083b58e192e, 9f7b5994fb8f877178974c7dd78ff27baf733fae08082e844540bacf2b8ed23e).

By substituting the variable public key P into the signature verification program, uG+vP=R was calculated to verify the feasibility of the experiment. The experimental running results are shown in Figure 6:
In order to avoid randomness, English case including digits, underscores, Spaces were selected as keys and information to verify the feasibility of the scheme. The data is shown in Table 1:

Table 1. Corresponding table of different signature verification

e content	generated e	Key e 'generated by OPT algorithm	Signature z content	z	The random number k	Verify validity
123456	115564964636320539531521438116064177340117575960192970697235805359596761035897	0x89bf38413ed1742501442b9c5deb819a81d389668bcd787fccb0c1fbc836ab18	54321	0x4556c961133d7a593d089834b9099654673d8ab3b8325123bb4f8a240af86d24	53712675124463753457704245564399574513522652948799933949709355191231716925281	True
xiaotian	155325743173117613215179879338624433473646122130804445313856864623 2380270216	0x6ad056cbf5bb2ce93e1bd9e1ff44f7dc7644194bb2db3d4b12a0fac50f8b26a1	give me money	0x8a72b7ca097bf6051af084aec507ecb30209951d6e21efa48329bc9f18d14e31	47830828368418485772249180832362749456599122105470023853307223682035763691561	True
yy@123_	54839313559401762638800234993338309829695745845529620161472800357585550554911	0x47132b12a22965e480818c29479645bb40fbae203da7cef88c0540c679e3bcf	I love you!	0xfbb344d409200cf36cc78e91fe34cbdd24dd23ed30bbffe53cfc5b4e4a11f5db	56674719528244637645379599233403503591405619527073202672617532259064677502160	True

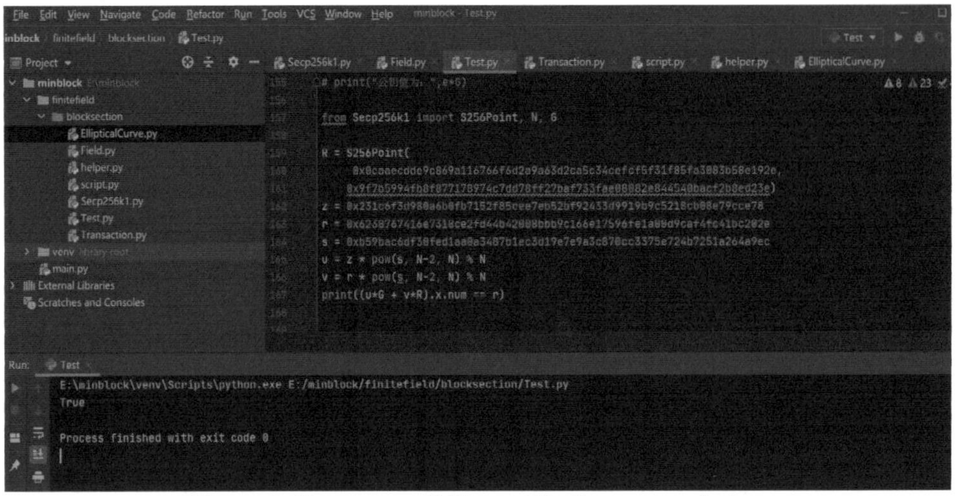

Figure 6. Signature validity verification screenshot

3.2. Security verification

1) Anonymity

Without the signer exposing the key, a third party cannot obtain the real key.Even if an eavesdropper can reverse solve the "private key" we use for information exchange with a mature quantum computer, it turns out to be a fake private key.Eavesdroppers also cannot obtain the user's real key, because the generation of fake private keys is random and one-time.In the algorithm, the pseudo-private key is generated by xor between the private key and the random number k, and k is a random number of 256 bits. The randomness of k ensures the variability of the user private key and the anonymity of the user in the communication process.The eavesdropper can't really get the identity of the correct signer to determine the message signature, so the scheme conforms to anonymity.

2) Security

Absolute security: given a ciphertext c, the attacker cannot tell whether the plaintext is m1 or m2.The encryption algorithm E(pk， m)=c,There is m1, m2 belongs to M(M is a set of plaintext) there is c belongs to C(c is a set of ciphertext).P (E(pk, m1)=c)= P (E(pk, m2)=c) that is, the algorithm that encrypts plaintext into ciphertext with equal probability (an attacker cannot distinguish m1 from m2) is absolutely secure.

Quantum attack resistant: traditional elliptic cryptography,According to the $eG=[(k-u)/v]G \rightarrow e=(k-u)/v$,Any combination of (u,v) can be proved to be the holder of e only if it satisfies the preceding equation,I don't know what e is by exhausting (u,v) until e is equal to (k-u)/v.In the condition of quantum computer is not mature, it is not practical to exhaust.But it is not safe in mature functional quantum computing devices.Some experts pointed out that Shor's algorithm takes advantage of the parallelism of quantum computing to quickly decompose common difactors and reverse calculate e in only 8 minutes, thus breaking the basis of RSA algorithm.But for the improved elliptic encryption algorithm,$e'G=[(k-u)/v]G \rightarrow e'=(k-u)/v$,$e=e' \oplus k \oplus k$.It is assumed that the inverse calculation of E 'through quantum equipment is 8 minutes, but it is impossible to calculate the random number K of 256 bits through enumerating, and the pseudo-key generation is random and one-time, which is constantly changing with the use of users.Assume that the user does not use the pseudo key after using it once but stays in the last use, and the eavesdropper steals the real private key after cracking the pseudo private key,assume that it takes 8 minutes to reverse the private key2**256*≈8*10**77 分 ≈1.33*10**76hours.It is not feasible to crack the private key in such a long time. It is theoretically verified that the pseudo-key algorithm is anti-quantum computer. By protecting the real key E, it can ensure that the user's private key is not obtained by the eavessteal, and it can be concluded that the pseudo-key algorithm is safe under the quantum computer.

4. Conclusion

The development of quantum computing has brought convenience as well as great hidden dangers,Anti-quantum encryption algorithms are imminent,The encryption algorithm proposed in this paper can effectively resist the attack of quantum computing, and the scheme protects the user private key from attack in terms of security.In addition to the high security of the above analysis algorithm, ECC characteristics are retained in terms of performance:less storage space, lower CPU overhead and less bandwidth,And the scheme is based on elliptic curve, sharing the same random number, can achieve password security with minimal changes, high practicability,Higher security at the cost of less computing power.Improved the disadvantages of OPT: convenient key sharing.It also provides a new idea for future key security.In the following study, we will further optimize the algorithm and verify the improved scheme from more aspects. There are still many scientific research work to be improved in the future.

References

[1] Neha, T and Ganesan, R. (2014) "Data Security in Cloud Architecture Based on Diffie Hellman and Elliptical Curve Cryptography"[J], IACR Cryptology ePrint Archive, pp. 49.
[2] Shor, P. W. (1997) Polynomial-time algorithms for prime factorization and discrete logarithms on a quantum computer[J]. SIAM Journal on Computing, 26(5):1484-1509.
[3] Grover, L. K. (1998) A framework for fast quantum mechanical algorithms [C]//Proceedings of the thirtieth annual ACM symposium on Theory of computing.53-62.
[4] Benioff, P. (1980) "The Computer as a Physical System: A Microscopic Quantum Mechanical Hamiltonian Model of Computers as Represented by Turing Machines"[J], Journal of Statistical Physics, 22(5), pp. 563-591.
[5] "IBM Makes Quantum Computing Available on IBM Cloud to Accelerate Innovation"[C], https://www-03.ibm.com/press/us/en/pressrelease/49661.wss, 2016-5-4.
[6] Shor, P W. (1994) "Algorithms for Quantum Computation: Discrete Logarithms and Factoring", Proceedings 35th Annual Symposium on Foundations of Computer Science[J].IEEE, pp. 124-134.
[7] Arute, F., Arya, K., Babbush, R., et al. (2019) Quantum Supremacy Using a Programmable Superconducting Processor[J]. Nature, 574(7779): 505-510.
[8] How to factor 2048 bit RSA integers in 8 hours using 20 million noisy qubits[C]https:// //arxiv.org/abs/1905.09749
[9] Bellovin, S. M. (2011) Frank Miller: Inventor of the one-time pad[J]. Cryptologia, 35(3): 203-222.
[10] Xiao, Y., Wang, B., Wang, Z., et al. (2022) One-time pad scheme based on polar code and OFDM for MMW-RoF system at W-band[J]. Optics Express, 30(3): 4412-4423.

Implementation of CCSDS lossless compression algorithm for geomagnetic data

Yue Zhang[1, 2, a], Li Li[3, 4, 5,*], Chengbin Huang[1, 2,b], Jinhua Wang[1, 2, c]

[1] China Telecom Research Institute; Shanghai, China
[2] National Engineering Research Center of Cloud and Network Infrastructure Security; Shanghai, China
[3] Innovation Academy for Microsatellites of CAS; Shanghai, China
[4] Shanghai Engineering Center for Microsatellites; Shanghai, China
[5] University of Chinese Academy of Sciences; Beijing, China

*lil@microsate.com; [a] zhangy138@chinatelecom.cn; [b] huangchb@chinatelecom.cn; [c]wangjinh@chinatelecom.cn

Abstract. Long-term geomagnetic data monitoring will generate a huge amount of data, and compression algorithms will be required to reduce the amount of data to save data storage and transmission expenses. To adapt to the application with limited resources, the CCSDS lossless compression algorithm is applied to geomagnetic data measurement in our work. This compression algorithm is made up of two parts: a preprocessor and an adaptive entropy coder. In this study, the unit-delay predictor is chosen as the preprocessor based on the properties of the geomagnetic signal, and the adaptive entropy coder chooses a suitable compression algorithm for geomagnetic data. A 14-hour continuous measurement was performed. The results reveal that when the block size is set to 64, the algorithm can obtain a compression ratio in the geomagnetic field of 0.269 to 0.318. Furthermore, the number of chosen encoding options for three-axis magnetic data was counted.

1. Introduction

The geomagnetic field is a vector field that changes with time and space and comes from the earth's interior [1]. The geomagnetic field follows some rules and is widely used in mineral exploration [2], navigation [3], earthquake prediction [4], and other applications. In the aforementioned applications, it is frequently important to know the geomagnetic distribution of a specific area at a specific moment ahead of time. Some geomagnetic models such as the international geomagnetic reference field (IGRF) [5] and the World Magnetic Model (WMM) [6] can estimate the magnetic field distribution at a certain time in a specific location in the world.

Accurate geomagnetic measurements serve as the foundation for the development of geomagnetic models. These measurements include satellite measurements, ocean measurements, ground measurements, etc. The measurement on the ground station has the benefit of convenience, high-precision magnetic field sensors, and a controllable measurement environment when compared to other measurement methods. It is an essential component of geomagnetic measurement. The time scale of geomagnetism can be many years, thus long-term observations of the geomagnetic field are necessary.

However, the long-term measurement puts strain on the storage and transmission of the measurement devices, particularly wireless geomagnetic measurement sensors with limited energy deployed in the field.

Lossless compression algorithms are commonly used to reduce the amount of geomagnetic data. The majority of lossless compression algorithms are derived from dictionary-based LZ77 and LZ78 [7, 8] as well as encoding-based Huffman encoding [9], Shannon encoding [10], and arithmetic encoding [11]. These algorithms, however, are complex and consume a large amount of processor and memory resources, making it difficult to meet the needs of wireless geomagnetic sensors with limited resources. The Consultative Committee for Space Data Systems (CCSDS) recommended an algorithm based on the Rice compression algorithm as the recommended standard for spatial data. This lossless compression algorithm has three editions, the most recent of which is CCSDS 121.0-B-3, which has been validated on a large amount of digital data [12]. The algorithm has the advantages of being simple to implement and consuming fewer resources, making it suitable for use in resource-constrained scenarios such as wireless nodes. As a result, we implement this lossless compression algorithm and applied it to the geomagnetic measurement. The feasibility of this compression algorithm with geomagnetic data is demonstrated.

2. CCSDS lossless compression algorithm

This lossless algorithm is made up of a preprocessor and an adaptive entropy coder. The preprocessor is used to minimize data correlation to produce the lowest entropy, while the adaptive entropy coder is used to represent the processed data with the fewest bits. Figure 1 depicts the composition of the lossless compression algorithm.

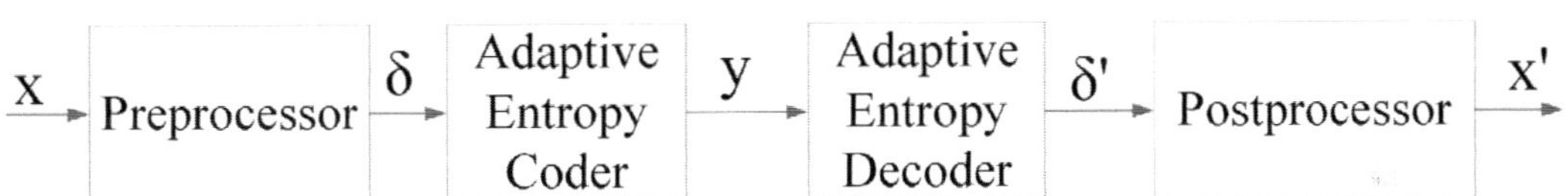

Figure 1. The composition of CCSDS lossless compression algorithm.

The data $x=x_1,x_2,\ldots,x_J$ in Figure 1 is the raw data in a block that needs to be compressed and J is the number of data samples called block size. $\delta=\delta_1, \delta_2,\ldots, \delta_J$ are the preprocessed data from the preprocessor. The data δ generated after the preprocessor needs to remove the correlation between the data as much as possible, and the preprocessing needs to be a reversible process for signal decoding. The adaptive entropy coder offers several algorithms for compression and the one that achieves the shortest code data y will be selected in each compression. Decoding is the reverse process of encoding. After the compression algorithm is identified in the adaptive entropy decoder, the corresponding analysis is performed to obtain the reference value and the corresponding mapped data. The mapped data will be calculated to get the prediction errors and then be recovered to the original data x'.

2.1. Preprocessor

The preprocessor can function as a predictor, with the output data representing prediction errors. Section III will present the characteristics of the geomagnetic data, which are slowly changing over time. As a result, the unit-delay predictor is used in our work. The first raw data in a block is considered the reference sample, and each subsequent value is represented by the difference between the current and previous values. To reduce the signal's entropy as much as possible, the data after the preprocessor should satisfy the following condition:

$$p_0 \geq p_1 \geq p_2 \geq \cdots p_j \geq \cdots p_{(2^n-1)}, \tag{1}$$

here p_j is the probability that δ_j equals integer j and n is the resolution in bits per sample. A mapper is adopted to satisfy Equation (1). The mapper operation is described as:

$$\delta_i = \begin{cases} 2\Delta_i, 0 \le \Delta_i \le \theta_i \\ 2|\Delta_i| - 1, \ -\theta_i \le \Delta_i \le 0 \\ \theta_i + |\Delta_i|, \text{ Others} \end{cases} \tag{2}$$

Where $\theta_i = \min(\hat{x}_i - x_{min}, x_{max} - \hat{x}_i)$ and Δ are the prediction errors. $x_{min} = -2^{n-1}$, $x_{max} = 2^{n-1}$ for signed values and $x_{min} = 0$, $x_{max} = 2^n - 1$ for unsigned values. The reference samples can also be shared by several blocks to compress the data further. However, the same effect can be achieved by changing the size of the block. Therefore, in our work, each block has a reference sample.

2.2. Adaptive Entropy Coder

For the mapped data, the coder provides several code options. In compression for a block, the option with the shortest code length will be chosen as the compressed method. Figure 2 depicts the adaptive entropy coder's options.

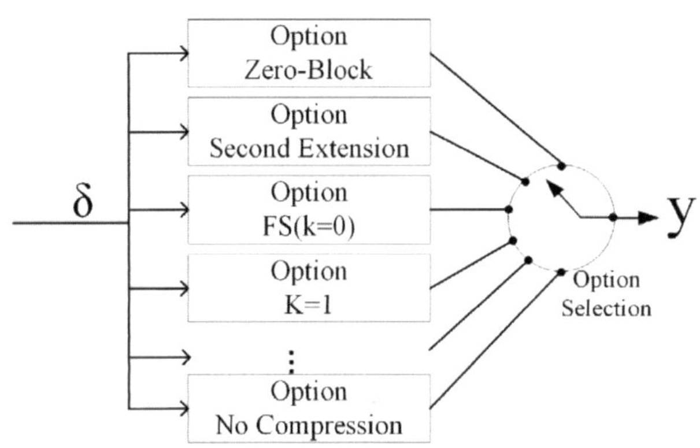

Figure 2. The composition of CCSDS lossless compression algorithm.

The coder's options are divided into four categories. The first is the Zero-Block, the second is the second extension, the third is the k split-sample options, and the fourth is no compression. If an option is selected, the corresponding unique ID is added to the encoding for later decoding.

1) Zero-block option

According to Section II A, the data in a block is made up of a reference sample and J-1 mapped predict errors. This block can be expressed as ID + Reference sample + 1b if the J-1 mapped predict errors are all zero.

2) Second extension option

This option is designed for data in the range of 0.5 bits/sample to 1.5 bits/sample. The implementation method is to use one value to represent two adjacent values δ_i and δ_{i+1} to reduce the amount of data. A zero value will be added to the sequence if J-1 is odd. The express of this option is

$$\gamma = (\delta_i + \delta_{i+1})(\delta_i + \delta_{i+1} + 1)/2 + \delta_{i+1} \cdot \tag{3}$$

3) The k split-sample options

This kind of option is based on the Fundamental Sequence (FS) code. The FS code can be regarded as a special case of k split-sample options when k is equal to zero. For an integer t encoded by FS code, it can be represented by t*'0' and a '1'. The k split-sample options mean that the low k bits of the signal remains unchanged, while the high n-k bits are represented by the FS code.

4) No compression option

In extreme cases, data encoded with FS code may produce long code. The CCSDS lossless compression algorithm includes this option to avoid this situation. If other algorithms are unable to achieve the compression effect, the data will not be compressed.

2.3. Adaptive Entropy Decoder

The decoder consists of two parts, one is to identify the compression algorithm according to the ID, and then decode according to the principle of the algorithm.

1) Zero-block option

The reference sample is identified at first and the rest J-1 mapped predict errors are all zeros.

2) Second extension option

There are only $J/2$ codewords that need to be extracted with this option. Because its encoding method is complicated, decoding is also a little troublesome. There are four steps to extract the data.

Step 1: obtain γ by counting the 0s of the FS codeword;

Step 2: obtain β and m_s according to Table 1;

Step 3: calculate $\delta_{i+1} = \gamma - m_s$;

Step 4: calculate $\delta_i = \beta - \delta_{i+1}$.

Table 1. Decoding method for the Second extension option

γ	β	m_s
0	0	0
1,2	1	1
3~5	2	3
6~9	3	6
10~14	4	10
15~20	5	15
21~27	6	21
28~35	7	28
36~44	8	36

It should be noted that the input of J is required when looking up the table. For $J = 16$, $\beta \leqslant 4$ and for $J = 64$, $\beta \leqslant 8$.

3) The k split-sample options

The codewords can be divided into two parts. The first part is the low k bits, which are coded with source bits. The second part is the high n-k bits, which are coded using FS code. After decoding the two parts separately, they are spliced together to obtain the data before encoding. Repeat the above process until all data in the frame is obtained.

4) No compression option

This option means that the original data is not compressed, and the data of J*n bits can be copied directly.

2.4. Postprocessor

The Postprocessor process is the inverse of the Preprocessor. Its decoding process consists of two steps. The first step is to use the mapped data to restore the predicted error. The second step is to recover the original data through the prediction error.

1) Mapped data decoding

According to Equation (2), the predicted data can be obtained by decoding the mapped data expressed as Equation (4) below.

$$
\begin{aligned}
&if\ \delta_i \le 2\theta_i \\
&\Delta_i = \begin{cases} \delta_i / 2 & (when\ \delta_i \text{ is even}) \\ -(\delta_i + 1)/2 & (when\ \delta_i \text{ is odd}) \end{cases} \\
&if\ \delta_i > 2\theta_i \\
&\Delta_i = \begin{cases} \delta_i - \theta_i & (when\ \theta_i\ \hat{x}_i - x_{\min}) \\ \theta_i - \delta_i & (when\ \theta_i\ x_{\max} - \hat{x}_i) \end{cases}
\end{aligned}
\tag{4}
$$

2) Predicted data decoding

The unit-delay predictor is used in our work. Therefore, the data after the reference value can be obtained by adding the previous value to the current error.

It can be seen from the above compression and decompression process that when CCSDS is used for data compression, the position of each bit of the data is sensitive. The key information such as ID is located according to the bit position, and the subsequent data is decoded accordingly. That is to say, as long as there is a data error of one bit, all subsequent data will be wrong. Therefore, the data length of each compression cannot be too long. However, the data compressed each time cannot be too short, otherwise redundant information such as ID will be frequently inserted, resulting in a poor compression effect.

3. Experiment

Geomagnetic measurements were taken in Shanghai, China, to test the feasibility of this lossless compression algorithm in geomagnetic data compression. A 3-axis magnetometer with a resolution of 0.01 nT was used to collect the geomagnetic data. The data is represented by a 32-bit signed integer, implying that the compression algorithm's resolution in bits per sample *n* is 32. The geomagnetic size is calculated by dividing the collected data by 100. The magnetometer is mounted on the bracket, and the magnitude of the magnetometer's geomagnetic field in three directions is measured over time. The magnetometer's sample rate is set to 4 Hz. The measurement devices are shown in Figure 3.

Figure 3. The measurement devices for geomagnetic data.

In our experiment, we record geomagnetic data for about 14 hours. Figure 4 depicts the data from one magnetometer.

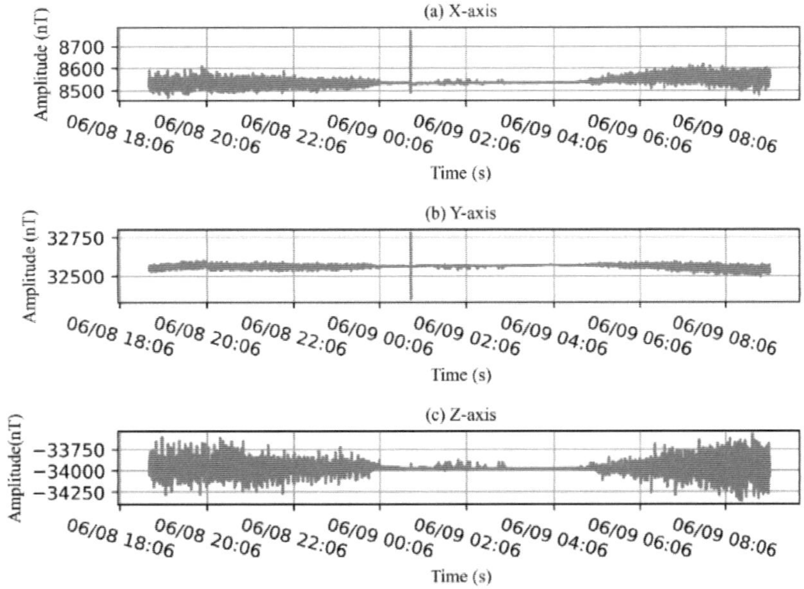

Figure 4. A magnetometer's geomagnetic data on three axes.

Figure 4 shows that the geomagnetic signal in one direction changes less in a short period. As a result, the unit-delay predictor described in Section II A can be applied to signal to preprocess. The CCSDS lossless compression algorithm was implemented. Different block sizes are set for the magnetic data on the three axes respectively, and the compression ratio is calculated accordingly. The compression ratio is expressed by dividing the compressed signal length by the original signal length. The results are depicted in Figure 5.

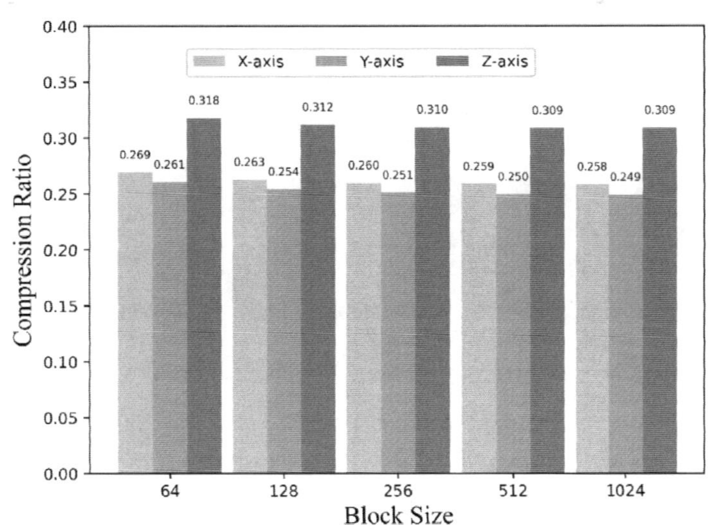

Figure 5. Compression ratio versus Block size.

Figure 5 shows that as the block size increases, the compressed signal length decreases slightly. When the block size is increased from 64 to 1024, the compression ratio of the x-axis decreases from 0.269 to

0.258, the compression ratio of the y-axis decreases from 0.261 to 0.249, and the compression ratio of the z-axis decreases from 0.318 to 0.30. The compression effect has improved but it is not significant. In addition, the mapped prediction errors need to be recovered sequentially from the reference samples in the decoding stage. When a bit error occurs, the samples in the block following the bit error cannot be decoded accurately. Block size should not be set too large from this standpoint. As a result, the block size in this experiment is set to 64. The measured data can be divided into 3408 blocks using this configuration. Each block's data is compressed, and the option used for each compression is counted. Figure 6 depicts the statistical findings.

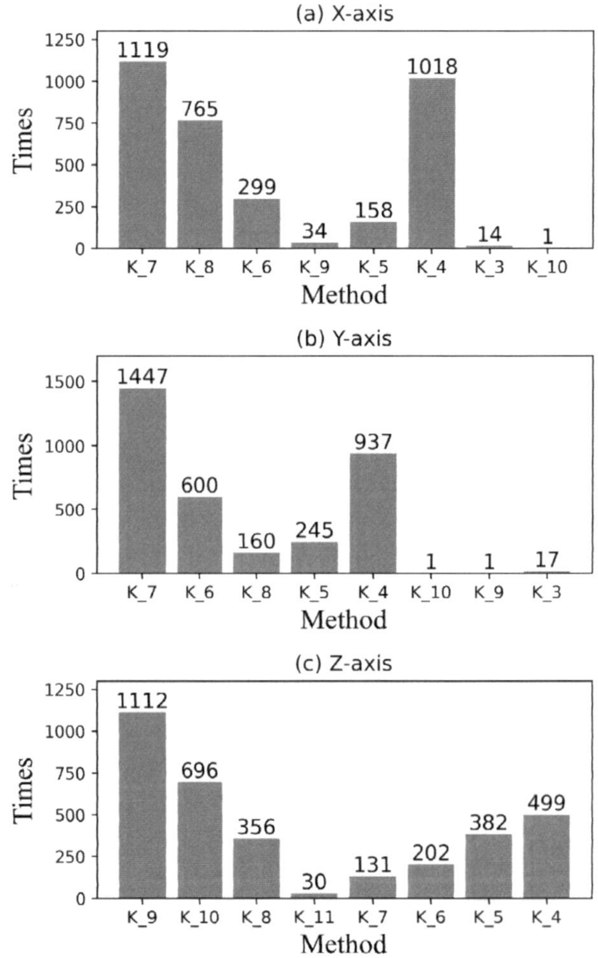

Figure 6. The number of options when the block size is 64.

Figure 6 shows that the k split-sample options are used for data in all three axes. Different options are chosen for different data types, which improves the performance of the adaptive entropy coder to different signals.

4. Conclusion

In this paper, the CCSDS lossless compression algorithm is applied to geomagnetic data measurement based on the characteristics of geomagnetic data. The CCSDS lossless compression algorithm is introduced and implemented, and the compression algorithm is applied to geomagnetic data measurement. The block size is set to 64 to achieve a balance between compression ratio and error propagation, and the compression options under this parameter set are counted. The results show that the algorithm can achieve a compression ratio of 0.2~0.3 on geomagnetic data, confirming the algorithm's feasibility for geomagnetic data compression.

References

[1] Mandea, M., Chambodut, A. (2020) Geomagnetic Field Processes and Their Implications for Space Weather[J]. Surveys in Geophysics, 41(6), 1611-1627.
[2] Gehrmann, R. A. S., North, L. J., Graber, S., et al. (2019) Marine mineral exploration with controlled source electromagnetics at the TAG hydrothermal field, 26° N Mid-Atlantic Ridge[J]. Geophysical Research Letters, 46(11): 5808-5816.
[3] Lohmann, K. J., Lohmann, C. M. F., Ehrhart, L. M., et al. (2004) Geomagnetic map used in sea-turtle navigation[J]. Nature, 428(6986): 909-910.
[4] Hayakawa, M., Hattori, K., Ohta, K. (2007) Monitoring of ULF (ultra-low-frequency) geomagnetic variations associated with earthquakes[J]. Sensors, 7(7): 1108-1122.
[5] Alken, P., Thébault, E., Beggan, C. D., et al. (2021) International geomagnetic reference field: the thirteenth generation[J]. Earth, Planets and Space, 73(1): 1-25.
[6] Chulliat, A., Brown, W., Alken, P., et al. (2020) The US/UK world magnetic model for 2020-2025.
[7] Ziv, J anda Lempel, A. (1977) A Universal Algorithm for Sequential Data Compression. IEEE Transactions on Information Theory. 23(3), 337–343.
[8] Ziv, J. and Lempel, A. (1978) Compression of Individual Sequences via Variable-Rate Coding. IEEE Transactions on Information Theory. 24(5), 530–536.
[9] Li, L., Liu, H., Zhu, Y., et al. (2020) A Lossless Compression Algorithm Based on Differential and Canonical Huffman Encoding for Spaceborne Magnetic Data[C]//Proceedings of the 2020 2nd International Conference on Image, Video and Signal Processing. 115-119.
[10] Shannon, C. E. (1948). A mathematical theory of communication. The Bell system technical journal, 27(3), 379-423.
[11] Rissanen, J. J and Langdon, G. G. (1979) Arithmetic Coding.IBM J Res Dev. 23 (2), 149–162.
[12] "C. C. for Space Data Systems", Lossless Data Compression CCSDS 121.0-B-3, vol. 1, August 2020.

AUTHOR INDEX

Institute of Physics
Dirac House, Temple Back
Bristol BS1 6BE UK

ISSN: 1742-6588
ISBN 978-1-7138-6492-9

9 781713 864929

CURRAN ASSOCIATES INC.
proceedings.com